Patrick Moore's Practical Astronomy Series

For further volumes:
http://www.springer.com/series/3192

Celestial Delights

The Best Astronomical Events through 2020

Francis Reddy

 Springer

Francis Reddy
Syneren Technologies Corp.
Lanham, MD, USA

ISSN 1431-9756
ISBN 978-1-4614-0609-9   e-ISBN 978-1-4614-0610-5
DOI 10.1007/978-1-4614-0610-5
Springer New York Dordrecht Heidelberg London

Library of Congress Control Number: 2011936392

Printed on acid-free paper

Springer is part of Springer Science+Business Media (www.springer.com)

*For Kari and Sharon, who eagerly
share the sky with a new generation.*

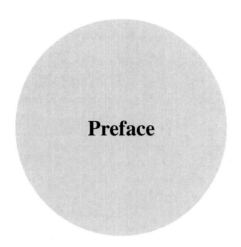

Preface

There is a widespread impression that the scientific appreciation of the universe must be left wholly to those who have had years of formal training or who devote a large part of their free time to science as a hobby. It isn't true, of course, and it's less true now than ever before. Thanks to the Web, volunteers with little specific training are now assisting scientists in tasks as diverse as classifying galaxies, searching for lunar features, and monitoring bird nesting sites.

But you don't need expensive equipment, a steep investment of free time or even a broadband connection to start observing the heavens. All you need are your eyes.

Everyone takes a moment to sky-gaze now and then – admiring the colors of a Sunset, pondering the Man in the Moon or playing connect-the-dots with bright stars. And while most of us are denied the awesome pleasure of a clear, dark country sky far from the tawny glow of city lights, we can still observe the solar system's brightest members – the Sun, the Moon and five planets. Their ever-changing configurations fascinated and puzzled sky watchers for the first few thousand years of human civilization, a time when the human eye was the primary observing tool. Tracking their wanderings through the sky requires nothing more than good weather and some guidance on when and where to look.

That's where *Celestial Delights* comes in. It's a resource that lays out the best, most observable events for years to come and supplies the background necessary to understand and appreciate them. Think of it as a *TV Guide* to the sky.

The first goal of this book is to share the simple beauties of the sky as seen by our ancestors, sights still available to all of us. By using the Moon and brighter planets as celestial signposts, the tables and charts in this book serve as a guide to every planet that can be seen easily with the unaided eye. Separate chapters detail upcoming eclipses, introduce the major constellations, describe the best meteor showers, and explore other more challenging, if less predictable, phenomena. The Timetable of Celestial Events (Appendix A) provides an almanac of easy-to-see astronomical occurrences from 2011 through 2020, all organized by date and time and cross-referenced with diagrams and tables found elsewhere in the book.

A secondary goal is to explore how we see the sky from the perspective of our spacefaring culture, one that has literally touched some of the worlds that human-kind has watched, prayed to and feared for ages. An exciting burst of planetary exploration is currently under way. Plans for the decade include continuing robotic exploration of the Moon, Mars, Mercury, and Saturn, and spacecraft are now en route to asteroids, comets and Pluto. I have attempted to share something of the excitement felt and the challenges faced by those who continue to delve into the long-observed phenomena that this book describes.

I hope you'll come to enjoy the beauty of the heavens that is, after all, everyone's heritage, and encourage you to participate in this centuries-old delight. All you need to get started is to occasionally look up and wonder: "What is that bright star?" With this book, the answer is literally within your grasp.

Bowie, MD.   Francis Reddy

Acknowledgments

Celestial Delights would not be possible without the assistance of many others. First and foremost is my friend and co-author of two earlier editions, Greg Walz-Chojnacki, who encouraged me to go solo this time out. Working with him on the previous editions and other projects was a terrific experience, and his input was sorely missed.

Fred (Mr. Eclipse) Espenak of NASA's Goddard Space Flight Center in Greenbelt, MD, makes an immense body of eclipse information publicly available, something appreciated by all eclipse-chasers and not a few science writers. Fred's work is ultimately the source for all eclipse and transit predictions in the book, and his maps long have inspired my own simpler creations. One new feature of the solar eclipse maps is the inclusion of city lights as a way to illustrate urban locations that are otherwise too numerous to label. This data, which is acquired by satellites of the U.S. Defense Meteorological Satellite Program, is made available by NOAA's National Geophysical Data Center.

Theodore Yapo, a doctoral student in the Computer Science Department at Rensselaer Polytechnic Institute in Troy, New York, graciously provided the book with another novel feature – detailed images of the Moon's appearance at mid-totality at upcoming total lunar eclipses. Ted rendered these "sneak peeks" using his own computational model that tracks how sunlight refracts, disperses and scatters in Earth's atmosphere.

The artistic eyes and obedient cameras of many photographers grace the pages of this book. If it's true that a picture is worth a thousand words, then they've saved me a great deal of work, but I doubt my words are even remotely as beautiful and compelling as their images. For the use of their photographs, heartfelt thanks go to Anthony Ayiomamitis, Juan Carlos Casado González del Castillo, António

Cidadao, Ben Cooper, Lars Christiansen, Miloslav Druckmüller, Iván Éder, Antonio Fernandez, Alfred Lee and Tunç Tezel.

My friend Robert Miller has long supported this project by creating custom software for it, and he reprised his role in the last edition by generously providing star maps showing the retrograde loops of Mars.

Most of the astronomical computation for the book was performed using the Multiyear Interactive Computer Almanac (MICA) developed by the U.S. Naval Observatory in Washington, D.C., and published by Willmann-Bell, Inc. of Richmond, Va. Additional helpful resources include the Horizons Web-based ephemeris computation service provided by NASA's Jet Propulsion Laboratory; Starry Night Pro Plus 6, sold by Simulation Curriculum Corp. of Edina, Minn.; Stardome Plus, a Java applet made available by *Astronomy* magazine; Runmap, a map-reprojection tool by Wesley Colley at the University of Alabama, Huntsville; and the Generic Mapping Tools software package created by Paul Wessel and Walter Smith at the University of Hawaii.

Many others have been generous with their time and knowledge during the writing of this and previous editions. They include: Pål Brekke at the Norwegian Space Center; Geoff Chester, U.S. Naval Observatory; Douglas Caswell, European Space Agency; Bill Cooke at NASA's Marshall Space Flight Center in Huntsville, AL; Daniel W. E. Green at the Harvard-Smithsonian Center for Astrophysics, Cambridge, MA; Michael Mumma and Michael Smith at NASA Goddard; Cary Oler at the Solar Terrestrial Dispatch in Stirling, Alberta, Canada; David Pankenier at Lehigh University, Bethlehem, PA; Michael Rappenglück, Institute for Interdisciplinary Science, Gilching-Geisenbrunn, Bavaria, Germany; Peter Schultz, Brown University, Providence, RI; Ewen Whitaker, retired from the U.S. Geological Survey, Flagstaff, AZ; Jon Giorgini, William Folkner and Donald Yeomans at NASA's Jet Propulsion Laboratory in Pasadena, CA, and James Young (W7FTT) of Wrightwood, CA. However, I am solely responsible for any errors in the text and illustrations.

Linda Connell, director of NASA'a Aviation Safety Reporting System, helped me locate a report where a pilot describes mistaking Venus for an oncoming jet. Diane Jarvi, a singer-songwriter based in Minneapolis, provided translated lyrics from *Revontulten Leikki* ("Northern Lights at Play"), and Judith Heymann translated the writings of French astronomer J. J. Lalande. Judy Young, director of The Sunwheel Project at the University of Massachusetts, has provided continued encouragement.

Last but not least, I'm grateful to my wife Cecelia for proofreading the text and assisting with the index and to everyone at Springer for their support and patience in creating this third edition.

About the Author

Francis Reddy is a senior science writer for Syneren Technologies Corporation on contract to the Astrophysics Science Division at NASA's Goddard Space Flight Center in Greenbelt, MD, where he writes science-related stories, scripts, press releases and other content. He previously served as a senior editor at *Astronomy* magazine, where his writing was honored by the Solar Physics Division of the American Astronomical Society and *FOLIO*: magazine. He has published several books for skygazers of all ages, including *Halley's Comet!* (AstroMedia, 1985), *The Rand McNally Children's Atlas of the Universe* (Rand McNally & Co., 1990), and two previous editions of *Celestial Delights* (Ten Speed Press, 1992 and 2002). He lives in Bowie, MD, with his wife, Cecelia.

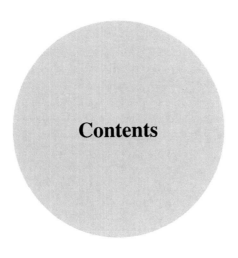

Contents

Chapter 1

The Meaning of the Sky

Humanity has been unraveling the basic workings of the solar system since some unknown skygazer first began tallying the Moon's phases, perhaps tens of thousands of years ago. In just the last two generations, we've gone from being passive observers of the Moon and planets to active participants in the solar system. We've witnessed the first tentative steps of humans on the Moon – an act we won't see repeated for decades. Spacecraft have flown past every planet and numerous comets and asteroids, and probes have landed on the Moon, Venus, Mars, the asteroids Eros and Itokawa, and Saturn's fascinating frozen moon, Titan. Other missions are now en route to the newly reclassified dwarf planets, Pluto and Ceres, and chasing comet Churyumov-Gerasimenko, which in a few years may make history as the location of the first comet landing.

In 2009, as we celebrated 400 years since Galileo first pointed an improved telescope skyward, astronomers had cataloged more than 400 planets orbiting around nearby stars. The current picture – of planets more massive than anything in our solar system in star-hugging orbits – represents a distorted view of what's out there, an artifact of the limitations of today's detection techniques. But astronomers are on the verge of finding large numbers of Earth-sized planets that orbit in their star's prime real estate. This is the so-called habitable zone, where liquid water can exist freely on the surface without instantly boiling or freezing – a necessity for life as we know it. In September 2010, astronomers claimed they'd found the first: Gliese 581g, a planet about three times more massive than Earth that orbits a star 20 light-years away – cosmically speaking, practically in our back yard. As the next generation of ground- and space-based telescopes begins work in the coming decade, astronomers will begin probing Gliese 581g and other "Goldilocks planets" for clues that will tell us whether life on Earth is truly unique.

F. Reddy, *Celestial Delights: The Best Astronomical Events through 2020*, Patrick Moore's Practical Astronomy Series, DOI 10.1007/978-1-4614-0610-5_1, © Springer Science+Business Media, LLC 2012

We now know basic facts about the cosmos that our ancient forebears could never have dreamed of, yet they shared our urge to understand humanity's place in the great scheme of things. We know the basic layout of our solar system and the physical laws that govern the universe, but most of us can't easily recognize how those laws express themselves in the sky above our heads.

This book is about matching our current knowledge of the cosmos to the practice of observing the sky. The basic motions that beguiled and bewildered our ancestors are still on display, waiting for us to enjoy, and we have the advantage of truly knowing what we're seeing.

Skywatching and Skywatchers

The single most striking observation of the sky is one so obvious that we might at first overlook it entirely – the daily motion of the Sun, Moon, planets and stars as they appear to rise in the eastern sky and set in the west. We know that this is caused by Earth's once-a-day spin on its axis, similar to how a rider on a merry-go-round sees the carnival crowd sweeping past with every rotation. Because we can't sense Earth's spin, it was natural to take the sky's apparent motion for a real one and imagine that the sky wheeled around us once each day.

We see the sky as if it were a vast sphere cut in two but a flat plane marked by the horizon (Fig. 1.1). At any moment, half of this celestial sphere is a great dome

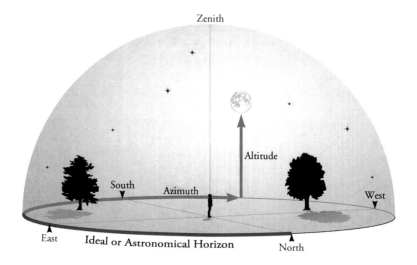

Fig. 1.1 We experience the sky as if it were a vast sphere cut in two by a flat plane represent- ing the horizon. At any one time, half of the sky is above the horizon and half is below it. An object's *altitude* is its angular elevation above an ideal horizon, such as the ocean. The *zenith* is the spot directly overhead, at an altitude of 90°. *Azimuth*, the angular distance to an object measured from north along the horizon, is always given in this book as a compass direction

above our heads, the other half a dome below our feet. As the day progresses, we turn eastward to face different parts of the celestial sphere. Objects below our horizon appear to rise in the east as we spin toward them, and objects above our horizon appear to set in the west as we turn away from them.

Sunrise and sunset define our day. Thanks to its vital powers of heat and light, the Sun was often the greatest deity of many ancient pantheons. But the Sun's energy isn't constant throughout the year, a fact we usually notice as longer days in summer, shorter ones in winter. If we tracked the direction where the Sun rises (or sets, a more practical option for city dwellers) throughout a year, we'd see that its position swings north and south between two extremes (Figs. 1.2 and 1.3).

The northern extreme occurs in summer around Jun. 21 – the summer solstice (literally, "standing Sun") – and the southern extreme occurs in winter around Dec. 21 at the winter solstice. (Unless specified otherwise, all references to seasons and their corresponding months apply to the Northern Hemisphere; they are, of course, reversed in the Southern Hemisphere.) Astronomically, a solstice isn't a full day long, but rather the instant in time when the Sun reaches its extreme north or south position on the celestial sphere. The longest day occurs at the summer solstice – the date when the Sun swings on its highest arc through the sky. The days then grow slowly shorter until the winter solstice, when the Sun traces its lowest arc through the sky. Then, the daylight hours gradually increase again until they reach another maximum at the following summer solstice.

Returning briefly to our previous analogy, the Sun's apparent motion occurs because our merry-go-round isn't level – it tips to one side by many degrees. Earth's axis isn't plumb to its orbit, but tipped in one direction. Picture putting this tipped merry-go-round on a flatbed truck and driving it around a circular racetrack. On one part of the track the merry-go-round tips away from the infield, while on the opposite side it tips toward the infield. From the perspective of a rider on the merry-go-round, a fixed spotlight in the infield will nod up and down in elevation

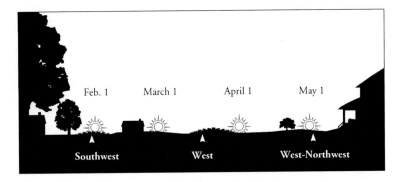

Fig. 1.2 Track the setting Sun for a few weeks and you'll see the direction of sunset move along the horizon. The Sun slides northward from late December to late June, then reverses course for the rest of the year

Fig. 1.3 Solar figure eight. Every year, the Sun executes a figure eight called the analemma, here seen from the Temple of Zeus in Athens, Greece. From Jan. 7 to Dec. 20, 2003, photographer Anthony Ayiomamitis imaged the Sun on 44 separate days with a single frame of film, then made one final exposure for the Temple. The pattern's altitude range results from Earth's tilted axis. Its width shows the effect of Earth's slightly elliptical orbit. Our planet moves faster when closest to the Sun and slower when farther away. This small difference causes the Sun to be early or late relative to a steady clock (Anthony Ayiomamitis)

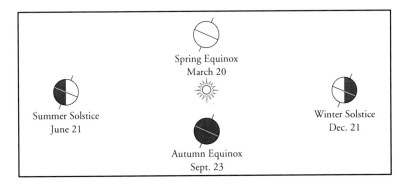

Fig. 1.4 The tilt of Earth's axis is the reason for seasons. On Jun. 21, the Northern Hemisphere angles into the Sun most directly, and this is when the Sun arcs highest in the sky, rising and setting at its northernmost points. Half a year later, on Dec. 21, northern observers see the Sun travel its lowest arc, rising and setting at its southernmost points

with every circuit. Similarly, as Earth orbits the Sun, its axis reaches a maximum tip toward the Sun (summer solstice), a maximum tip away from the Sun (winter solstice), and two points (equinoxes) midway between them where the planet's tip doesn't favor either hemisphere (Fig. 1.4) The solar motion we perceive is the cumulative effect of our living on a spinning, tilted rock in orbit around the Sun.

The directional change to the sunset/sunrise point is easiest to spot in the spring and fall, when the day-to-day shift is greatest and when the Sun passes through points precisely midway between its solstice extremes. These midpoints are called equinoxes (from "equal night") because at these times the nighttime hours nearly equals the daytime hours. Astronomically, the equinoxes mark the instant when the center of the Sun's disk crosses the celestial equator moving north (spring) or south (autumn). The spring or vernal equinox occurs around Mar. 20; the autumnal equinox occurs about Sept. 22. At these times, an observer on the equator would see the Sun directly overhead.

Armed with this fundamental knowledge of the Sun's behavior, ancient Sun-watchers could track the march of the seasons and guide essential activities – such as planting and harvesting – and the festivals and rituals associated with them. This knowledge is embodied in some of the world's most famous ancient constructions, from England's Stonehenge, to the passage tombs of Ireland, to the pyramids of Mexico. By building solar movements into their most important structures, ancient peoples connected the mundane to the cosmic, the seasonal to the celestial, and the all-too-brief human experience to the infinite and eternal.

We find echoes of the Sun's journey in the dates of many holidays. The rules for determining Easter and Passover explicitly tie them to the spring equinox. Christmas falls near the winter solstice, when the Sun's life-giving energy ebbs to its minimum and begins to strengthen again. In the U.S., the solstices and equinoxes are regarded as the official start of each season, but they can just as easily be considered mid-season markers. That interpretation is explicitly acknowledged in cultural traditions like the late-June European holiday of Midsummer, which falls near the summer solstice.

Many observances occur midway between the solstices and equinoxes, on so-called cross-quarter days. February opens with Candlemas and Groundhog Day, derived from the Celtic festival of Imbolc. May brings us both the bonfire festival of Beltane and its descendant May Day, as well as the Bavarian witch-vigil of Walpurgisnacht. The English feast of Lammas and the Celtic festival of Lughnasa occur at the start of August, and the end of October sees the Celtic Samhain festival celebrating summer's end and its descendant Halloween, as well as All Saints' Day immediately following. Day of the Dead, a Mexican celebration held Nov. 1 and 2, descends from a festival Aztecs held in early August on our calendar. It was relocated from near one cross-quarter day to another by Spanish conquerors eager to replace an indigenous tradition with their own, the same process of assimilation that links Celtic festivals and holy days in Western Christianity. Ultimately, what connects these dates is a firmly rooted notion that the Sun's well-being is essential to us all and that humankind – indeed, all life on Earth – has a stake in the Sun's continued annual journey around the sky.

Seen from the tropics, the Sun attains another position of importance that simply isn't possible for those located elsewhere. On certain dates specific to each locale, the Sun passes through the zenith – directly overhead – at noon. Historical and archaeological evidence from Mexico to South America shows that many peoples observed and celebrated one of the Sun's zenith passages.

The Moon, a lesser light, often draws modern attention to the night sky. Early calendars were based on lunar motions, mainly because its cycle is shorter, more obvious, and therefore more accessible than the Sun's. The priest-astronomers of ancient Mesopotamia and Greece determined the start of a new lunar month by watching for the first appearance of the Moon's slender crescent in the evening twilight. In Egypt the month began with the disappearance of the waning crescent Moon from the predawn sky. Systems for marking time by the Moon doubtless go back millennia. Suggestions of some sort of lunar time-reckoning can be found among the 17,500-year-old cave paintings of Lascaux in France, and Paleolithic bone artifacts found throughout Europe bear carved tally marks that may represent even older lunar timecards.

But the Moon's apparent convenience as a short-term timekeeper is complicated by a thorny mathematical problem. Twelve lunar cycles fall short of the seasonal year by about 11 days, so a purely lunar calendar slowly drifts through the seasons: Imagine celebrating New Year's Day on the Fourth of July. Ancient calendar-keepers needed some way to reboot the two cycles if a pure lunar calendar was to keep step with the seasons, and there are three possible fixes, all of them slightly messy. They could ignore the problem and let the seasons drift. They could insert whole days at culturally acceptable times – a solution in the same vein as the extra day in February we add every leap year. Or they could use observations of the Sun and stars to explicitly sync up the cycles every year.

The Lure of the Planets

Long before skywatching became part of a formalized time-keeping system, ancient observers realized that the Sun and Moon weren't the only objects moving through the heavens. Ancient priest-astronomers noticed five peculiar "stars" that seem to wander independently, each moving at different speeds through the constellations. These strange objects adhere to roughly the same sky path as the Sun and Moon and usually drive eastward relative to the stars. But now and then these lights slow down, stop, loop west for a time, stop again, and then resume their eastward journey. Ancient astronomers knew five of these *planetes* ("wanderers," the Greek origin for our word "planets"): Mercury, Venus, Mars, Jupiter and Saturn. Although under ideal conditions Uranus can be seen with the naked eye, it isn't bright enough to call attention to itself (see Table 1.1). Uranus and the much fainter Neptune would remain undiscovered until long after the invention of the telescope.

In their wanderings through the sky, one or more planets may appear to pass close to each other or to the Moon, forming captivating and sometimes striking arrangements. The diagrams in this book illustrate many of these events through 2020 and enable those new to skygazing to be more certain about they're seeing.

The greater the number of planets involved in these groupings, the longer the time between successive displays. For instance, Jupiter and Saturn only appear close about every 20 years – with an excellent example closing out 2020 – and gatherings involving all five visible planets are much less common.

The political hierarchy of ancient societies, which typically gave the ruler the role of chief priest or shaman, was reflected in their view of the celestial realm: What went on in the sky was imbued with meaning for the state, direct communiqués intended for the ruler. Sometimes these were regarded as blessings, but often the more unusual or unexpected the event was, the more sinister its message.

Nowhere was this more true than in early China, where the planets were viewed as the ministers of Shangdi, the Lord on High, and their gatherings were seen as a meeting of celestial administrators called to deliberate major policy changes. As the chief shaman, the emperor's role on Earth was a heavenly appointment, and early in the second millennium B.C. observation of the heavens became his greatest responsibility. It was the emperor's divine obligation to ensure that all human activity conformed as closely as possible to Shangdi's desires. This precept, which became more formalized over time and was first explicitly stated in early writings of the Zhou Dynasty (1046–256 B.C.), is referred to as the "Mandate of Heaven" (*tian ming*). Chinese rulers believed that detailed skywatching could reveal the will of Shangdi, and astronomical and weather events could indicate displeasure with the way earthly affairs were being conducted.

David Pankenier of Lehigh University has shown that tight five-planet conjunctions are closely associated with Chinese dynastic transitions and with the development of the Mandate of Heaven. A meeting of Shangdi's ministers suggested that the current ruler had fallen out of favor, sending a signal throughout the land to

Table 1.1 Understanding visual magnitudes. Astronomers describe the apparent visual brightness of stars, planets and other astronomical objects using a special scale where a difference of five magnitudes represents a change in brightness of 100 times. The system takes some getting used to because the brightest objects have negative magnitudes. The notes below give a general sense of the faintest objects visible in various environments; some magnitudes have been rounded

Mag.	Example object	Visibility notes
−26.7	Sun	Visible to the naked-eye, even from large cities
−20	Brightest meteors	
−12.7	Full Moon	
−6	Crescent Moon	
−4.8	Venus at its brightest	
−4.0	Faintest objects visible in full daylight	
−3.8	Venus at its faintest	
−2.9	Mars and Jupiter at their brightest	
−1.5	Sirius, the brightest star	
−0.5	Saturn at its brightest	
0	Arcturus and Vega	Generally visible to the naked eye from large cities and striking when seen from dark, rural locations
0.9	Antares	
1.8	Dubhe, brightest star of the Big Dipper Mars at its faintest	
2.0	Polaris	Just visible to the naked eye from small cities and suburbs
3.3	Megrez, faintest star of the Big Dipper	
4		Faintest naked-eye stars visible from small cities
5		Binoculars needed from cities and suburbs
5.3	Uranus at its brightest	
6	Milky Way is obvious	Faintest stars visible to the naked eye from rural sites within 100 miles (150 km) of major cities; binocular objects from the suburbs
		Binoculars required in rural areas; telescope needed from suburbs
7.8	Neptune at its brightest	
10		Faintest stars visible in 20 × 80 binoculars from a dark rural site; telescope needed from brighter sites

would-be political challengers. Official records credit close planetary groupings as heralds of the Zhou, Han, and Song dynasties, and astronomical and historical evidence suggest that the Shang and Xia dynasties also began with tight planetary conjunctions. Driven by their heavenly mandate, Chinese rulers created the world's longest continuous record of astronomical observation, one that remains an important source of historical data today.

A Case of Cosmophobia

Planetary conjunctions were harbingers of political change in China, but in medieval Europe they were increasingly seen as signs of coming apocalypse. In the 12th and 13th centuries, as Latin translations of Greek and Arabic works became more widely available, ideas connecting astronomical events to world affairs and widespread religious change gained ground. One of the strongest supporters of these notions was the Franciscan friar Roger Bacon (ca. 1214–1294). Based on theories put forward by a 9th-century Arabic astrologer named Abu Ma'shar, Bacon argued that the world would see only six major religions, and each of these was symbolized by a conjunction between Jupiter and one of the six other celestial bodies (the visible planets, plus the Sun and Moon). Bacon wrote in 1266 that only one of these conjunctions – involving Jupiter and the Moon – had not yet come to pass and that the final religion it symbolized would bring about the world's end as detailed in the biblical Book of Revelation. He suggested that by studying astronomy and ancient prophecies, one could gain an improved understanding of when this change would come about.

At the time, Bacon's proposition struck most scholars as terribly fatalistic, and for decades they successfully pushed back against such approaches. But less than a century later, a series of devastating crises shook Europe – the arrival of bubonic plague in the 1330s, the start of the Hundred Years' War in 1337, and the east-west split of the Church in 1378. As the cultural fabric disintegrated, some turned to the planets for clues about what was coming next.

As the Black Death swept toward Paris, King Philip VI asked university physicians to report on the origin of the terrible pestilence. At the time, epidemics were viewed as having local causes associated with a corruption of one of the four basic elements: air, water, earth and fire. But bubonic plague was too widespread to be understood this way. Doctors and scholars looked for a cause that could corrupt the elements in many places at the same time – and the planets obliged. In the *Paris Consilium*, published in October 1348, the French physicians concluded that a principle cause of the Black Death was the March 1345 conjunction of Jupiter, Saturn and Mars. With a two-and-a-half-year lag between the conjunction and the appearance of the plague in France, the authors fully realized that this explanation was incomplete and suggested other complicating factors, such as a recent earthquake, also were involved.

Jupiter-Saturn conjunctions emerged as the go-to events for medieval astrologers looking for an eschatological trigger. In 1444, Jean de Bruges wrote about the global implications of Jupiter–Saturn conjunctions for the coming century – and he found nothing good to say about any of them. Points for specificity go to the German mathematician Johannes Stöffler, who in 1499 predicted that a great flood would cover the world on February 20, 1524, when all five planets gathered in the constellation Pisces. The event was promoted so heavily as the time approached that one German nobleman even built a three-story-tall ark for insurance. Later in the 16th century, the French astrologer Pierre Turrel hedged his bets by computing the world's end four different ways, yielding four dates between 1537 and 1814.

And ... we're still here. Our latest appointment with doom is in December 2012, when what would normally be considered a page-turn of the Mayan calendar, merged with the approach of a *completely fictional* planet, is supposed to bring about catastrophe. Astronomer David Morrison at NASA's Ames Research Center writes a Web-based question-and-answer column called "Ask an Astrobiologist" and says he's logged more than 2,500 questions about the 2012 issue. Some people, he notes, clearly fear this phony prophecy. In response, Morrison posted more than 200 answers and even a video summary explaining why the prediction will amount to nothing. He has coined the term "cosmophobia" to describe the gnawing fear that the universe is somehow out to get us.

Astronomy and astrology began parting company in the 17th century. Now, hundreds of years later, we know that other bodies in the solar system truly can affect life on Earth – through impacts or the long-distance interactions of gravity – in ways explainable by physical laws determined from observation and mathematics. Ancient observers recorded data that provide important insights in many astronomical investigations today, but their motivations were very different from those of modern scientists and their interpretations are part of another world. In short, we've moved on.

Star Light, Star Bright

We see the travels of the Sun, Moon and planets against the backdrop of the starry sky, so it should not be surprising that some of the world's oldest surviving literature contains references to the same objects and patterns we see there today. Homer's *Iliad*, the epic Greek poem about the Trojan War, can be traced back to at least 740 B.C. In Book 18, Hephaestus fashions and decorates the shield of Achilles with emblems of our winter sky:

> *and there the constellations, all that crown the heavens,*
> *the Pleiades and the Hyades, Orion in all his power too*
> *and the Great Bear that mankind calls the Wagon:*
> *she wheels on her axis always fixed, watching the Hunter,*
> *and she alone is denied a plunge in the Ocean's baths.*

We can still see the inspirations for these adornments: the familiar stars of Orion the Hunter, the bright clusters of the Pleiades and the Hyades in Taurus, and Ursa Major (the Great Bear, whose brightest stars form what we call the Big Dipper). Homer's observation that the Great Bear "wheels on her axis always fixed" and is "denied a plunge in the Ocean's baths," is a clear reference to the fact that these stars circled about the northern sky and never set below the horizon – a fact of middle-northern latitudes as true for us today as it was for the ancient Greeks.

A few generations later the Greek poet Hesiod wrote *Works and Days* – a kind of *Farmer's Almanac* from 700 B.C. *Works and Days* is a poem about peasant life that weaves an astro-agricultural calendar into a framework of mythology and the virtues of labor. The key dates for the calendar were determined by the rising and setting of certain stars – Arcturus, Sirius, the constellation Orion, and the Pleiades cluster in Taurus – just before sunrise or just after sunset.

"When the Pleiades, the daughters of Atlas, are rising," Hesiod wrote, "start the reaping; start the plowing when they are setting. Indeed they lie hidden for forty nights and days, and when the year has gone round they first appear again when the iron is being sharpened." In his time, the Pleiades could be seen rising just before the Sun in what on our calendar would be May, heralding the proper time to begin the harvest of winter wheat. As November opened, these same stars could be seen setting in the west at sunrise – and it was time for farmers to prepare their fields and sow their grain. Five months later, in early April, the cluster became lost in the setting Sun's glare, but in May the Pleiades emerged before dawn, commencing another yearly cycle. This pattern still holds true, but the star dates now fall about a month later and their link to the growing season in Ascra, Greece, where Hesiod lived, is much weaker.

Hesiod relates other aspects of everyday life to his stellar schedule. The first appearance of Arcturus in the evening, together with certain animal signs, announced the coming of spring. When the golden thistle bloomed and Sirius rose just before dawn, it was the height of summer, when goats were fattest, "wine at its best, women at their lewdest, and men at their weakest." The morning rising of Arcturus announced the start of autumn and the best time for picking grapes.

We can watch the motion Hesiod describes by locating a bright star in the south or west as evening twilight fades. A good example is Aldebaran, the brightest star in the winter constellation of Taurus the Bull, gleaming low in the western sky in late spring. If we check on the star after a week or two, we'll see that it has drifted slowly westward from its original position, gradually appearing lower in the sky. Continued observation reveals that it moves ever closer to the Sun, sliding westward until it is eventually lost in the glow of evening twilight. This westward drift of the stars can be interpreted in another way: as an eastward drift of the Sun through the starry background. Figure 1.5 illustrates how the movement of the Earth along its yearly orbit places the Sun closer to our line of sight with Aldebaran. Eventually both Sun and star lie roughly along the same line of sight and the star is lost in the Sun's brilliant glare. When Aldebaran emerges from twilight on the opposite side of the Sun, it will be visible just before sunrise.

As Hesiod's descriptions of the Pleiades bear out, ancient observers had noticed that the Sun makes a complete circuit through the stars once each year. The path traveled by the Sun through the stars became known as the ecliptic; the twelve constellations along the ecliptic gave rise to the famous astrological "signs of the zodiac." In Chap. 7, we'll discuss these and other constellations in more detail.

While there can be no question that the sky played an important role in the great cultures of the past, we take for granted the vestiges of its significance remaining in today's societies. Our week has 7 days and there are seven independently moving objects in the sky – the Sun, Moon, and five planets. Sunday derives from the Roman *dies Solis*, "day of the Sun," Monday comes from *dies Lunae*, "day of the

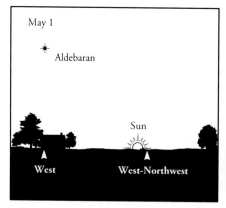

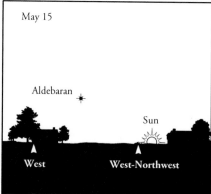

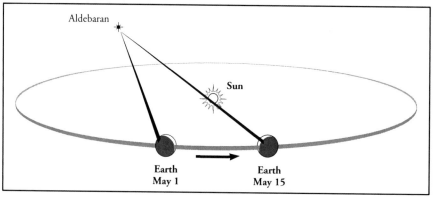

Fig. 1.5 *Top*: Throughout May, the bright star Aldebaran appears to slide ever closer to the setting Sun. We see different constellations at different times of the year because the Sun appears to move eastward relative to the stars. *Bottom*: What's really happening is that the Sun's apparent motion reflects Earth's real motion around it. Once each year, Earth's motion in space brings the Sun close to our line of sight with distant Aldebaran

Moon," and Saturday is Saturn's Day. The remaining celestial connections come to us obscured by the names of figures from the Nordic pantheon: Tuesday is named for Mars, Wednesday for Mercury, Thursday for Jupiter, and Friday for Venus. We'll revisit the mythology associated with each planet in later chapters.

One obvious aspect of celestial objects we haven't yet mentioned is their extraordinary range of brightness. The maps and diagrams throughout this book indicate the brightness of the stars and planets by the size of the black dots representing them – the bigger the dot, the brighter they are. This follows a numerical scale of brightness that owes its essential form to the Greek astronomer Hipparchus. Chief among his contributions (129 B.C.) is a catalog of stars visible from his home with notes on their positions and brightness. He classed the brightest stars as first magnitude, the next brightest were second magnitude, and so on, down to sixth magnitude, into which he grouped the faintest stars visible to the unaided eye. Although the brightness scheme makes sense when described this way, those new to astronomy are often puzzled to learn that a star with a higher magnitude number is actually fainter than one with a lower value. It helps to think of the word "magnitude" as meaning "class" (first-class stars being brighter than second-class stars).

Later, when the brightness scale was given a more precise footing, it became clear that some of the stars Hipparchus had classified as first magnitude were actually considerably brighter than others. Astronomers therefore found it necessary to extend the magnitude scale down to zero and even into negative numbers; Sirius, the brightest star, has a magnitude of -1.5. Applied to objects within the solar system, things get strange indeed, with the Sun topping the scale at -26.7. Table 1.1 contains additional examples.

Reclaim the Night

For much of the world's population, the night sky literally isn't what it used to be. Anyone who has taken an evening flight from a large city airport has seen the lattice of orange and green streetlights and the gleam of decorative lighting on buildings far below – patterns that are easily visible from space (Fig. 1.6). Streetlights are intended to illuminate the ground directly below them, but most of them spill light sideways and upward as well. Worse yet is decorative illumination used to light building exteriors, much of which goes straight into the sky. The skyglow within cities and suburbs is so bright that all but the brightest stars may be impossible to find, and Mars and Saturn may be not be visible most of the time. Even from the North Rim of the Grand Canyon, the dome of scattered light above Las Vegas, 175 miles away, affects the visibility of stars.

The night sky is now so foreign to many city dwellers that its sudden appearance can be cause for alarm. Power outages following the 1994 Northridge, CA, earthquake produced the darkest Los Angeles sky in decades. Residents running outside as the shaking stopped were met with a sky filled with brilliant stars and with the Milky Way – a glow from innumerable stars too faint to see individually – arching across the sky. City emergency services, radio stations and Griffith Observatory logged hundreds of calls from people asking why the stars were so bright and concerned that the unusual "silver cloud" (the Milky Way) had something to do with the quake. When told that this was the normal appearance of the night sky, many simply didn't believe it.

Although astronauts have the best view of both heaven and Earth, when given the opportunity to sightsee, most of them focus on watching the planet below drift past their window. But some, like Sandra Magnus, who flew on Expedition 18 to the International Space Station between Nov. 2008 and March 2009, spent some time taking in the stars. Her evocative description of the experience gives a sense of what most of the rest of us are increasingly missing. "Even though the Earth's

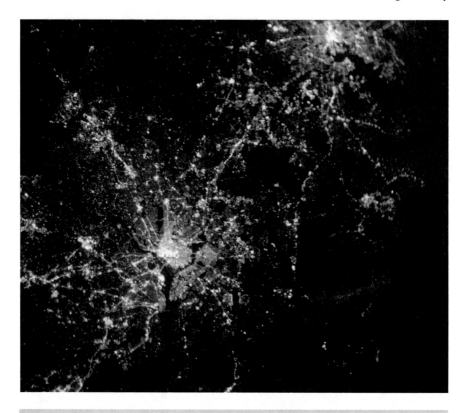

Fig. 1.6 Seen from space, our planet sports its own constellations made by city lights. This picture, taken by astronaut Don Pettit from the International Space Station, shows Washington, D.C., (*left center*) and Baltimore, MD (*upper right*). A clear boundary surrounds the District of Columbia because the density of lighting drops beyond it (NASA)

horizon is dark, light provided by the clouds and the city lights reflecting off of the clouds provides enough illumination to discern the difference between the Earth and space," she wrote. "The night sky is inky black against the night horizon of the Earth. In the night sky, though, sparkle uncounted points of light, some white, some red, some orange, all of different sizes. They are everywhere. The Milky Way is clearly evident. It rises up from behind the Earth like a glowing white path leading off into the distance, inviting you to follow. … You are swimming in a sea of beautiful lights that can only be seen in the dark. … They are so remote and seem so tiny. The vastness of space is truly evident as you watch the Earth turn slowly beneath. It is awe inspiring and overwhelming all at once and oh, so beautiful!"

Light going where it is neither useful nor wanted is called "light pollution." Astronomers were the first to be affected as rising city populations encroached on what were once isolated mountaintop observatories. Light pollution is only beginning to be recognized as a serious environmental problem and a risk to human health and safety. Nocturnal animals depend on darkness to hunt or hide from predators and migratory animals rely on natural light sources easily overwhelmed by artificial ones. Glaring light can scatter within the eye, reducing contrast and creating unsafe conditions, especially for vision-impaired and older drivers. In 2009, for these reasons and others, the American Medical Association unanimously voted to support efforts to control light pollution. In addition, excessive nighttime illumination penetrating the eyelids of sleeping people can affect the very hormones that orchestrate the human sleep-wake cycle.

Plus, it's incredibly wasteful. Light unintentionally thrown directly into the sky wastes that portion of the energy used to make it. The International Dark Sky Association, a group that supports legislative efforts to limit light pollution and approves lighting fixtures, now estimates that the cost of wasted light is at least $2.2 billion a year in the U.S. alone.

Satellites give this story a global perspective. In 2001, Pierantonio Cinzano and Fabio Falchi of the University of Padova in Italy and Chris Elvidge of the U.S. National Oceanic and Atmospheric Administration used satellite images to create the first world atlas of artificial night sky brightness. According to their study, 99% of the U.S. population and two-thirds of the world's population live in areas where the night sky can be considered polluted. For about half of the world's population, including all but a few percent of U.S. residents, light pollution brightens a clear moonless night more than a first-quarter Moon brightens the sky at the darkest observatory sites. For two-thirds of the U.S. and one-fifth of the world, the faint band of the Milky Way is completely lost in the artificial skyglow. More troubling still, the night sky remains so bright for one-sixth of the world's people that their eyes cannot complete the dark adaptation required for full night vision. The authors concluded: "Mankind is proceeding to envelop itself in a luminous fog."

Light pollution is not an inevitable price of progress, and we can bring back the night sky through a combination of technology, legislation and common sense. The sky-brightness atlas revealed Venice as the only city in Italy with a population greater than 250,000 where observers in the city's center could see the Milky Way. The reason? Venice strives for a romantic image and preserves its ambiance by using unobtrusive outdoor lighting. By emphasizing quality rather than quantity, Venice meets public needs without excising the stars.

Solutions that work do exist and cities all over the world are beginning to address the issue. The Czech Republic included a declaration about light-pollution in its 2002 Clean Air Act. The statement obligates citizens to "take measures to prevent the occurrence of light pollution," which it defines as "every form of illumination by artificial light which is dispersed outside the areas it is dedicated to, particularly if directed above the level of the horizon." But this language was never followed up with specific governmental rules. So in 2007, Solvenia became the first country to pass a national law that actually mandated limitations on the brightness and shielding of outdoor lighting; while enforcement remains a problem, it's a good start. Elsewhere, "dark-sky preserves" – regions protected by unusually strong

Table 1.2 The sky on the Web

Astronomy Picture of the Day
apod.nasa.gov

Cities at Night: The View from Space
earthobservatory.nasa.gov/Features/CitiesAtNight

GLOBE at Night
www.globeatnight.org

Google Sky
www.google.com/sky

International Dark Sky Association
www.darksky.org

The Sunwheel Project
www.umass.edu/Sunwheel

The World at Night
www.twanight.org

WorldWide Telescope
www.worldwidetelescope.org

lighting ordinances – have popped up in Canada, Chile, the Czech Republic, Hungary, Italy, Poland, the U.S. and Scotland. Even local businesses are finding that reducing the glare from their lighting actually improves security and makes them better, less intrusive neighbors. For more about the topic of light pollution, see the resources in Table 1.2.

All ancient peoples made some sort of connection with the sky. To them it was a place where powerful beings worked and played – at times helping, at times hindering humanity. To us, it's important as a new frontier, a place where a few of us now live and work but all of our minds can explore. Representing both our collective heritage and our collective destiny, it would be a shame to lose the night sky in a fog of accidental light.

References

Aveni AF (2001) Skywatchers. University of Texas Press, Austin

Aveni AF (1994) Conversing with the planets. Kodanasha America, Inc., New York

Cain P (2010) Slovenia takes dim view of light pollution. BBC. http://www.bbc.co.uk/news/world-europe-11220636. Accessed 10 Sept. 2010

Carmichael AG (2008) Universal and particular: the language of plague, 1348–1500. Medical History Suppl., 27: 17–52

Carmichael E, Sayer C (1991) The skeleton at the feast: The Day of the Dead in Mexico. University of Texas Press, Austin

Chepesiuk R (2009) Missing the dark: health effects of light pollution. Environ. Health Perspectives, 117:A20–A27. doi:10.1289/ehp.117-a20

Cinzano P, Falchi F, Elvidge CD (2001) The first world atlas of the artificial night sky brightness. Mon. Not. Royal Astron. Soc. 328:689–707. See also www.inquinamentoluminoso.it/worldatlas/pages/fig1.htm

Green DWE (2010) The astronomical magnitude scale. http://www.cfa.harvard.edu/icq/MagScale.html. Accessed 10 Sept. 2010

Homer, Fagles R (trans) (1990) The Iliad. Viking, New York

Krupp EC (1983) Echoes of the ancient skies: The astronomy of lost civilizations. Harper and Row, New York

Magnus S (2009) The night pass. http://www.nasa.gov/mission_pages/station/expeditions/expedition18/journal_sandra_magnus_10.html. Accessed 10 Sept. 2010

Morrison D (2009) Doomsday 2012, the planet Nibiru, and cosmophobia. Astron. Beat. http://www.astrosociety.org/2012. Accessed 10 Sept. 2010

Motta M (2009) U.S. physicians join light-pollution fight. Sky & Telesc, http://www.skyandtelescope.com/news/48814012.html. Accessed 10 Sept. 2010

Owen D (2007) The dark side: making war on light pollution. The New Yorker. http://www.newyorker.com/reporting/2007/08/20/070820fa_fact_owen. Accessed 10 Sept. 2010

Pankenier DW (1998) The mandate of heaven. Archaeol, March/April, 26–34

Randi J (1997) Appendix III. Forty-four end-of-the-world prophecies–that failed. In: An encyclopedia of claims, frauds, and hoaxes of the occult and supernatural. St. Martin's Griffin, New York

Sharkey J (2008) Helping the stars take back the night. New York Times. http://www.nytimes.com/2008/08/31/business/31essay.html. Accessed 10 Sept. 2010

Smoller L (1999) Apocalyptic calculators of the later middle ages. In: Proceedings of the third annual conference of the center for millennial studies. http://www.bu.edu/mille/publications/Confpro98/SMOLLER.PDF. Accessed 10 Sept. 2010

Tandy DW, Neale WC (1996) Hesiod's Works and Days: a translation and commentary for the social sciences. Univ. of California Press, Berkeley

Chapter 2

Moon Dance

The Moon is our nearest neighbor in space, Earth's partner in a never-ending dance around the Sun, and for much of each month it's also the most prominent object in the nighttime sky. The Moon whirls around the celestial vault in just over 27 days and each day appears slightly different – a sliver of light at the beginning and end, a brilliant silvery disk in between. Compared to every other naked-eye object in the sky, the Moon is a flagrant show-off. Its regular waxing-waning cycle is so obvious that it formed the basis of early calendars, providing a convenient interval of time between the day and the more subtle annual cycle of the Sun and seasons. While there is only one purely lunar calendar in wide use today, the Moon remains an important cultural symbol. The flags of more than half a dozen nations feature its crescent, and some of the world's most important religious festivals are linked to its phases. As both a signpost of the cosmic and a symbol of inaccessibility, the Moon has been a source of inspiration and mystery throughout the ages.

Today, however, the Moon also reminds us of a remarkable technological milestone. A dozen men have walked on its dusty gray surface, placing scientific instruments, retrieving rock samples, taking photographs – and in one case even hitting a golf ball. A renaissance of lunar exploration is now taking place – with China, Europe, Japan, India and the U.S. currently taking part – but the Moon remains the first and only alien world that humanity has visited in person. So it's fitting that this space-age stepping stone begin everyone's personal exploration of the solar system.

F. Reddy, *Celestial Delights: The Best Astronomical Events through 2020*,
Patrick Moore's Practical Astronomy Series, DOI 10.1007/978-1-4614-0610-5_2,
© Springer Science+Business Media, LLC 2012

Light of the Silvery Moon

We see the Moon because it bounces back to us a small portion of the Sun's light. The darkest parts of the lunar surface reflect just 5% of the sunlight they receive, similar to the rich black asphalt of a newly paved road. Even the brightest regions reflect only about three times this amount, much like the lighter gray of a well-used road that hasn't been paved in years. Our impression of the Moon as a brilliant disk stems in part from its contrast with the dark night sky and in part from the sensitivity of the human eye once it adapts to darkness. The ghostly daytime Moon, faintly visible in the bright blue sky, gives a true picture of our satellite's brightness. Yes, that's right, the Moon doesn't only come out at night. You can easily catch it while the Sun is up during afternoons in the days before full Moon and during mornings in the days after.

The Moon sweeps quickly through the sky, completing one orbit around Earth every 27.32 days on average. This period, called the sidereal month, represents the time it takes the Moon to make one full pass through the sky with respect to the stars. So if one night we find the Moon in the vicinity of a bright star, a sidereal month must pass before the Moon returns to the same location. The Moon's orbital motion carries it from west to east by about 13° each day, a slightly greater angle than the apparent width of a fist held at arm's length. When the Moon lies near a bright star or planet, this eastward motion may be apparent in as little as an hour. Because the Moon moves eastward so quickly, Earth must spin about 50 min longer each day to turn us toward its new position. As a result, the Moon rises an average of 50 min later every day.

That's hardly the first thing we notice about the Moon, of course. What we notice first is its changing appearance, from a slender crescent to a fat silvery orb and back, in a repeating sequence of phases called a "lunation." Lunar phases are merely a trick of lighting – no matter how the Moon looks, half of it is always in sunlight. Only when the Moon is opposite the Sun is its sunlit side completely facing us – a full Moon. The reason we see phases at all is because the Moon's motion around Earth changes how much of its sunlit side we see (Fig. 2.1).

Ever hear the phrase "dark side of the Moon"? Sometimes people use this in reference to the part of the Moon that faces us but isn't yet in sunlight – in other words, where it's still night on the visible side of the Moon. More often, the phrase crops up in reference to the Moon's far side, the lunar landscape we can never see from Earth ("dark" being used in the sense of "unknown"). Of course, both the near and far sides of the Moon see equal amounts of sunlight during a lunar month.

Each lunation begins at new Moon, when the Moon is positioned almost exactly between Earth and the Sun. Because the side facing away from Earth receives all of the sunlight, we can't see the new Moon – unless it passes directly in front of the Sun and produces a solar eclipse. The Moon's motion carries it far enough east of the Sun that attentive observers can find its hair-thin crescent low in the west just after sunset, usually about 25 h after the moment of new Moon (sometimes, less than 15). As the Moon moves eastward away from the Sun, the time between sunset and moonset steadily increases and the Moon shows us a greater portion of its

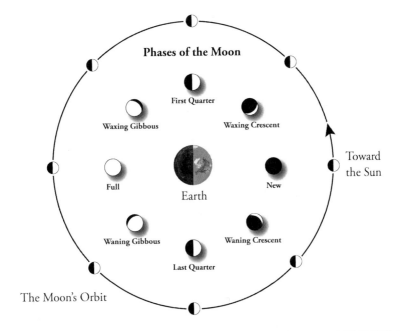

Fig. 2.1 Sunlight (streaming from the right in this picture) always illuminates exactly half of Earth and the Moon. The Moon's motion along its orbit (*outer circle*) allows us to see different fractions of its sunlit hemisphere, which creates the cycle of lunar phases we see (*inner circle*)

daylight side – that is, it's waxing. After about a week, that slim crescent has grown to a half-lit disk. This is the phase called "first quarter" because it occurs one-fourth of the way through the monthly cycle. The first quarter Moon rises around noon, sets near midnight, and is highest in the sky at sunset.

Each quarter phase represents the precise time and date when the Moon reaches an additional 90° east of the Sun, so we start at 0° for new, 90° for first quarter, etc. In between the quarter and full phases, the bulging Moon is said to be "gibbous" (from a Latin word for "humpbacked"). It's a waxing gibbous Moon after first quarter, and a waning gibbous Moon after full.

A week after first quarter, the Moon again aligns with Earth the Sun – but this time Earth is in the middle. Now opposite the Sun in the sky, the full Moon rises in the east as the Sun sets in the west and is visible all night long. After this milestone, the Moon is said to be waning. In a week, the Moon shrinks back to a half-disk – its last quarter phase, when it rises around midnight, stands highest in the sky at dawn, and sets around noon. Throughout the last week of a lunation, the waning crescent appears ever thinner and closes in on the Sun. The cycle begins again when Earth, Moon and Sun realign 29.53 days later, at new Moon. Figure 2.2 shows how the Moon's shape changed almost every day over a single lunation.

For the Yolngu people in northern Australia, lunar phases represented the trials of the fat and lazy Ngalindi, who corresponded to the full Moon. His wives

Fig. 2.2 These images, which were acquired over the course of a single month, illustrate the Moon's changing face. First visible in the western sky at dusk as a slender crescent at dusk (*top row*), the Moon fattens up through first quarter (*second row*) and gibbous phases (*third row*) as its angle from the Sun increases. When opposite the Sun, the Moon is full (*fourth row*) and rises around sunset. The Moon's disk then becomes slimmer (wanes) with each following day, eventually becoming a thin crescent in the predawn sky (*bottom row*). A few days later, it appears again in evening twilight (António Cidadao)

punished his loafing by chopping off pieces of him with their axes, producing the waning Moon. Ngalindi escaped them by climbing a tree to follow the Sun, but his wounds were too severe and he died (new Moon). After 3 days, however, Ngalindi is resurrected. He grows fatter again until, 2 weeks later, his outraged wives go at him again and the lunar cycle begins anew. When this cycle played its first time, humans and animals were immortal beings. But as Ngalindi died, he cursed everyone on Earth with mortality so that only he could return to life.

Now check out Fig. 2.1 again. The Sun lies to the right and Earth's orbital motion carries it down the page. When we gaze at the first quarter Moon at sunset, or at the last quarter Moon at dawn, we're looking roughly along the direction of Earth's orbital motion. At first quarter, the Moon marks the place in space that Earth occupied about three and a half hours earlier; at last quarter, the Moon shows us where Earth will be at roughly the same time in the future.

You may have noticed an apparent discrepancy: The Moon takes 27.32 days to circle Earth, but the phase cycle lasts an average of 29.53 days. Here's the trick: The Moon completes an orbit in one sidereal month, but in that time both Earth and the Moon have moved about one-twelfth of their way around the Sun. Looked at another way, the Sun appears to have moved about 27° east of its location relative to the background stars since the start of each lunation. So before the line-up with Earth, Sun, and the Moon is reset to produce the same lunar phase, the Moon must travel a little bit farther in its orbit – and this takes it a couple of extra days.

This longer period between repeating phases is called the synodic month and it forms the basis of most lunar calendars.

Earthlight Becomes Her

Give the crescent Moon some extra scrutiny the next time you see it and you'll notice something both strange and lovely. You'll actually see more of the Moon than its thin, sunlit crescent – the rest of the disk also glows faintly. Sometimes called the "ashen glow" or, more romantically, "the old Moon in the new Moon's arms," what causes this secondary illumination is sunlight first reflected from our own planet and then bounced back to us by the Moon. This twice-reflected sunlight, called "earthshine," is best seen when the ecliptic cuts the horizon most steeply and the Moon stands high. This happens in evenings during spring for the waxing crescent (Fig. 2.3) and in mornings during autumn for the waning crescent.

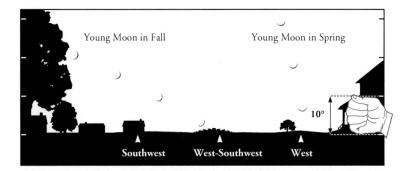

Fig. 2.3 The angle of the Moon's path changes throughout the year. The young Moon can be seen more easily in spring evenings than in the fall; the reverse is true for the waning crescent seen in the predawn sky. Here, for clarity, the Moon is shown greatly enlarged. Horizon scenes like these occur throughout the book. Arrows along the horizon show compass directions; lines along the sides show altitude intervals of 10° – about the width of your fist at arm's length

Box 2.1 Illusive Moon

Most people judge the rising or setting Moon to be about twice as big as the Moon they see higher in the sky on the same night. Yet, regardless of its altitude, the Moon's apparent size is the same – about half a degree across. This paradox, called the "Moon illusion," isn't even limited to the Moon. It also enhances sunrise and sunset and makes constellations near the horizon seem larger and more prominent.

The illusion has been known since antiquity and as yet there is no complete explanation. Like any optical illusion, it's interesting because it underscores a breakdown between reality and the way our brains process information about the outside world. For most people, all it takes to break the illusion is to view the enlarged Moon upside-down, or between pinched fingers or through a cardboard tube. By subtracting the visual cues that somehow create the misperception, the Moon shrinks back to its rightful size.

The first to offer the correct explanation seems to have been none other than Leonardo da Vinci (1452–1519), who understood the parallel between a full Moon illuminating an earthly night and what a "full Earth" must do for the Moon. Writing in his *Codex Leicester* around 1510, a full century before the telescope, da Vinci's lunar knowledge shows some incredible hits as well as a couple of misses. He understood that the phases of Earth and Moon complement each other: At new Moon, a lunar astronaut would see Earth at full phase. So when we gaze at the waxing crescent Moon and blue-gray light fills the lunar night, that light is coming from a waning crescent Earth. Da Vinci also suggested that clouds from terrestrial storms could brighten or dim earthshine, and such changes can be detected in modern measurements. Leonardo then states his incorrect belief that the Moon possesses an atmosphere and oceans and that its watery surface is what makes it such a good reflector.

Modern interest in earthshine is that it provides a way to see our planet in a new light. The amount of sunlight Earth reflects is an important quantity in climate studies. It is usually measured by satellite observations, which are expensive, challenging to calibrate, and often cover only a small fraction of the planet at any given time. Earth's ever-changing weather affects daily cloud cover, and seasonal changes affect, for example, the amount of snow on the ground. Pulling all of these intertwined factors apart can be difficult.

By contrast, earthshine blurs the complex details of the entire sunlit side of our planet into a single, globally averaged measurement that can be made from robotic observatories on the ground. For more than a decade now, Philip Goode at Big Bear Solar Observatory in California has led Project Earthshine to do just this. The project now includes a second robotic observatory in the Canary Islands, and Goode eventually hopes to grow it into a worldwide network. What they've found so far is that

Earth's reflectance, also called its albedo, is much more complex and variable light than traditionally thought. The instruments can even detect the increase in earthshine as the Sun rises over Asia and the massive land mass rotates into daylight.

And earthshine may help us understand planets astronomers are just beginning to locate. By splitting earthshine into its component wavelengths, scientists can view a spectrum of our own planet as if it were seen from afar. Because earthshine smears out details, it essentially reduces Earth to a single point of light. When we find earth-size planets around other stars – and astronomers are getting closer each day – even the most powerful telescopes will see them only as points of light. Earthshine lets astronomers use our planet as a stand-in for investigating what we can learn from these new worlds once we find them.

Enric Pallé at the Institute of Astrophysics on the Canary Islands, who leads research with the Project Earthshine station on Tenerife, has been exploring just what kind of information an alien astronomer with capabilities similar to ours could determine about Earth – and what future astronomers might glean about newfound planets. He has shown that the changing amounts of cloud cover over land and sea can be detected, which suggests dynamic weather, and because time-averaged cloud patterns are closely tied to fixed surface features, such as continents and ocean currents, the planet's rotation period can be determined, too. Earthshine also reveals the glint of sunlight from oceans and other water bodies during crescent phases. Earthshine fades out toward the ultraviolet, which is caused by the absorption of sunlight from ozone, an atmospheric gas consisting of three oxygen atoms. This ozone-caused dimming has a significant implication.

Left to itself, oxygen rapidly combines with other molecules, so its presence in Earth's atmosphere means that some surface process must maintain the gas. That process, of course, is plant life. Other gases, like carbon dioxide, water vapor, and methane, are similarly kept out of chemical equilibrium by life on Earth, but their spectral signatures occur at near-infrared wavelengths. Because the Moon emits these wavelengths and Earth's atmosphere strongly absorbs them, earthshine data can't uncover these gases – but a future space telescope could. Detection of all of these gases in a planet's atmosphere would be a strong indication that the planet hosts life as we know it.

Calendar Moon

Humans have been using the Moon as a timekeeper for millennia. The first hints of lunar records can be found among Upper Paleolithic carvings and cave paintings in Europe and Asia. Many stone and bone artifacts from this time are carved with circular or serpentine designs, although no one knows their meanings. Guesses range from the practical – such as tallies of kills in a given hunt – to the purely artistic. In the 1970s, Alexander Marshack of the Peabody Museum of Archaeology and Ethnology found intriguing evidence that some of these designs had a practical side. He proposed that they represented the oldest human records of the passage of time and involved lunar phases. For example, a 30,000-year-old bone plate from

---

Box 2.2 Moon cheats

If you know the lunar phase for a date of interest, such as a birthday, anniversary or meteor shower, you can estimate past or future phases pretty closely just by knowing a simple rule, no apps needed. For each succeeding year, the lunar phase on a given calendar date slips backward by about half a week. So with every 2-year advance, the Moon shows the *preceding* lunar phase on the same date. For example, a full Moon occurs on June 4, 2012; a first quarter Moon on Jun. 5, 2014; and a new Moon on Jun. 4, 2016. Keep backing through the Moon's phases this way and in 8 years you come full circle – to a full Moon on Jun. 5, 2020.

Although the usefulness of this trick is minimal today, with the possible exception of bar bets, such patterns were immensely helpful to ancient calendar makers. For lunar phases, the gold standard is the famous 19-year Metonic cycle, where after 235 lunations the Moon returns to the same phase on the same calendar date. The two periods agree to within about 2 h, but the extra days we add during leap years can advance the calendar date by a day or two. Full Moons occur on Jun. 4 in 1993 and 2012 and on Jun. 5 in 2031 and 2050.

Dordogne, France, seems to show a 2-month-long record of lunar phases, with each turn in a winding design representing a new or full Moon. Researcher Michael Rappenglüeck in Bavaria, Germany, finds similar meaning in sequences of dots appearing in the Paleolithic paintings that adorn walls at the Lascaux and La Tête du Lion caves in France. These markings bear a resemblance to lunar tallies that have been documented among various indigenous peoples, including Native Americans.

Civilization first arose in the valleys fertilized by major rivers – the Indus, the Nile, and the Tigris and Euphrates. The land between these last two rivers, Mesopotamia, saw the rise of the earliest Sumerian cities sometime in the fourth millennium B.C., and by the second millennium a calendar based on lunar phases had taken shape. Each month held either 29 or 30 days, alternating irregularly – the basis of the month in today's Western calendar. Days began at sunset and religious authorities declared a new month to have begun when they first saw the young crescent Moon low in the evening twilight. If weather interfered and the length of the month exceeded 30 days without a sighting of the young Moon, rules called for a new month to be declared anyway. By the fifth century B.C., the calendar lost its reliance on observation and the new month's start was instead determined by computation alone.

While the Moon served as a convenient short-term timepiece, it's problematic for longer periods. Twelve complete lunations typically take 354 days – 11 days short of a seasonal year. This means that each succeeding calendar year would begin 11 days earlier, progressively drifting back through the seasons. This is exactly the case with the Islamic calendar, the only true lunar calendar still in wide use. The beginning of each calendar year slides through the seasons for about 33 years, after which things are back where they started.

The Western calendar, established by Pope Gregory XIII in 1582, abandons the Moon altogether. We ensure that the calendar stays in sync with the seasons by adding 1 day every few years according to some relatively straightforward rules. We add (or intercalate) a 366th day (Feb. 29) to every year that is exactly divisible by four except in century years, but if a century year is evenly divisible by 400 then it too becomes a leap year.

Calendars are complicated because the relevant cycles – the day, the lunar month, the seasonal year – do not mesh perfectly together and even change over time themselves. Most cultures found that the seasonal drift of a lunar calendar was too difficult to live with because the seasonal, or solar, year controlled agricultural activities. Early astronomers spent considerable effort devising methods of intercalation to pad out a lunar calendar for a better fit with the average solar year. A calendar that merges features of a lunar calendar with the solar cycle is called lunisolar.

In Egypt, no fewer than three calendars were in use by the fourth century B.C. The survival of Egypt depended on the annual flood of the Nile, which fertilized the land. Egyptian priests noticed early on that the first appearance of the star Sopdet – the celestial guise of the goddess Isis, known to us as Sirius, the brightest star in the sky – in the east just before dawn coincided with the flood season. By 3500 B.C. they had devised a lunisolar calendar that kept in step with the seasons by intercalating a month whenever Sopdet's appearance occurred on certain days. This calendar was used in timing religious events related to agricultural and seasonal activities.

But the inconvenience of a year with either 12 or 13 months was not lost on the Egyptians, and by 2800 B.C. a civil calendar appeared for governmental and administrative purposes. The months were fixed at 30 days each, resulting in a year 360 days long: 5 days and a fraction short of the seasonal year. The 5 extra days, tacked on at year's end, were considered very unlucky. The Egyptians made no attempt to keep this calendar in step with either the seasons or the Moon – the whole point was to strip out natural cycles that could not be made to agree. Each year contained exactly the same number of days. Administrators and businessmen alike could perform calculations over several years without having to determine when intercalations had occurred. For the same reason, the Egyptian civil calendar also found a home with astronomers – including Nicolaus Copernicus – even into the 1500s.

When it became apparent that their original lunar calendar was drifting with respect to the civil calendar, Egyptians devised a second lunar calendar to be used for scheduling festivities tied to the Moon. Whenever the first day of this new lunar calendar fell before the first day of the civil calendar, religious authorities inserted a month to bring the two back into agreement.

Corrections to the earliest lunar calendars were probably first made simply by watching the harvest or other natural indicators and then adding whatever length of time seemed necessary. Over time, and with continued observation and record keeping, other cycles were recognized that could facilitate the process. One very useful cycle, credited to Meton and Euctemon of Athens in 432 B.C., was already in use in Mesopotamia and was also known in India and China. Meton and Euctemon

Box 2.3 Do full Moons make us loony?

People working in police departments, hospitals and other emergency services routinely credit the full Moon with busier, crazier shifts. Some statistical studies seem to show suggestive lunar links, while others show no effect at all.

Astronomers and statisticians in various countries have looked for a lunar modulation in events ranging from dog bites and emergency room admissions to violent crimes. Usually, there's no significant correlation with full Moons, and when there is, it isn't especially strong. When a signal stubbornly refuses to separate from background noise no matter how much data are used, the odds favor there being no signal at all. In 1996, James Rotton of Florida International University and Ivan Kelly of the University of Saskatchewan combined data from 37 published and unpublished studies and put all of this data through a statistical technique called meta-analysis – in essence, a study of studies. They showed that lunar phases could account for no more than 0.03% of the rise and fall of crimes, crisis calls, psychiatric admissions and other activities supposedly influenced by the Moon.

The researchers concluded that some studies used inappropriate analytical techniques and failed to properly account for other kinds of cycles, like the recurrence of weekends. These problems can create the appearance of relationships between behavior and lunar phase where none exists. Which seems more likely: A mysterious influence from the full Moon? Or poor decisions on a Saturday night?

Studies show that most people making the association between human behavior and a full Moon are usually completely unaware of the Moon's actual phase. And when we do notice a full or nearly full Moon after an eventful day, it just validates a connection we've already heard. It's human nature to seek patterns, especially in emotionally demanding situations. We also tend to remember when these patterns hold true and to forget when they misfire. So no, full Moons don't bring out our loony side – but definitely watch out for weekends.

recognized that 235 synodic months occurred in almost exactly 19 seasonal years – the error is less than 2 h per cycle, but the leap years in our calendar can advance the date by a couple of days. This means that lunar phases recur on essentially the same date every 19 years. Earth's annual motion around the Sun gives rise to the seasons, the monthly relationship between the Sun and Moon causes phases – and through the Metonic cycle, they can be locked together.

Not everyone based their month on lunar phases, though. In India, an early calendar gave more importance to the Moon's motion through the stars (its sidereal period) and accordingly had months of 27 or 28 days. The Inca civilization of South America also worked the sidereal month into its calendar. The Maya of Mesoamerica perfected a unique calendar system inspired by astronomical cycles but, like the Egyptian civil year, never explicitly referenced to them, thus avoiding clumsy intercalation schemes.

Vestiges of the Moon's importance as a timekeeper survive in the timing of religious festivals. Rosh Hashanah, the first day of the Jewish calendar, falls on the new Moon after the September equinox, although calendar rules may cause it to be postponed for a couple of days. The Christian festival of Easter falls on the first Sunday following the first full Moon that falls on or after the vernal equinox. Church rules fix the equinox date as March 21, so Easter can occur as early as Mar. 22 and as late as Apr. 25. In truth, neither the Jewish calendar nor the Christian calculation of Easter actually refers to the real Moon, employing instead an abstraction – a "mean Moon" that behaves with much greater consistency. The Quran, however, does specifically instruct Islamic religious authorities to see the first visible crescent after new Moon before beginning and ending Ramadan, the holy month of daytime fasting.

Princess of Tides

Nearly everywhere the Moon was correctly linked to a primary natural rhythm – the ebb and flow of the ocean tides. Among the Maori of New Zealand the Moon's responsibility for the tides became a part of its name: Rona-whakamau-tai, Rona the Tide Controller. Along the southeast coast of what is now Alaska, the Tlingit, Haida and Tsimshian tribes imagined the Moon as an old woman who governed the tides. They tell the story of how Raven and Mink tricked the old woman into making the sea flood and ebb twice each day, thereby providing people with a seafood buffet at every low tide.

Yet even into the 17th century the link between the Moon and the tides was not fully accepted. Galileo Galilei (1564–1642), whose observations revolutionized knowledge about the Moon, sharply criticized his contemporary Johannes Kepler for suggesting that the Moon has an influence over the sea. Galileo insisted that tides were proof that Earth moved. Just as a vase holding water sloshes when the vase is moved, he argued, so the tides resulted from Earth's combined rotation and motion around the Sun.

The issue was finally settled in 1687 by a man born the year Galileo died. That's when English physicist Isaac Newton (1642–1727) published what is considered to be one of the greatest works in the history of science. In his *Mathematical Principles of Natural Philosophy* – better known as the *Principia*, from a short form of its Latin title – Newton describes the theory of universal gravitation that allowed him to "deduce the motions of the planets, the comets, the Moon, and the sea." Just as the gravitational attraction of the Sun prevents the planets from flying off into space, so Earth's gravity keeps the Moon in its orbit. But gravity is a mutual attraction. The Moon also pulls back on Earth – and its strength changes over distance, which means that the force of the Moon's pull varies slightly across Earth's diameter. It's this differential force that matters.

The Moon's pull creates a large bulge of matter on the side of Earth directly beneath it and another bulge on Earth's opposite side. Think of these tidal bulges as the crests of enormous waves that rise only a couple of feet in open ocean but extend halfway across Earth's circumference. Each day, Earth spins beneath these giant waves. In addition, the waves progress slowly across Earth's surface as the Moon proceeds in its monthly orbit. So a given location on Earth passes under each tidal bulge once every 24 h and 50 min (because the Moon rises 50 min later each day). Since there are two bulges, we see a high tide every 12 h and 25 min. This is greatly simplified picture: The positions of continents, friction, the irregular shapes of ocean basins, and the fact that the Moon's orbit is not in the same plane as Earth's equator all influence the timing and water range of high and low tides. The greatest water ranges occur where tidal periods closely match the natural frequency of the underwater terrain. Earth's top three tidal ranges: the Bay of Fundy in Nova Scotia and Ungava Bay in Quebec, both in Canada, and the Severn Estuary between Wales and England in the United Kingdom. In these places, the difference between the water level at high and low tide can reach or exceed 49 ft (14.9 m).

The Sun also raises tidal bulges – although these are less than half the height of the Moon's – and the two sets of tidal bulges interact. When Earth, Sun, and Moon are aligned during full and new Moon, the solar and lunar tides combine, resulting in a higher-than-normal high tide (called a spring tide). When solar and lunar tides work against each other at the Moon's quarter phases, a lower-than-normal high tide (neap tide) results. These extremes may change the water level by up to 20% above or below normal tidal limits.

The Sun and Moon also produce tides in the atmosphere and, more surprising, in the solid body of Earth. The rocks rise and fall less than about a foot under tidal influences; this oscillation goes unnoticed because it occurs over enormous horizontal distances. For more than a century, researchers have looked for a connection between tides and the timing of earthquakes. Since about 2000, geologists have shown clear correlations in areas near volcanoes or below hydrothermal vents in the ocean floor. In 2004, a study of 2,000 earthquakes along subduction zones – places where one giant plate of Earth's crust rides over another, such as along the coasts of Alaska, Japan, New Zealand and western South America – found a strong tidal correlation for shallow earthquakes (less than 25 miles or 40 km deep). Imagine a stressed earthquake fault near a coast, flexing under the changing weight of water above it. In this way, the Moon's influence can provide the final destabilizing nudge that pushes stressed rock too far, beginning the fracture that starts a quake.

Earth has a similar effect on the Moon – one that's a bit more obvious. Seismometers placed on the lunar surface by Apollo astronauts transmitted information on the depth, energy, and frequency of moonquakes until they were turned off in 1977. This gear showed that most of the seismic energy released on the Moon comes from a few modest, random quakes in the upper part of the lunar crust. However, there are also more numerous but much weaker moonquakes that occur in about 100 discrete locations at greater depth, more than 500 miles (800 km) down. These locations experience a moonquake every 27 days, or about the time it takes for the Moon to circle Earth. Each orbit, tides deform the Moon – flexing its surface, changing its shape, and increasing the internal stress that triggers moonquakes.

Lunar tides also have a long-term effect on Earth: They slow down its rotation and gradually increase the length of the day. Modern calculations and historical astronomical observations indicate that Earth has been slowing down by less than 2 s every 100,000 years. Paleontologists confirm that Earth's day was indeed shorter in the past. Some of the best evidence comes from thin, stacked beds of sandstone, siltstone and mudstone – collectively known as rhythmites – that display periodic thickness variations. These changes are linked to the different rates at which sediment was deposited, a process modulated by the tides. George Williams at the University of Adelaide in South Australia, who has analyzed the tidal patterns of rhythmites dating to about 620 million years ago, found that way back then the day was just 22 h long.

As you might expect, Earth's tides have had a much more significant impact on the Moon. In fact, it's the reason the Moon always keeps the same side facing Earth. Whatever the Moon's rotation may once have been, tides have locked it to its orbital period. The Moon is spinning on its axis, but it's doing so at exactly the same rate that it orbits us. Astronomers call this phenomenon synchronous rotation or tidal locking. We can actually see a little more than half of the lunar surface from Earth. That's because, over time, the tilt of the Moon's orbit and slight variations in its motion turn it a bit in the east–west direction and a bit in the north–south direction. These changes, called librations, allow us to see about 57% of the lunar surface.

Tens of billions of years from now, the steady brake of lunar tides will slow Earth's spin to the point that it, too, will permanently keep one side facing the Moon. (Pluto and its big moon Charon enjoy exactly this relationship right now.) Long before this happens, though, physical changes on the Sun will remodel the inner solar system: The boiling off of Earth's oceans will be one of the lesser effects. Today, no one knows how that change will alter the decelerating dance of Earth and Moon.

There is another consequence of Earth–Moon interactions that we can measure directly: The Moon is slipping farther away. Earlier, we noted that one of the tidal bulges raised by the Moon sits directly beneath it, but this isn't quite true. As Earth spins beneath the tidal bulge, friction with the ocean floor tends to drive the bulge forward, ahead of the Moon. This offset means that the mass of the bulge now pulls on the Moon, which then accordingly speeds up in its orbit by a small amount. Per the dictates of celestial mechanics, the accelerated Moon must then increase its distance from Earth.

How do we know this is really happening? Back in the Moon-landing heyday, from 1969 to 1972, special reflectors were placed there by Apollo astronauts and carried by two unmanned Soviet rovers. The reflectors were designed to bounce back laser light beamed from Earth – and that's exactly what scientists have been doing for 40 years. Laser pulses allow scientists to determine the distance to the reflectors with astonishing precision, equivalent to knowing the distance between Los Angeles and New York to within half the thickness of a U.S. dollar bill. These measurements tell us that the Moon's average distance is currently retreating from Earth by 1.5 in. (3.8 cm) every year. Naturally, this means the Moon has been closer in the past – something we'll explore a bit later.

Apart from its connection with tides, the Moon has been worshipped and personified in many different ways, although it was rarely a chief deity. Among the Fon people of Benin, the Sun and Moon form a powerful pair of twins that express the dualistic aspects of nature. Mawu holds the female principle – Earth, Moon, coolness, fertility – and Liza holds the male principle of sky, Sun, warmth and power. Together they created the universe and their balanced union suffuses it with order.

The Moon's role in Egyptian mythology is more diffuse. In Thebes, it was known as Khonsu, a god with healing powers, and in Hermopolis it was worshipped as Thoth, the god of wisdom, inventor of writing and language, and the one who recorded the verdict as the dead were tried before Osiris in the afterlife. As god of the dead, vegetation and fertility, Osiris is himself associated with the Moon in the late period of Egyptian history.

Osiris and his sister-wife Isis aided humanity by sharing knowledge of agriculture and the arts. Osiris' brother, Set, was so jealous over their successes that he murdered Osiris, suffocating him in a box and then dumping the box into the Nile. Informed of the betrayal, Isis quickly tracked down the body and arranged to have it returned to her, but she took precautions to keep it hidden. Nevertheless, Set one day came upon the body of his brother. He cut it into 14 pieces and scattered the remains along the Nile. Isis then searched for the pieces and recovered all but one. She embalmed Osiris, creating the first mummy and devising rites that would give him everlasting life in the underworld. According to scholars, the 14 pieces of Osiris represent the waning Moon of the last half of the lunar cycle – a full silvery disk that progressively shrinks before disappearing in the Sun's fire. But the Moon, like Osiris, is resurrected. Because the Nile retained one piece of Osiris, his gift of agriculture endured in the river floods that fertilize the desert each year. This event was always heralded by the star Sopdet (Sirius), which was the starry guise of his loyal wife Isis.

To be fair to the Egyptians, it should be noted that this story dates from the period of heavy Greek and Roman influence. The Egyptians were eminently practical, and although the foundations of their religion were inspired by the sky, they showed little interest in actual observation beyond the needs of reckoning time and orienting structures. There is nothing in purely Egyptian records that reveals any interest in detailed observation of lunar and solar eclipses or planetary movement.

The Babylonian Moon god was Sin, controller of the night and of the calendar; his earlier Sumerian counterpart was Nanna. He was a wise and generous god who marked off time and whose advice was sought by other deities each month. Sin's bright light kept evil forces at bay during the night, but one day his daughter Ishtar conspired with these forces to snuff out Sin's light. Marduk, the chief Babylonian deity, fought them off and restored Sin to his former glory.

In ancient Greece, Artemis ruled the Moon, the hunt, and all nature, brought fair weather for travelers, and protected young girls. She later became identified with Selene, who fell in love with the sleeping shepherd Endymion and stopped in her passage through the sky to visit him nightly. She bore him 50 daughters, representing the 50 synodic months between the Olympic Games. The Roman Moon goddess Diana, identified with Artemis, became the patroness of witches in medieval Europe.

In the surviving books of the Maya, the Moon was Ix Chel, Lady Rainbow, sometimes depicted as an old, toothless woman sitting in the glyph representing the Moon sign and holding a rabbit. She was the patron of childbirth, healing and the art of weaving, and there was a popular shrine to her on the island of Cozumel off the Yucatán coast.

We often picture the Moon's dark markings as forming the face we identify as the Man in the Moon; according to one story, he was caught working on a Sunday

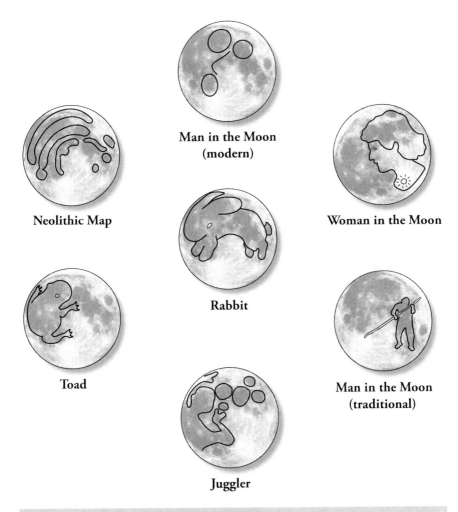

Fig. 2.4 The full Moon's most prominent features form patterns recognized throughout history and across cultures. One common pattern is a face, the best known being the modern Man in the Moon. The Rabbit is associated with the Moon in China and the Americas. The traditional Man in the Moon was known in Europe in Shakespeare's time. The Neolithic view is based on the appearance of pecked bands on the "Moon stone" in the Irish passage-tomb of Knowth, which may be the earliest representation of lunar markings (After Stooke). The Juggler, first recognized in 2008, proves we can still find something new on the Moon's face; the figure can easily accommodate interpretation as a basketball, volleyball or soccer player (After Murray). The Man in the Moon is most easily seen around moonrise; the Woman, when the Moon is high in the sky; and the Rabbit is easiest to find around moonset (Illustration by the author; photo by T. Rector/NOAO/AURA)

and cast there as punishment. But in Shakespeare's day, people referred to another pattern – the outline of a figure, sometimes viewed as bearing a long stick – as the Man in the Moon. Other commonly recognized patterns include the profile of a girl or old woman and a rabbit (Fig. 2.4). The rabbit was a lunar symbol in Europe,

China and the ancient Americas typically associated with fertility, although different cultures didn't always construct the rabbit from the same lunar markings.

In an Aztec myth, the gods gathered in Teotihuacán to determine which of them would be sacrificed in flame to illuminate the world. The proud, rich god Tecuciztecatl (to-koo-sis-te-CA-tl) immediately stepped up to the task, and after some deliberation the gods selected the poor, humble, pimply god Nanahuatl (nana-WA-tl) to be his companion. When the time came for the sacrifice and each was to throw himself into a bonfire, it was Tecuciztecatl's right as a wealthy god to go first. Four times he tried, but the heat of the flames always forced him back. The gods then invited sickly Nanahuatl to jump into the fire. When he managed it on the first try, the shamed Tecuciztecatl quickly followed. Nanahuatl was transformed into the Sun, Tecuciztecatl the Moon. But the remaining gods were troubled because both Sun and Moon shone with equal light. How would day and night be different? So the gods dimmed the Moon's light by striking it with a rabbit, a mark it still bears today and a final indignity to a haughty god.

In a curious mix of the ancient and the modern, a Chinese myth about the lunar rabbit was mentioned to Apollo 11 crewmembers as they approached the Moon. "Among the large headlines concerning Apollo this morning," said Mission Control in Houston, "there's one asking that you watch for a lovely girl with a big rabbit." According to the legend, the beautiful Chang-o has been living on the Moon for 4,000 years, banished there when she stole the pill of immortality from her husband. "You might also look for her companion, a large Chinese rabbit, who is easy to spot since he is always standing on his hind feet in the shade of a cinnamon tree. The name of the rabbit is not reported," Houston concluded.

Buzz Aldrin, the lunar module pilot, replied: "OK. We'll keep a close eye out for the bunny girl."

Box 2.4 Full Moon full of love

For us, the Moon is a symbol of romance, but for some species it's a literal love light. Every year, a full Moon triggers coral reefs across broad patches of the ocean to dissolve in a synchronized release of trillions of eggs and sperm. This mass spawning strategy is a kind of shotgun approach designed to overwhelm the impact of predators. Divers who have seen it liken it to swimming in a snow globe.

In the Caribbean, corals spawn over a couple of nights after the July/August full Moon. Along Australia's Great Barrier Reef, the corals "get busy" after the full Moon of October/November. Marine biologists have known about this lunar timing for decades, but until 2007 they had no idea how these massive annual spawnings – Earth's biggest sex events – could be coordinated across hundreds of different coral species on thousands of individual reefs.

Corals are actually colonies of small multicellular organisms called polyps that together create a communal skeleton. In 2007, Oren Levy at the University

(continued)

Box 2.4 (continued)

of Queensland, Australia, and his coworkers discovered that coral polyps contained two light-sensitive proteins called cryptochromes. These proteins enable polyps to detect both sunlight and moonlight. Changes in sunlight track the seasons, and changes in moonlight set the spawning time within the correct season.

Cryptochromes are also found in plants, insects and mammals, where they may play roles in migration and in regulating "biological clocks," the daily activity cycles known as circadian rhythms. Because they're so widespread, these proteins must be incredibly ancient. It's clear that the ability to sense cyclic light changes already must have been part of the blueprint of life some 450 million years ago, when corals first emerged.

Science Shoots the Moon

By the fifth century B.C., Greek astronomy and mathematics had become sophisticated enough for some imaginative scientists to explore the Moon's true nature. Anaxagoras (ca. 500–428 B.C.) correctly explained the causes of eclipses and was banished for considering that the Moon was at least partly made of the same stuff as Earth. Both he and the influential philosopher Aristotle (384–322 B.C.) said that the Moon was a solid sphere illuminated by sunlight and thereby explained its phases. Using simple geometry, Aristarchus (ca. 320–250 B.C.) showed that the Moon was much closer than the Sun, and Hipparchus (ca. 190–125 B.C.) later determined the Moon's distance to within 10% of the correct value. Nevertheless, the Moon came to be regarded as a sphere of crystalline smoothness and perfection, a member of the flawless realm beyond our mundane one.

Who was the first to draw the Moon as we actually see it, rather than in some highly stylized form? Leonardo da Vinci generally gets credit as being the first to attempt to portray lunar markings realistically, although a small naturalistic Moon appears in several earlier works by Jan van Eyk (1390–1440). Da Vinci's chalk drawings, only one of which is known today, were made between 1505 and 1514 – surprisingly recent for a depiction of an object so easily visible and culturally prominent. Apart from these efforts, all known previous depictions of the Moon appear to be symbolic rather than representational — except, perhaps, one.

In 1994, Philip Stooke of the University of Western Ontario described a contender for the oldest lunar map, one that would extend the history of lunar mapmaking back to the same period as the earliest maps of Earth and sky. It appears on a stone that archaeologists have dubbed Orthostat 47 inside the passage tomb of Knowth, located in the Boyne Valley of Ireland. Constructed some 5,000 years ago – predating both Stonehenge in England and the great pyramid of Giza – the

Irish passage tombs are places where Neolithic peoples placed the cremated remains of their dead so that their spirits could be reborn in the afterlife.

The "Moon stone" of Knowth occupies the center recess at the end of the tomb's eastern passage. Pecked onto the stone are three long arcs, a short arc and several circular patches, markings that Stooke believes represent the face of the setting Moon. These decorations match the relative positions of lunar surface features well enough that Stooke feels justified in saying that the Moon stone is a primitive lunar map. Other patterns pecked onto the same stone seem to illustrate how lunar features appear to change orientation through the night, rotating as the Moon rises and sets.

Similar pecked designs showing nested arcs appear all over Knowth. Kerbstone 52, for instance, bears a complex design that some suggest functioned as a lunar calendar. Investigations into astronomical alignments at Knowth have shown that at certain times moonlight could stream down the entire eastern passage. It's difficult to resist the romantic vision of the pale Moon occasionally illuminating a map of itself.

The view of the Moon as a perfect crystalline sphere changed forever in 1609, when Galileo Galilei became one of the first to examine the Moon through his improved version of a new Dutch invention – the spyglass, later named the telescope. Although only as powerful as a good pair of modern binoculars, Galileo's spyglass revealed features resembling jagged peaks, valleys, and pock-marked plains. Although he was not the first to sketch the Moon as seen through a telescope–the Englishman Thomas Harriot beat him by 4 months – Galileo was the first to publish. His drawings and observations appeared in March 1610 in a short treatise called *The Starry Messenger*. Galileo studied surface features through the Moon's changing phases and recognized their three-dimensional nature by the shadows they cast; the changing Sun angle revealed mountain chains and bowl-shaped depressions. Galileo noted that the brighter regions of the Moon were "uneven, rough and full of cavities and prominences, being not unlike the face of the Earth, relieved by chains of mountains and deep valleys" and the dark areas, which he referred to as "large lunar spots" are "not seen to be broken in the above manner … rather they are even and uniform and brighter patches crop up only here or there."

An existing tradition held that the features on the Moon's face reflected an image of Earth and the dark regions were our seas, although Galileo carefully avoids any decisive statement about water on the Moon. Nevertheless, each of his lunar spots came to be called a mare (Latin for "sea," pronounced MAH-ray; plural MAH-ria). Naturally enough, the rest of the Moon's surface became known as terrae (lands). Today the terrae are more commonly known as the lunar highlands, but the term mare remains in the names of the dusky regions.

Galileo's observations were an important step in proving the Copernican view that Earth was an ordinary member of the solar system. Isaac Newton, similarly inspired by the Moon, extended this line of reasoning to conclude that the force that brought apples to the ground was the same force that kept the Moon circling Earth and the planets revolving around the Sun. As astronomers discovered satellites orbiting other planets they began to appreciate the uniqueness of our own Moon. With the exception of distant Pluto and its oversized moon, Charon, our Moon is the largest in the solar system relative to the planet it orbits. The Moon is just over one-fourth Earth's diameter, with a total surface area slightly larger than Africa, and holds less than one-eightieth of Earth's mass. The force of gravity on the lunar surface is only one-sixth what we experience on Earth. This is too weak for the Moon to hold onto gases escaping from its interior; as a result, the Moon has no atmosphere to speak of. Table 2.1 lists some basic facts about Earth and the Moon (Fig. 2.5).

After Galileo announced his telescopic discoveries, cartographers set to work mapping lunar features. The first real map of the entire visible hemisphere was published in 1645 by Michiel Van Langren (1600–1675), an astronomer in the court of Philip IV of Spain. Van Langren identified several hundred features, most of them craters, and he demonstrated some political acumen by naming them for assorted kings and noblemen. Van Langren also honored philosophers, explorers, saints and scientists, including himself. He applied his moniker to both a prominent crater and the mare near it, but today only the crater name, Langrenus, still stands.

Table 2.1 Facts about Earth and Moon

Earth	
Diameter	7,926.4 miles
	12,756.3 km
Surface temperature	
Maximum	136° F (58° C)
Minimum	−126° F (−88° C)
Atmospheric surface pressure	1.013 bar
Atmospheric composition	78% nitrogen (N_2)
	21% oxygen (O_2)
	1% water vapor (H_2O)
Sidereal rotation period (length of day)	23.934 h
Obliquity (tilt of spin axis with respect to the orbital plane)	23.45°
Sidereal year (true period of revolution around the Sun)	365.26 days
Tropical year (time between successive vernal equinoxes; i.e., the seasonal year)	365.24 days
Average distance from Sun	92.96 million miles
Light takes 8.3 min to travel this far.	149.60 million km
	1 Astronomical Unit (AU)
Moon	
Diameter	2,159.2 miles
	3,468.8 km
	27% of Earth's
Average surface temperature	
Day	253° F (123° C)
Night	−387° F (−233° C)
Sidereal period	27.32 days
Synodic period (time between repeating phases, e.g. new Moon to new Moon)	29.53 days
Mean distance from Earth	238,855 miles
Light takes 1.3 s to travel this far.	384,400 km
Orbit inclined to Earth's	5.15°

Fig. 2.5 Earth and the Moon compared (NASA and NOAO photos; montage by the author)

Modern lunar nomenclature originated with Giovanni Riccioli (1598–1671), who employed it on the lunar map in his two-volume *New Almagest*, published in 1651. He gave the maria Latin names reflecting qualities or characteristics (Sea of Tranquility, Sea of Serenity, Sea of Cold, Sea of Nectar) and early ideas that connected the Moon with weather changes (Sea of Rains, Sea of Vapors, Sea of Clouds, Ocean of Storms). Craters were named for scholars and scientists (Copernicus, Tycho, Kepler) and mountain ranges were named after famous terrestrial ranges, such as the Alps or Apennines. For the next 300 years, as ever larger telescopes revealed finer details on the battered lunar surface and mapmakers strained to keep up with the flood of new features and names, Riccioli's nomenclature proved flexible enough to endure (Fig. 2.6).

The known lunar territory began to double in 1959 with the success of the Soviet Luna 3 mission, which returned 29 grainy images revealing 70% of the lunar farside, the side that never turns toward Earth. The Space Age had begun and the Moon was center stage. The Soviet Luna and Zond programs provided the first close-up views and the first robotic landing, but the numerous American Ranger, Surveyor, and Lunar Orbiter probes supplied unprecedented detail that paved the way for the highly successful piloted lunar missions of the Apollo program.

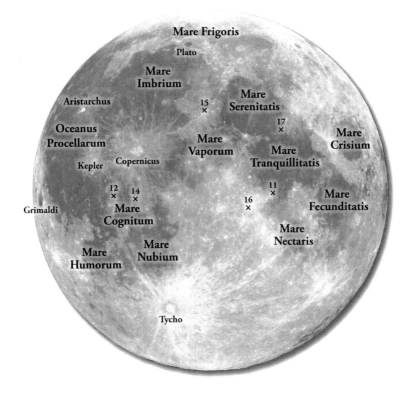

Fig. 2.6 Lunar features easily visible with the naked-eye or binoculars are identified in this image. The numbers indicate the mission names for each of the Apollo landings and an × marks the site. The feature known as Tycho is among the Moon's newest major craters, formed by an asteroid crash 108 million years ago. Debris thrown out during impacts creates bright rays that extend far from the craters. These are especially noticeable around Tycho, Copernicus, and Kepler (Photo by T. Rector/NOAO/AURA)

In 1990, 18 years after the last Apollo landing (Figs. 2.7–2.8) and 14 years after the last Luna mission, Japan became the third nation to successfully fly by, orbit, and impact the Moon with a modest probe named Hiten.

In 2003 the European Space Agency reached for the Moon with its successful SMART-1 probe, and Japan returned to the Moon in 2007 with Kaguya, a space-craft with a high-definition color camera that returned remarkable video from orbit. Both China – with Chang'e 1 and 2, launched in 2007 and 2010, respectively – and India, with Chandrayaan 1 in 2008, have joined the "Moon club" by successfully placing spacecraft in lunar orbit.

But what's the reason for all this activity? Despite having direct samples, for more than a decade lunar scientists have had the creeping suspicion that they didn't know the Moon as well as they thought.

Fig. 2.7 About an hour after Apollo 11 astronaut Neil Armstrong first stepped onto the Moon on July 20, 1969, his companion Buzz Aldrin photographed his own boot print in the dust. The surface consists of rock pulverized into powder by eons of meteorite impacts (NASA)

Water from the Moon

The recent discovery of water on the lunar surface forces an astonishing about-face in our understanding of the Moon. During the Apollo era, chemical studies of returned lunar rock samples quickly established that the Moon was bone dry. Indeed, this result became one of the foundations for any explanation of the Moon's origin. Forty years down the road, with much more sensitive instrumentation, water seems to be everywhere.

The idea that water could exist on the Moon was first suggested in 1961. Deep craters located near the Moon's north and south poles can provide permanent protection for ice because the Sun can never shine into them. The same objects that made lunar craters, asteroids and comets, also brought water to the Moon. Sunlight quickly breaks down water into hydrogen, hydroxyl (OH) and oxygen, and these byproducts then disperse into space. But if water molecules drifted into a permanently shadowed crater before this happened, they would find a "cold trap," a place where they could mix with lunar soil and remain stable for billions of years.

Fig. 2.8 *Top*: Apollo 12 astronaut Charles "Pete" Conrad Jr. visits Surveyor 3, which landed on the Moon 2 years before him. In November 1969, Conrad and Alan Bean set down their lunar module, visible in the background, just 600 feet from the probe. NASA *Bottom*: The Apollo 12 landing site as it looks today, first imaged in 2009 by NASA's Lunar Reconnaissance Orbiter. Subtle dark lines radiating from the lander show where the astronauts' activities disturbed surface material (NASA/GSFC/Arizona State Univ)

The next watery surprise came 30 years later, with radar studies of Mercury in 1991. Radio waves reflected from craters near both of the planet's poles were unusually bright and strongly polarized. The returned signals were similar, in fact, to radio waves reflected by the Greenland ice sheet, the large frozen moons of Jupiter and the polar caps of Mars. Scientists concluded that the permanently shadowed polar craters on Mercury likely contain ice deposits.

In 1994, a U.S. mission lunar mission named Clementine detected a similar radar reflection at a location near the Moon's south pole, providing the first indication of lunar water there. Another spacecraft, Lunar Prospector, provided even greater detail about the mineral composition of the lunar surface. Shortly after its arrival at the Moon in 1998, scientists examining data from an instrument designed to detect subsurface hydrogen announced they had found some – presumably in the form of frozen water – at both lunar poles. At the end of the mission, the Lunar Prospector science team attempted to excavate some of this ice by slamming the spacecraft into one of the permanently shadowed craters. The impact was expected to launch a plume of lunar material high enough to reach sunlight, where it could be analyzed by telescopes on Earth, but in the end no debris plume was detected.

In 2009, scientists announced that NASA's Moon Mineralogy Mapper, an instrument flown on the Indian lunar mission Chandrayaan 1, had discovered spectral features associated with water and hydroxyl on the Moon's surface. However, this water is different from the water ice scientists were seeking. The Sun produces an outflow of particles called the solar wind, and most of those particles are protons, the nuclei of hydrogen atoms. Protons striking lunar minerals containing oxygen can initiate a reaction that produces H_2O. This so-called adsorbed water adheres to the surfaces of soil minerals as films just a molecule thick. Perhaps some of the water molecules produced, say, on the sunlit rim of a polar crater, can go through cycles of evaporation and condensation that allow the molecules to migrate into the crater's eternal night, where they would be stable in the cold trap's deep freeze. The findings show that the Moon not only retains water but also, in a limited way, even manufactures it.

Just weeks later, a NASA mission provided evidence of what was really in one of those permanently shadowed craters by using a more ambitious version of the final Lunar Prospector experiment. The Lunar CRater Observation and Sensing Satellite (LCROSS) mission crashed a rocket body and a probe into Cabeus crater near the Moon's south pole. The rocket crashed first, which allowed a spacecraft following it to observe the impact and fly through the plume before hitting the Moon itself. According to Anthony Colaprete, the chief scientist for LCROSS at NASA's Ames Research Center in California, the mission appears to have thrown up "a range of fine-grained particulates of near pure crystalline water-ice." The debris cloud, which represented the upper meter or two of lunar soil at the impact site, contained more than 5% water by mass, or about twice as wet as the Sahara Desert. Not exactly a water world, but definitely not bone dry, either.

In March 2010, scientists using data from a NASA radar instrument flown on Chandrayaan 1 announced that some 40 craters ranging in size from 1 to 9 miles

(2–15 km) wide contained nearly pure deposits of ice. Scientists estimate that these deposits may be several yards deep and, if melted, would provide about 160 billion gallons (600 billion liters) of water, or about the volume in Australia's Sydney Harbor. "How that would come about I haven't a clue," said Paul Spudis, a scientist at the Lunar and Planetary Institute in Houston and the principal investigator of the Chandrayaan 1 radar experiment.

This is exciting for a number of reasons. First, it tells us something new about the Moon and provides potential clues to its natural history, which is closely bound to Earth's. Second, it means that proposals to return humans to the Moon and even establish residence there may not be so far-fetched. With local water resources, astronauts could manufacture oxygen to breathe and hydrogen for rocket fuel – necessities that would otherwise require expensive resupply flights from Earth. According to Spudis, the Chandrayaan 1 ice deposits could provide enough hydrogen fuel to launch one space shuttle per day – for 2,200 years.

In an ironic twist, those initial Clementine and Lunar Prospector results that first piqued interest in lunar water turn out to have been misinterpreted. No radar could detect as little ice as LCROSS found, and while subsurface hydrogen is there, it isn't all locked up in the form of water. LCROSS showed that the debris plume contained as much of the element in the form of molecular hydrogen (H_2) as was bound up in water. Just what processes formed the molecular hydrogen? Well, that's another lunar mystery.

Pummeled Moon

Although we clearly still have much to learn, we've come to understand much about the Moon's complex history and the violent origins of the solar system. From 1969 to 1972, six American Apollo missions brought a dozen men to the Moon's surface and 843 lb (382 kg) of lunar soil and rocks to the Earth's; between 1970 and 1977 Soviet Luna probes provided another 11 oz (300 g) of surface material from three additional locations on the nearside, the side that faces Earth. Since then, geologists have identified 71 lunar rocks that, incredibly, came to us from the Moon all by themselves. Blasted off the lunar surface as part of the debris thrown out by an impact, the rocks cruised through space and finally fell to Earth where they eventually could be collected by scientists. To date, geologists have found at least 108 lb (49 kg) of lunar meteorites and, by comparing them with known Moon rocks, confirmed them as having originated there.

The main feature of the Moon's geologic history is bombardment by rocks large and small. Seen through a telescope or from orbiting spacecraft, the lunar highlands break up into an endless series of overlapping meteorite craters. These regions took the brunt of a bombardment that formed the solar system's moons and planets through powerful collisions. The top few kilometers of the Moon's surface have been repeatedly mixed and pulverized. The lunar highlands, the Moon's most ancient terrain, contain rocks that solidified within a global ocean of molten rock some 4.5 billion years ago. Even after the young Moon developed a thin crust, molten rock seethed below its surface. At this time the solar system was filled with debris, leftovers from building the planets. Numerous impacts blasted the crust, eroding and mixing the uppermost layers, destroying the oldest lava flows, and at the same time throwing blocks of debris from the deep crust out onto the surface.

The maria represent more recent terrain. They cover about 16% of the Moon's surface, mostly on the nearside. As the Moon's surface cooled and its crust solidified, several massive impacts formed huge multi-ringed basins around 3.9 billion years ago. These giant impacts occurred even as the amount of debris striking the Moon began to slacken. Dense basalt magma from the lunar interior oozed its way through the fractured crust and flooded onto the basin floors. The maria are the frozen remains of ancient dark lava flows that erased older craters. This explains why the maria have far fewer craters than the neighboring highlands. The rate of impacts leveled off around 3 billion years ago, and by then only small amounts of magma could find its way to the surface.

The impacts of the last billion years, such as the one that formed Copernicus (57 miles or 91 km across), have gouged the lunar crust and excavated subsurface layers, throwing out blankets of brighter ejecta that highlight the Moon's most recent wounds. Blocks of debris thrown hundreds of miles struck as a multitude of smaller impacts, revealing brighter soil and creating the linear rays that radiate away from many craters. Apart from these last few large impacts and many smaller ones, the Moon's face has changed little since.

The prominent crater Tycho (54 miles or 87 km across) and its bright ray system may well represent the most significant change to the Moon's face since dinosaurs walked the Earth. Scientists estimate the impact occurred 108 million years ago, which means the crash may have been witnessed by the likes of Iguanodon, Utahraptor and other dinosaurs of the early Cretaceous Period. Fittingly, the Tycho impact foreshadowed how the Cretaceous would end, when another large space rock struck what is now the Yucatán Peninsula in Mexico. Geologists now generally agree that this impact produced an environmental catastrophe that forever ended the reign of the dinosaurs – and gave mammals their chance for dominion.

Improbable Moon

Accounting for the origin of the Moon was a problem for planetary scientists. It was hoped that analysis of actual lunar rocks would eventually favor some theories and disprove others, but in fact no pre-Apollo theory of the Moon's birth adequately fits our knowledge of its orbital characteristics and chemical and geological makeup. As chemist Harold Urey once summed up the situation, "All explanations for the origin of the Moon are improbable."

Any proposal for the origin of the Moon must address several facts: the strange inclination of the lunar orbit, the Moon's low density compared to Earth's, geochemical information gleaned from lunar samples, and the high angular momentum contained in the Earth's spin and the Moon's orbit. Angular momentum is a property of rotating systems that includes both the speed of rotation and revolution and the masses of the bodies involved. Earth and Moon together possess more angular momentum per unit mass than Mercury, Venus or Mars.

One early model pictured the Moon as Earth's "sister," a body that had formed alongside our planet that has orbited it ever since. This view requires a Moon that is a miniature version of Earth, made of the same ratio of rock and metal. We now know that lunar rocks contain unexpectedly small amounts of elements such as cobalt and nickel that normally accompany iron-containing minerals. Most lunar samples also lack materials like water that vaporize at low temperatures (so-called volatiles). The "sister" model cannot account for the different densities of Earth and the Moon or the high angular momentum of the Earth-Moon system. Conclusion: Fail.

We know that the Moon is moving away from us as it slows down Earth's spin. Extrapolating backward in time means that in the distant past the Moon must have been closer – and the Earth spinning faster. If we could somehow reel the Moon into Earth today, our planet would "spin up" until a day became just 5 h long. George Darwin, a son of English biologist Charles Darwin, suggested in 1879 that when the Earth was molten it spun so fast that it threw off a chunk, which became the Moon. In this scenario, the Moon is Earth's "daughter." Although the geochemical aspects are on the right track, the details don't match what we know from lunar samples, such as the Moon's depletion of volatiles, like water. Plus, even an Earth with a 5-h day isn't spinning fast enough to do what Darwin proposed. Fail.

Perhaps the Moon formed elsewhere in the solar system and was captured into orbit as it wandered by, thus becoming Earth's "spouse." We know that lunar rocks formed without the presence of water and volatile elements; we also know that the Moon is otherwise similar to rocks in Earth's mantle. The large satellites of other planets, on the other hand, are composed of mixtures of ice and rock. The likelihood of a capture event is very low to begin with, but for the Earth to have snared a unique body like the Moon seems very improbable indeed. It also fails to explain why Earth and Moon – two bodies that, according to this scenario, were created in different parts of the solar system – share what compositional similarities they do. Another fail.

In the mid-1970s two groups of scientists independently offered a new scenario. They argued that the Moon was made from material blasted from Earth by a giant off-center impact shortly after it formed – that it was, in essence, a "chip off the old block." A decade later, computer simulations provided an experimental laboratory where scientists could watch the event unfold and see the effects of slight differences in important parameters, such as the mass, speed and composition of the impactor. Although heavily criticized when first proposed – in part because there was a prevailing view that planetary formation was a gentler process – the impact model accounts for diverse aspects of the Moon's chemistry and dynamics better than any other explanation so far, and is now widely accepted. While many details remain unclear, in broad outline this is a win (possibly even an epic one).

Box 2.5 What's a blue Moon

This popular term refers to the second full Moon occurring in a given calendar month, a meaning first introduced in 1946. Blue Moons are an astronomical curiosity, a fun but inevitable result of the interplay between lunar and calendar cycles.

Blue Moons occur about every 2.5 years and naturally tend to fall in the longest months. Two blue Moons can occur in a year when February goes without a full Moon. The next time this happens is in 2018, when, conveniently, the year's first blue Moon occurs with a total lunar eclipse.

Table 2.2 below lists upcoming blue Moon dates in Eastern and Universal Time. Because correcting for other times zones can push one of the full Moon dates into either the preceding or the following month, it's important to remember that a blue Moon for you may not be one for somebody else.

Table 2.2 Blue Moons through 2020

2012	Aug. 31	9:58 A.M. EDT (13:58 UT)
2015	July 31	6:43 A.M. EDT (10:43 UT)
2018	Jan. 31	8:27 A.M. EST (13:27 UT)
	March 31	8:37 A.M. EDT (12:37 UT)
2020	Oct. 31	10:49 A.M. EDT (14:49 UT)

In its current form, the scenario begins 4.45 billion years ago, about 50 million years after the start of Earth's formation and very near its completion. Another body, one about 10% the mass of Earth and about the size of Mars, had formed in the same part of the solar system and was on a collision course. The two worlds struck one another with a blinding flash; jets of vaporized rock shot into space. The collision completely melted the impactor, remelted Earth's surface and blasted away its atmosphere. Our planet shuddered as the two bodies merged and Earth absorbed the impactor's momentum, ramping up the planet's original modest rotation to a brisk 5-h day. Much of the ejected material, most of which came from the mantle of the impactor, either fell back to the glowing, wounded Earth or escaped into the solar system. But some of it – less than 2% of Earth's mass – went into orbit, settling as a disk of debris in our planet's equatorial plane. In a few decades, at the disk's outer edge some 20,000 miles (32,000 km) away, about half of this mass coalesced to form the Moon. At this distance, the Moon that first rose over the ancient Earth looked nearly 12 times larger than it does today.

The Moon's gravity created waves in what was left of the debris disk, which only remained for another few hundred years, and it was this interaction that rapidly cranked up the Moon's orbital tilt. Gradually, the rest of the debris disk fell back to Earth, leaving the Moon alone as impacts and volcanism remodeled its surface into the pale pummeled disk we see today.

One of the scientists who proposed the giant impact scenario, William Ward of Harvard University, showed in 1974 that the Moon's presence helps stabilize our planet. The angle Earth's axis makes with respect to the ecliptic is called its obliquity; it's this tilt that gives us seasons. The angle varies slightly over a period of 41,000 years and this cycle, working together with shorter-term variations in the shape of the Earth's orbit and a slow wobble of the planet's spin axis known as precession, directly affects how much sunlight a given locale receives each season. These oscillations are the major players in the climate swings of Earth's past. They are ultimately powered by the gravitational influence of the Sun, the Moon and the planets – especially Jupiter. But with the Moon's strong, steady pull acting like a giant flywheel, Earth is able to resist the most dramatic obliquity swings. In 1993, Jacques Laskar and colleagues at the Bureau of Longitudes in Paris found that without the Moon, Earth's obliquity changes were, over tens of millions of years, non-linear, unpredictable, and dramatic. They concluded:

> It can thus be claimed that the Moon is a climate regulator for the Earth. If it were not present, or if it were much smaller ... the obliquity values of the Earth would be chaotic with very large variations, reaching more than 50° in a few million years and even, in the long term, more than 85°. This would probably have drastically changed the climate on the Earth.

Picture Earth tipped on its side, seemingly rolling along its orbit at each solstice much as the planet Uranus does. Consider that the North and South Poles would take turns being baked each year when the Sun passed overhead, as it now does in the tropics. And at mid-winter an entire hemisphere, as opposed to just the Arctic or Antarctic, would never see the Sun. Drastic climate change, indeed.

Long an inspiration to poets and lovers, the Moon remains a symbol of mystery and an eerie beacon of otherworldliness. It's the only celestial body humans have walked upon, and it's the only one whose geography we can explore with binoculars. The story this battered landscape tells is one of violence we can hardly imagine. Gazing at the Moon's pockmarked surface, it's sobering to realize that the intense cosmic bombardment that hammered it into shape also must have taken place here on Earth.

The Moon steadies Earth's spin to make our home climate more stable and temperate. It rules the seas by powering the tides and, perhaps, important ocean circulation patterns, and it provides a lighting cue for the spawning of some ocean species. The next time you see a slender crescent Moon, its nightside bathed in the bluish light of our own planet, take a moment to reflect on a partnership that began shortly after the solar system formed – and long before there was life on Earth to appreciate the view. Table 2.3 gives a taste of the selenolgical surprises that await you on the Web, but don't neglect the subtler rewards of viewing our planet's biggest satellite with your own eyes.

Table 2.3 The Moon on the Web

Active lunar missions

Lunar Reconnaissance Orbiter (LRO)
lunar.gsfc.nasa.gov
www.Moonzoo.org

Past lunar missions

Apollo Lunar Surface Journal
history.nasa.gov/alsj

Chandrayaan 1 (India)
www.chandrayaan-i.com

Kaguya (Japan)
www.kaguya.jaxa.jp/en
www.youtube.com/jaxachannel

Lunar CRater Observation and Sensing Satellite (LCROSS)
lcross.arc.nasa.gov

Lunar Orbiter Image Recovery Project
www.Moonviews.com

SMART-1 (ESA)
www.esa.int/esaMI/SMART-1

Lunar images and maps

Lunar Photo of the Day
lpod.wikispaces.com

The Consolidated Lunar Atlas
www.lpi.usra.edu/resources/cla

Google Moon
www.google.com/Moon

World's largest ground-based digital lunar mosaic
www.lunarworldrecord.com

References

Austin AL, Ortiz de Montellano BR, Ortiz de Montellano, T. (trans) (1996) The rabbit on the face of the Moon. Univ. of Utah Press, Salt Lake City

Canup RM, Asphaug E (2001) Origin of the Moon in a giant impact near the Earth's formation. Nat. 412:708–712. See also http://www.swri.org/press/impact.htm

Cochran ES, Vidale JE, Tanaka S (2004) Earth tides can trigger shallow thrust fault earthquakes. Sci. 306:1164–1166. doi: 10.1126/science.1103961

Drake S (1957) Discoveries and opinions of Galileo. Doubleday and Co., Inc. Garden City, New York

French BM (1977) The Moon book: exploring the mysteries of the lunar world. Penguin Books, New York

Goldstein DB et al (1999) Impacting Lunar Prospector in a cold trap to detect water ice. Geophys. Res. Lett. 26:1653–1656

Goode PR et al (2001) Earthshine observations of the Earth's reflectance. Geophys. Res. Lett. 28: 1671–1674

Kaufman L, Kaufman JH (2000) Explaining the Moon illusion. Proc. Natl. Acad. Sci. 97: 500–505

Kelly IW, Rotton J, Culver R (1996) The Moon was full and nothing happened. In: Nickell J, Karr B, Genoni T (eds.) The Outer Edge. Prometheus, Amherst, New York

Kerr R (2010) How Wet the Moon? Just Damp Enough to Be Interesting. Sci. 330: 434

Korotev RL (2010) List of lunar meteorites. http://meteorites.wustl.edu/lunar/moon_meteorites.htm. Accessed 10 Sept. 2010

Laskar J, Joutel F, Robutel P (1993) Stabilization of the Earth's obliquity by the Moon. Nat. 361:615–617

Lemonick MD (2010) Wetter Than the Sahara! (On the Moon, That's Good). Time. Fri, Oct. 22. http://www.time.com/time/health/article/0,8599,2027057,00.htm. Accessed 22 Oct. 2010

Levy O, Appelbaum L, Leggat W et al (2007) Light-responsive cryptochromes from a simple multicellular animal, the coral Acropora millepora. Sci. 318:467–470

Murray PE (2008) Basketball player on the Moon. In: Dembowski WM (ed.) The Lunar Observer, Aug. 2008. http://www.zone-vx.com/TLO200808.pdf. Accessed 10 Sept. 2010

NASA (1969) Apollo 11 technical air-to-voice transcription. Manned Spacecraft Center: Houston. http://www.hq.nasa.gov/alsj/a11/a11transcript_tec.pdf. Accessed 10 Sept. 2010

Norris R (2009) Searching for the astronomy of aboriginal Australians. In: Jonas, Vaiskunas, (ed.) Astronomy & cosmology in folk traditions and cultural heritage. Klaipėda Univ Press, Klaipėda, Lithuania

Pallé E (2010) Earthshine observations of an inhabited planet. In: Montmerle T, Ehrenreich D, Lagrange AM (eds.) Physics and Astrophysics of Planetary Systems, EAS Pub Series, 41: 505–516. doi: 10.1051/eas/1041041

Palmer JD (1996) Time, tide and the living clocks of marine organisms. Am. Sci. 84:570–578

Richards EG (1998) Mapping time: The calendar and its history. Oxford Univ. Press, New York

Rincon P (2010) Ice deposits found at Moon's pole. http://news.bbc.co.uk/2/hi/science/nature/8544635.stm. Accessed 10 Sept. 2010

Sinnott RW, Olson DW, Fienberg RT (1999) What's a blue Moon? Sky Telesc. May 1999, 37–38

Spudis P (1996) The once and future Moon. Smithsonian Inst. Press, Washington

Stooke PJ (1994) Neolithic lunar maps at Knowth and Baltinglass, Ireland. J. Hist. Astron. 15: 39–55. See also: The lunar maps of Knowth, publish.uwo.ca/~pjstooke/knowth.htm

Vasavada AR, Paige DA, Wood SE (1999) Near-surface temperatures on Mercury and the Moon and the stability of polar ice deposits. Icarus 141:179–193

Ward WR, Canup RM (2000) Origin of the Moon's orbital inclination from resonant disk interactions. Nat. 403:741–743. See also: New theory links Moon's current orbit to its formation via a giant impact, www.swri.org/9what/releases/incline.htm

Whitaker EA (1999) Mapping and Naming the Moon. Cambridge Univ. Press, New York

Williams GE (2000) Geological constraints on the Precambrian history of Earth's rotation and the Moon's orbit. Rev. of Geophys. 38: 37–59

Chapter 3

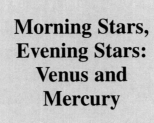

Morning Stars, Evening Stars: Venus and Mercury

Track the Moon long enough and it will guide you to new celestial sights. Some morning just before new Moon, or some evening just after it, the slender crescent will share the twilight with a star that outshines all others. If the timing is just right, a second speck of light – fainter, a bit redder – will hover in the unsteady air close to the horizon. With that observation your grasp of the universe swells over 200-fold, reaching far beyond the orbit of the Moon to encompass the planets Venus and Mercury. These two "stars" never appear more than a few hours ahead of or behind the Sun. They emerge into the morning or evening twilight only to reverse course and slip back into the Sun's glow. The terms "morning star" and "evening star" usually refer to any planet bright enough to stand out in the twilight glow of dusk or dawn, a condition that all of the classical planets satisfy sooner or later. But only two planets spend all or most of their visibility periods in twilight, so these terms best apply to Mercury and Venus, the planets whose progress through the sky are most closely tied to the Sun's.

Actually, Venus so outshines every other planet that it deserves a title all its own. It's the third brightest object in the sky and can be seen in full daylight under ideal observing conditions. At its best, Venus swings far enough from the Sun to leave the twilight behind: For weeks on end at its best appearances, sky watchers will see it as a brilliant jewel high in a darkened sky. The orbit of Venus also brings it within about 25 million miles (40 million km) of Earth, or only about 100 times the distance to the Moon – the closest approach of any planet. All this adds up to one simple fact: Anyone can find Venus armed with only the most basic information about when and where to look for it.

Mercury does everything Venus does, only faster, more faintly, and at lower altitudes. Even at its best, Mercury hugs the horizon, shining weakly through the haze and twilight. With Mercury, timing is everything. It pops above the predawn

F. Reddy, *Celestial Delights: The Best Astronomical Events through 2020*,
Patrick Moore's Practical Astronomy Series, DOI 10.1007/978-1-4614-0610-5_3,
© Springer Science+Business Media, LLC 2012

horizon for a week or so, retreats back toward the Sun, and then shows itself briefly in the west after sunset for another week before diving back into the Sun's glow and repeating the cycle. To make matters worse, the time of year also critically affects the planet's visibility. Taken together, Mercury presents an observational challenge unmatched by any of the planets known since antiquity.

When the two planets shine together in morning or evening twilight, Venus acts as a celestial beacon lighting the way to the more elusive Mercury. The crescent Moon and other planets join them too, creating beautiful arrangements that make wonderful astronomical photo opportunities.

Queen of Heaven

As the brightest, closest, and most visible of all the planets, Venus has fascinated stargazers and scientists alike for centuries. Venus was probably the first planet noticed by ancient cultures. The astronomers of imperial China new it as as Taipo, the Great White One, around 300 B.C. and later as Jinxing (Star of Metal). Our name for the planet comes from the Romans, who identified it with their goddess of love and beauty, but it's the Norse fertility goddess Freya who lends her name to the day ruled by Venus. "Freya's day" is Friday.

Venus was an obscure deity in ancient Rome until she became associated with the powerful Greek goddess Aphrodite. Julius Caesar himself enshrined her as Venus Genetrix, the ancestor of his own family. Because the planet Venus alternates between evening and morning appearances without remaining visible all night long, some cultures knew the planet by one name when it appeared in the west and by another when it shone in the east. Among the Greeks, both Pythagoras (c. 570–c. 495 B.C.) and his contemporary Parmenides of Elea are credited with realizing that the evening star Hesperos and the morning star Phosphoros ("bringer of light") – the equivalent Roman names were Vesper and Lucifer – were in fact one and the same object.

Among the Yolngu people in northern Australia, Venus is Banumbirr, an important ancestral spirit. During the Dreaming, when aborigines believe spirits created the universe, Banumbirr created and named animals, plants and landforms as she traveled westward across Australia. Ray Norris, an astronomer at the Australia Telescope National Facility, has studied astronomical beliefs among aboriginal peoples and notes that Banumbirr plays a central role in a Yolngu ceremony designed to communicate with deceased ancestors. This "Morning Star Ceremony" starts at dusk, continues through the night, and climaxes when Venus rises a few hours before dawn. Banumbirr is said to trail behind her a faint rope that serves both as a telegraph line for transmitting messages from the dead and as a tether that prevents her from ever moving away from the Sun. Venus rises a few hours before dawn at different times from year to year, so the Morning Star Ceremony requires a certain level of advance planning and observation to carry out.

Half a world away, among the Maasai of Kenya and Tanzania, Venus is Kileken, the Orphan Boy, a star that became incarnate. According to a popular folktale, a mysterious young boy one day appeared outside the hut of an old, poor, lonely tribesman. The boy called himself Kileken, said he was an orphan, and offered to help the old man look after his cattle in the morning and evening. His only request was that the old man must never see him work. Kileken completed his chores with superhuman speed and brought good fortune and riches to his host, but the old man sensed magic afoot and demanded an explanation. Kileken warned that the old man would lose all he had gained if he understood the boy's secret. Overcome by suspicion and curiosity, one day the tribesman attempted to spy on the boy as he worked. He was quickly discovered and in a blinding flash Kileken disappeared and returned to the sky. Unable to trust in a gift from the gods, the old man returned to solitude and poverty.

African peoples generally did not recognize that the evening and morning stars were the same object, although the Karanga and Xhosa tribes in the southern part of the continent did connect the two. Among some of the more imaginative names for Venus in Africa were "evening fugitive," "the watching one" and, with the planet high in twilight at dinnertime, "peeper into pots." Unlike Mesopotamians and the Maya, the early Egyptians were not inclined to systematic observation. Still, they referred to Venus as "the crosser" or "the star that crosses" and symbolized it as Bennu, a legendary heron-like bird that carried the reincarnated soul of Osiris. Like another legendary bird, the phoenix, Bennu would die in flames and be reborn from its own ashes – a suggestion that the Egyptians too knew the true identity of the morning and evening aspects of Venus, perhaps by the second millennium B.C.

Associating the "wandering stars" with the most important gods originated in Mesopotamia. There, Venus was considered a manifestation of Ishtar, a powerful goddess responsible for the fertility of land and beast. It was known earlier by the Sumerian name Ninsianna and was associated with the goddess Inanna (literally "Queen of Heaven"), the forerunner of Ishtar. Hymns to Inanna indicate that by 2000 B.C. she was associated with both the morning and evening apparitions of Venus:

> *Mighty, majestic and radiant,*
> *You shine brilliantly in the evening.*
> *You brighten the day at dawn.*
> *You stand in the heavens like the Sun and Moon.*

The tale of Inanna's descent into the underworld, and her death and rebirth, was inspired by the movements of Venus and clearly resonates with the Egyptian concept of Bennu. Records dating back nearly 4,000 years show that Babylonian observers were thoroughly familiar with the details of the planet's visibility cycle, which is what gives rise to its connection to fertility. During a single apparition from first visibility to last. Venus can be seen 260 days; this is about 20 days shy of the human gestation cycle, from conception to birth. Evidence for early Babylonian knowledge of the planet's movements comes from a list of dates for the beginning and end of each of the planet's morning and evening appearances during the 21-year reign of Ammisaduqa (1702–1682 B.C.). Appended to the observations are astrological predictions – for example, "a king will send messages of peace to

Fig. 3.1 A view of the Caracol, a building designed for observing the Sun, Moon and Venus at the Maya site of Chichén Itzá on Mexico's Yucatán Peninsula. Behind it stands the imposing Pyramid of Kukulcan (Photo by the author)

another king" and "the heart of the land will be happy." One can think of these portents as emails from the gods to the king to aid him in the smooth operation of earthly affairs.

Over two millennia later, the Maya civilization in what is now Belize, Guatemala, and southeastern Mexico independently discerned the pattern of Venus's motions. They knew Venus by several names – Great Star, Wasp Star, Bright Star – and associated it with Kukulcan, Feathered Serpent, god of the wind and inventor of the calendar (which included a 260-day cycle). Its movements were of such astrological importance that the Maya oriented at least three major buildings with Venus in mind – the Caracol at Chichén Itzá (Fig. 3.1) and the Governor's Palace at Uxmal, both on the plains of Mexico's Yucatán Peninsula, and the Temple of Venus at Copan in Guatemala. They devoted six pages to Venus in the *Dresden Codex*, one of a handful of surviving documents, and include detailed astronomical and religious information. They greatly feared the planet in the days just after its switch to the morning sky. Illustrations in the *Codex* show Venus deities flinging spears at various earthly victims, with singularly unpleasant prognostications nearby – "Woe to the turtle; woe to the warrior and pregnant woman"; "evil excessive Sun; the misery of maize seed." Maya knowledge of the movements of Venus was eminently practical but was driven by religious and even military necessity.

For centuries, warfare between Maya cities consisted mainly of raiding parties in which nobles sought personal glory and captives for religious sacrifice.

That changed around A.D. 400, when war imagery began to reflect an influence from the central Mexican civilization of Teotihuacán, where Tlaloc, a rain and fertility god, had been associated with Venus for centuries. The Maya at Tikal in Guatemala embraced this new influence and war took on a new character, as seen in the conquest of the neighboring city Uaxactún in A.D. 378. What scholars call the Tlaloc-Venus concept became formalized as the standard image of the Maya conqueror king, and within a couple of generations the Maya began timing their battles to coincide with the actual appearance of Venus and other planets.

Elements of the Maya pantheon worked their way back to central Mexico and were adopted by the Aztecs, who rose to prominence after the Mayan culture had declined. The Aztec god of wind and fertility was Quetzalcoatl, who represented the predawn appearance of Venus and whose name also means "feathered serpent." His twin, Xolotl, the deity of magicians, shone as the evening star.

For the Inca in Peru, Venus was identified with Chasca, a fertility goddess, and was worshipped as a consort of the Sun god Inti, from whom the Inca believed they were descended.

Queen of Deception

On the astronomical magnitude scale, where smaller numbers indicate greater brightness, Venus hits −4.8 at its best – only the Sun and Moon are brighter. The planet figures prominently among reports of "unidentified flying objects," and there are numerous stories about airport tower controllers who, mistaking Venus for an arriving flight, radioed the planet its landing instructions. Similarly, pilots have been known to mistake the planet for aircraft position lights. Such incidents are difficult to document because no one really wants to talk about them, but a single anonymous report in the Web-accessible database of NASA's Aviation Safety Reporting System illustrates how real these admittedly rare events are.

One night in April 1996, as a passenger flight made its initial approach to Ohio's Port Columbus International Airport, air traffic control had bumped the reporting pilot to number two in the landing sequence to accommodate a faster jet. He was instructed to maintain visual separation, which was unusual at night, but the pilot confirmed that he had the jet in sight. Unfortunately, what he took to be the lights of the approaching jet was in fact Venus. "Only when the lights of the jet were shining down on me did I realize that what I had been watching in the dark night" was not the approaching jet, but rather "a very bright planet or star." The jet, which at this point was just 500 ft away, was descending "right on top of me." Approach control issued an immediate 180° left turn for evasive action, and both flights landed without further incident. Later, the pilot recorded, he spoke with the air traffic control supervisor, who said that they had seen the problem developing and had held the jet at a steady altitude until it passed over him. At the time, Venus was near its best and brightest and setting near midnight, far from the twilight glow where one might typically expect to find it.

The planet's brilliance still takes us by surprise and occasionally fuels episodes of popular delusion. What follows are examples from the past three centuries.

On December 10, 1797, while on his way to a victory banquet at Luxembourg Palace in celebration of his successful Italian campaign, Napoleon noticed that the throng lining the streets was looking at the sky instead of paying attention to his procession. He asked what was going on and was told that a daytime star (Venus) could be seen shining above Luxembourg Palace, something which might be looked upon as a timely indication of celestial endorsement. But as Venus reached greatest eastern elongation and gained brightness, the public perception strangely changed. Astronomer J. J. Lalande wrote that the evening star came to be regarded as a comet – and naturally, one on a collision course with Earth.

> On Jan. 16th, people were on the Pont-Neuf claiming the existence of a new comet and many were frightened. However it was just Venus, which was seen in daylight in Luxembourg, the very day when 20,000 people, waiting for General Bonaparte, were looking in that direction. We could see it in the same way every 19 months if we paid attention to it, but we rarely take the opportunity. This time people were overcome with a peculiar terror; all of the shows and clubs talked about comets. *The Comet* and *The End of the World* were performed in the Vaudeville.

A century later, an evening apparition of Venus stirred popular imagination throughout the United States. In the Midwest, it created a sensation as the signal light of some mysterious airship, but New Englanders instead attributed it to experiments by an East Coast celebrity. An editorial in the *Boston Evening Transcript* of Apr. 15, 1897, begins:

> A local astronomer was heard to remark the other day in a joking way that he had not been able to work lately, so busy was he kept answering questions about the new Edison experimental star in the western sky.... It is two or three months since the story was spread that Edison was experimenting with electric lights, and that it was his desire to learn how far he could signal from a balloon with an electric light. Sober people of Massachusetts have watched the decline in the western sky of this marvellous and brilliant light, and known it for the planet Venus, glowing with unwonted brilliancy, its thin crescent being discerned easily by any possessor of a small telescope.

The American physicist J. Robert Oppenheimer related in a 1950 letter to Eleanor Roosevelt a similar case of mistaken identity that occurred in Los Alamos, New Mexico, in the tense days before the test of the first nuclear bomb.

> I remember one morning when almost the whole project was out of doors staring at a bright object in the sky through glasses, binoculars and whatever else they could find; and nearby Kirtland Field reported to us that they had no interceptors which had enabled them to come within range of the object. Our director of personnel was an astronomer and a man of some human wisdom; and he finally came to my office and asked whether we would stop trying to shoot down Venus.

Writing in a climate of rising cold-war political frenzy stemming from the Soviet Union's development of nuclear weapons, its Berlin Blockade, and the start of the Korean War, Oppenheimer saw the Venus incident as a cautionary tale. He concluded: "I tell this story only to indicate that even a group of scientists is not proof against the errors of suggestion and hysteria."

Beneath a Cloudy Veil

While ancient cultures followed and catalogued the movements of Venus, scientists have come to understand its physical nature only with agonizing slowness. Although Venus is the brightest and nearest of the planets, observed over four millennia, little was definitively known about actual conditions there until the middle of the twentieth century. There is a very good reason for the slow progress: Venus is totally enshrouded in a reflective mantle of bright, pale, nearly featureless yellow clouds (Fig. 3.2). Those clouds at once reveal the planet to the stargazer and veil it from the inspection of planetary scientists.

Fig. 3.2 Venus conceals its true character beneath a reflective veil of thick, sulfurous clouds. Although featureless in visible light, the clouds show substantial detail only in images taken through violet and ultraviolet filters. NASA's Pioneer Venus orbiter returned this ultraviolet image in February 1979 (NASA)

Fig. 3.3 Twin crescents. Earth's nearest neighbors in space are the Moon and, only a hundred times more distant at its best, the planet Venus. Both can be seen during the day. On May 21, 2004, the two-day-old crescent Moon passed directly in front of Venus. When the planet emerged about an hour later, a photographer in Budapest, Hungary, recorded this dramatic daylight image of complementary crescents. The large ghostly arc of the Moon results from the small fraction of its disk — just 5% — then being illuminated by the Sun (Iván Éder)

Near the part of its orbit that brings it closest to Earth, Venus appears as a brilliant, featureless crescent when viewed through a telescope. Galileo Galilei recognized in 1610 that Venus mimics the Moon by going through a cycle of phases (Fig. 3.3). This was hard evidence that the planet revolved around the Sun, evidence that supported the heliocentric ideas proposed by the Polish monk Nicolaus Copernicus in 1543. Although Venus phases were expected under the older geocentric system, the geometry of that system demanded that Venus must always be less than half illuminated, which meant that a cycle similar to the Moon's was impossible.

While watching Venus enter and depart the Sun's disk during the 1761 transit, the Russian astronomer Mikhail Lomonosov took note of some unusual phenomena. As Venus crept toward the Sun's limb, he wrote, "the solar edge in the place of Venus ingress, which was clearly seen before…became unexpectedly vague and obscured."

I thought initially that my tired eye caused an obscuration, but when looking again in a few seconds I found a black indentation from the coming Venus, which replaced the former vague spot. I continued to look attentively how the trailing side of the planet approached the Sun; suddenly, a hair-thin bright radiance (luminescence) between Venus' trailed side and solar edge appeared that lasted only less than a second.

He observed a similar obscuration along the Sun's edge as Venus approached the solar limb toward the end of the transit. Lomonosov correctly interpreted what he saw as evidence "that the planet Venus is surrounded by a distinguished air atmosphere similar (or even possibly larger) than that is poured over our Earth." Both the flash and the obscurations of the Sun's limb were the result of sunlight passing through and refracted by Venus' dense atmosphere.

Yet even by 1960, such basic information as the planet's rotation period remained a subject for lively speculation. What little was known to scientists suggested that Venus was a slightly smaller version of our own planet. Although its proximity to the Sun gives it twice the sunlight we receive, its planet-wide cloud layer reflects nearly all of that light back into space. The clouds give the planet its brilliance, but they also ensure that Venus absorbs less solar energy than Earth (and about the same as Mars) despite its nearness to the Sun. Some observers believed the thick, featureless cloud deck was formed by water droplets in an atmosphere much like our own. Many thought the length of the Venus day was about the same as Earth's. In short, Venus appeared to be Earth's twin.

A very different picture emerges from Table 3.1, which summarizes the important physical and orbital data for Venus. In size and composition Venus does resemble Earth, but there the similarity ends (Fig. 3.4). The temperature on the planet's surface averages 856°F (458°C) – hot enough to melt tin, lead, and zinc – and shows little variation from equator to pole or between day and night. The bulk of the atmosphere consists of carbon dioxide, which explains the high surface temperature as "global warming" gone wild. The bright clouds reflect away all but 25% of the Sun's energy and only a small fraction of this actually reaches the surface. But the dense atmosphere swaddling Venus makes this small fraction count. Short-wavelength energy (light) reaching the surface is absorbed and then reradiated as heat at longer (infrared) wavelengths. Atmospheric carbon dioxide repeatedly absorbs and reradiates this energy, preventing the heat from quickly escaping into space. By the time the outgoing heat radiation balances the incoming solar energy, there's already enough thermal energy bouncing between the surface and the atmosphere to keep them both plenty hot.

This process is the familiar "greenhouse effect" that also operates in our atmosphere, where water vapor plays the chief greenhouse role. It's reasonable to assume that Venus and Earth were born with similar amounts of water, but there is no water on Venus now and very little in the atmosphere. Planetary scientists speculate that a combination of the greenhouse effects of water vapor, extra solar energy – currently about twice what Earth receives – and a slow increase in the Sun's energy output worked to keep Venus much hotter than Earth throughout most of its early history. This would have kept lots of water vapor in the atmosphere, which at higher altitudes ultraviolet sunlight would have broken down into hydrogen and oxygen. Over the eons, the hydrogen simply escaped into space, and you can't have water without hydrogen.

Table 3.1 Facts about Venus and Mercury

Venus

Diameter	7,521 miles
	12,104 km
	94.9% of Earth's
Surface temperature	856°F (458°C)
Surface atmospheric pressure	92 bars, or 92 times that of Earth's atmosphere. This is equivalent to the ocean pressure at a depth of 3,078 ft (938 m).
Atmospheric composition	96.5% carbon dioxide (CO_2)
	3.5% nitrogen (N_2)
Moons	None
Rotation period	243 days, in the direction opposite Earth's
Obliquity	177.4°
Sidereal period (time for one full orbit around the Sun)	224.7 days
Synodic period (time between successive conjunctions with the Sun)	583.92 days
Average distance from Sun	67.2 million miles
Light takes 6 min to travel this far.	108.2 million km
	0.723 Astronomical Unit (AU)
Orbit inclined to Earth's	3.39°

Mercury

Diameter	3,032 miles
	4,879.4 km
	38.3% of Earth's
Surface temperature	
Maximum	800°F (427°C)
Minimum	−280°F (−173°C)
Mercury has the greatest day/night temperature range of any planet	
Moons	None
Rotation period	58.65 days
Obliquity	0.1°
Sidereal period	87.97 days
Synodic period	115.88 days
Average distance from Sun	35.9 million miles
Light takes 3.4 min to travel this far.	57.9 million km
	0.387 Astronomical Unit (AU)
Orbit inclined to Earth's	7°

The atmosphere of Venus held still more surprises. In the late 1960s the Soviet Union attempted to land instrumented capsules on the planet's surface as part of their on-going Venera series of space missions. The first three attempts that successfully entered the atmosphere ceased operating before they reached the surface. When the next probe, Venerea 7, survived its descent and returned surface pressure

Fig. 3.4 Earth, the Moon, Mercury and Venus compared (NASA and NOAO photos; montage by the author)

measurements, it became clear why the earlier Venus landers had failed: They had been crushed as they descended. The dense atmosphere presses onto the surface with a force more than 90 times greater than Earth's, equivalent to the pressure experienced by a submarine at a depth of more than 3,000 ft (914 m) beneath the sea.

The Venusian catalog of atmospheric horrors would not be complete without mentioning the contents of those perpetual yellow clouds. They can't be made of water droplets or ice crystals like earthly clouds because water in any form is scarce on Venus, less than 0.01% in the atmosphere. Instead, the clouds form from droplets of concentrated sulfuric acid, with traces of hydrochloric and possibly hydrofluoric acids thrown in as well. So in addition to its stifling heat and crushing pressure, the atmosphere of Venus holds the record as being the most corrosive in the solar system.

Venus has seen more terrestrial hardware than any other planet, with 30 Soviet, American, European and Japanese space missions returning data from the planet's vicinity since attempts began in 1961. The Soviet Venera and VEGA missions included flybys, orbiters, atmospheric probes and landers. Successful U.S. missions included the Mariner 2 and 5 flyby spacecraft, the Pioneer Venus Orbiter and

Fig. 3.5 Our closest views of Venus' surface come from Soviet landers. This view from the Venera 13 lander, which touched down March 1, 1982, shows flat rock slabs and soil as well as part of the spacecraft and the camera's lens cover (*center*). The lander survived the hostile environment for only 2 h, 7 min (NSSDC)

Multiprobe, and the radar-mapping Magellan orbiter. In addition, several missions to other planets – Mariner 10 and MESSENGER as they headed for Mercury, Galileo on its way to Jupiter, and Saturn-bound Cassini – observed Venus as a target of opportunity; they were in the neighborhood to change their speeds by using the planet's gravity as a fuel-saving technique. In total, ten spacecraft have returned data from the surface of Venus, but most survived less than 2 h in the hostile environment. Four Soviet landers have even transmitted images from their landing sites, revealing a sterile, rocky, arid landscape bathed in a diffuse orange light (Fig. 3.5). Only about 2% of incident sunlight actually reaches the surface, so Venus has about the same illumination as a heavily overcast day here on Earth. Images of the surface and chemical analyses of the rocks indicate that all areas explored by the landers were volcanic plains.

Since the perpetual clouds block all visible light from the surface, any global analysis of Venus had to be done by radar. The first crude maps were made in the 1960s with large ground-based radio telescopes, which pinged the planet with radar pulses whenever it was nearest Earth. The returned signal contained information about the way radio waves interacted with the surface and allowed the first identification of surface features. It was through these studies in 1961 that scientists were at last able to pinpoint the planet's rotation rate. Venus spins east to west, opposite to the direction of Earth and most other planets, and takes 243 Earth days to make one complete rotation. Since it completes one orbit around the Sun in just 225 Earth days, the "Venus day" is actually longer than the "Venus year." No one knows why the planet spins so slowly. The entire atmosphere rotates as well in the same direction, but takes just 4 days to do so at the altitude of the cloud deck.

Mapping of the planet from orbit began with Pioneer Venus Orbiter, which determined its topography with a radar altimeter, and continued with Venera 15 and 16, which carried radar imagers to produce photograph-like pictures of about 25% of the surface. The culmination of Venus mapping came with Magellan, which between 1990 and 1994 used a radar imager to map 98% of the planet in the greatest detail yet. Magellan revealed that landforms on Venus consisted of two main types – smooth volcanic plains (80%) and intensely deformed elevated plateaus called tesserae (8%) – and revealed extensive meandering rift valleys, over 1,700 volcanic landforms or deposits and more than 1,000 impact craters.

The results from Magellan raised many more questions. Like other bodies in the solar system, Venus must have undergone intensive bombardment from comets, asteroids and smaller debris over the past 4 billion years, but the density and number of craters, plus the fact that only a few of them are modified by lava flows or geologic faults, indicates that none of the planet's rock units are older than between 300 million and 500 million years. Left with no better alternatives, planetary scientists concluded that the crater distribution was most consistent with a nearly complete volcanic repaving of the planet somewhere in that period. The lava flows wiped the surface clean of larger older craters and all ancient landforms except the highest tesserae. When this extraordinary episode ended, they suggested, Venusian volcanism petered out and impact craters began piling up.

Even in its basic geological framework, Venus seems to bear little kinship with Earth. It shows no evidence of plate tectonics, the horizontal motion of large units of Earth's crust that drives earthquakes and volcanoes, builds mountains, and moves continents. Instead, the surface of Venus appears to have been repeatedly extended and compressed, warped by the successive rise and fall of magma plumes from deep within the planet. But if plumes of hot magma are still rising within Venus, then they could create active volcanoes today.

And apparently they do. Researchers using data from the European Space Agency's Venus Express spacecraft, which has been orbiting the planet since 2006, detected three regions where infrared emissions, which can penetrate the clouds, indicate different rock compositions. On Earth, cooled lava flows rapidly react with the atmosphere and quickly change their chemical composition. The Venus Express data indicate that rocks in regions called Imdr Regio, Dione Regio and Themis Regio appear unusually bright in the infrared. Scientists say this indicates the rocks are unweathered lava no more than 2.5 million years old – and likely much younger than this. To planetary scientists, this is a strong indication that Venusian volcanoes are still at work shaping the planet's surface.

The View from Earth

Observers had long known that the motions of Venus and Mercury were intimately tied to the Sun, but explaining just why this was so proved difficult. To fully appreciate why we see Venus and Mercury more easily at some times than at others, it's necessary to understand just how they move.

If we could view the solar system from a far-flung point high above the Sun's north pole, we would see the planets moving around the Sun in their elliptical orbits much like cars on a racetrack (Fig. 3.6). Those in the innermost lanes circle fastest, those on the outermost the slowest. Turn the solar system on its "side" – that is, view it along the plane of Earth's orbit, the "ground level" of the racetrack – and the planets seem to slide from one side of the Sun to the other. From behind the Sun

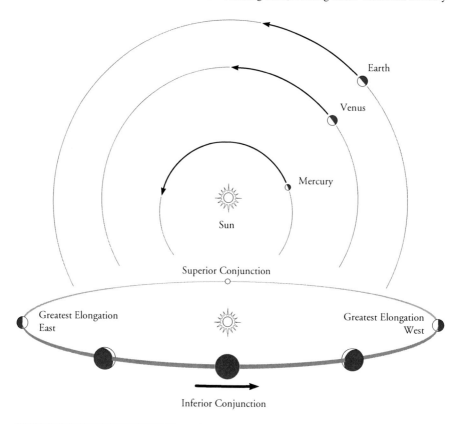

Fig. 3.6 *Top*: A view from high above the "racetrack" of the inner solar system. Arrows indicate the average distance each planet travels in a month. Both Venus and Mercury move faster than Earth. *Bottom*: How we see Venus or Mercury depends on their positions relative to Earth and the Sun. Throughout their visibility cycle, the planets undergo changes in apparent size and illumination

each planet moves to the left until it appears farthest from the Sun, at which point it doubles back. Proceeding now to the right, the planet's apparent distance from the Sun shrinks. It eventually crosses our line of sight to the Sun and continues until it again reaches a maximum angle, then it reverses course and heads to the far side of its orbit.

From our perspective on Earth we view the solar system not only along the plane of the racetrack, but from one of the innermost lanes. The planets on the outer orbits, from Mars on out, arc away from the Sun and pass behind us, where we can watch them throughout the night. Only Mercury and Venus exhibit the back-and-forth motion described above because only they follow orbits inside Earth's. For skygazers, both planets put on their best shows near the time of greatest elongation, the point where they reach their widest angle east or west of the Sun. For Venus this is between 45° and 47°; Mercury, more variable, can reach angles between 18° and 28° from the Sun.

With this in mind, we can track the relative positions of Earth and Venus through one complete cycle of visibility, or apparition. We'll begin as before, with Venus located on the far side of the Sun as seen from Earth at the point termed superior conjunction. Venus, now at its greatest distance from us, is lost in the glare of the Sun. If the planet could be seen, a telescope would reveal a very small but fully illuminated disk. Venus moves farther east each day, setting progressively later than the Sun. After a few weeks we can see it low in the west, glimmering in the twilight of dusk before following the Sun below the horizon. The planet continues its eastward slide and pulls farther away from the Sun, each day appearing higher above the horizon after sunset and becoming more noticeable in the evening sky. About 7 months after superior conjunction, Venus reaches greatest eastern elongation and blazes in evening sky, setting a couple of hours after the Sun. To telescopic observers the planet's disk has grown larger, but now only half of its sunlit side faces Earth.

Venus now runs on the near side of its orbit, simultaneously approaching us and reversing course in a westward rush. Its disk continues to grow, but a telescope reveals an ever brighter – though ever slimmer – crescent. The planet grows brighter until about 5 weeks after greatest elongation, when the fading light from its ever-shrinking crescent finally offsets its increasing angular size. Venus falls quickly from its summit in the evening sky, taking just 10 weeks to plunge back into the Sun's glare. The evening apparition ends just before Venus passes between Earth and the Sun (inferior conjunction), when it laps us on the solar system racetrack.

Venus isn't lost in the Sun's radiance for long, though. It quickly pulls west of the Sun and begins its morning apparition. Venus climbs higher into the predawn sky as rapidly as it fell from the evening sky, glowing brightest about 5 weeks after inferior conjunction and reaching greatest western elongation 5 weeks after that. The planet then reverses course and, now on the far side of its orbit, begins a lazy, 7-month-long descent toward the horizon. Venus completes the cycle when its slow eastward slide to the Sun brings it back to superior conjunction. Figure 3.7 shows how Venus moves through the sky during typical evening and morning apparitions.

While Venus completes an orbit around the Sun in 225 days (its sidereal period), the planet's visibility cycle takes 584 days, or 1.6 years (its synodic period). We noted a similar apparent discrepancy in Chap. 2's discussion of the Moon's orbit, and the solution here is related. Because our viewpoint within the solar system's racetrack also moves, it takes some extra time for Venus to catch up so that the same geometry recurs between Earth, Venus, and the Sun. Table 3.2 lists the date ranges when Venus can be seen as a conspicuous evening and morning star for observers at middle northern latitudes. The table dates are limited to times when the planet appears at least 10° above the horizon about 45 min after sunset or before sunrise. Looking for Venus closer to dusk or dawn will extend these dates somewhat.

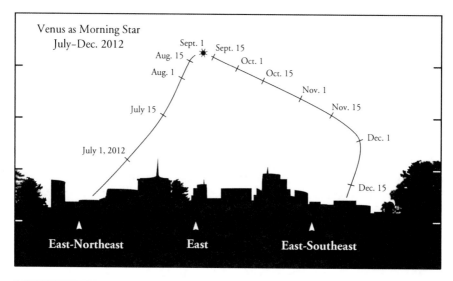

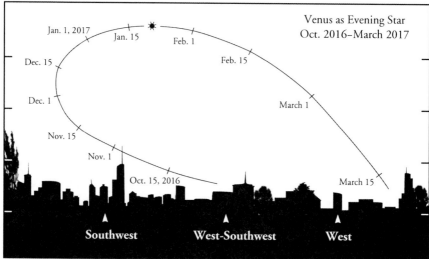

Fig. 3.7 *Top*: Venus arcs through the morning sky from late June to mid-December 2012, as seen 45 min before sunrise. The planet reaches its greatest elongation from the Sun on Aug. 15 but keeps climbing above the horizon into early September. *Bottom*: Venus as it tracks through the evening sky from Oct. 2016 to March 2017, as seen 45 min after sunset. The planet attains its greatest angle from the Sun on Jan. 12, but it gains a little additional altitude in the following weeks

Although Venus is always rather easy to find, it only commands attention when its greatest elongation occurs in the right season. At its best, Venus rises nearly 4 h before dawn or lingers in the evening sky for as long after sunset. All of the planets stay near the ground-level of the solar-system racetrack, which is known as the

Table 3.2 Visibility of Venus through 2020. Each entry gives the approximate range of dates for which an observer at mid-northern latitudes will find Venus at least 10° above the horizon 45 min before sunrise or after sunset

Year	As evening star (in the west after sunset)	As morning star (in the east before sunrise)
2011	–	Mid-November 2010 to early March. Best in late December 2010, when it rises 3¾ h before dawn.
2012	Early December 2011 to late May. Best in early April, setting 4 h after sundown.	Late June to late December. Best in early September, when it rises 3¾ h before dawn.
2013	Early October to late December. Best in late November, setting nearly 3 h after the Sun.	–
2014	–	Late January to early August. Best in late February, when it rises little more than 2½ h before dawn and remains low in the southeast.
2015	Late January to mid-July. Best in early May, when Venus sets 3½ h after the Sun.	Early September to early February 2016. Best in early November, rising 3¾ h ahead of the Sun.
2017	Late October 2016 to mid-March. Best in mid-January, when it sets 4 h after the Sun.	Early May to late October. Best in early August, rising 3 h before the Sun.
2018	Early April to early September. Best in early June, but it remains low in the west and sets about 2½ h after sundown.	–
2019	–	Mid-November 2018 to mid-March. Best in late December, rising about 3¾ hours before dawn
2020	Early December 2019 to mid-May. Best in early April, setting 4 h after sundown.	Late June to late December. Best in early September, when it rises 3¾ h before dawn.

ecliptic. The ecliptic marks the Sun's apparent annual course through the starry sky, but Earth is really doing the moving, so the ecliptic also marks the plane of Earth's orbit. Throughout the year, the ecliptic's angle to the horizon changes, intersecting it most steeply near the start of spring and fall. In the northern hemisphere, Venus gives its best evening displays when greatest eastern elongation occurs near the start of spring, and the worst when it occurs near the end of summer. (The opposite holds true for morning displays.) In the span covered by this book, the best evening apparition occurs in March 2012, with Venus 35° high 45 min after sunset. The worst evening appearance occurs in August 2018, when Venus stands at less than a quarter of the altitude at the same time after sunset (Fig. 3.8). Yet on both occasions the planet is 46° from the Sun.

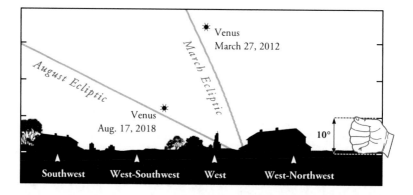

Fig. 3.8 Venus gives its best evening shows when its greatest angle east of the Sun occurs near the spring equinox, which happens in 2012. Its worst evening apparitions occur when it reaches greatest eastern elongation near the autumnal equinox, as in 2018. The steepness of the ecliptic accounts for the difference in altitude even though the planet is 46° east of the Sun on both occasions

Earth and Venus share a curious relationship, one that contributed to Maya and Mesopotamian fascination with the planet. Venus completes five synodic cycles (2,919.6 days) in almost exactly 8 years (2,922 days); the difference is less than 60 h. So after every set of five evening and morning apparitions, Venus returns to the same relationship with Earth and Sun and replays its performance. For example, the planet's fine morning show at the end of 2004 repeats almost exactly in 2012 and again in 2020.

Messenger of the Gods

The meaning of Mercury's Egyptian name, Sebeg, is unknown, but an early text dating to the reign of Ramses VI (1145–1137 B.C.), refers to the planet as "Set in the evening twilight, a god in the morning twilight." This tells us that the Egyptians likely understood that Mercury was the same object seen before dawn and after sunset, just as they apparently did for Venus. It also suggests that they viewed Mercury as having a dual personality. In the evening sky, it took on the malevolent nature of Set, Osiris' treacherous brother; in morning twilight, as an unidentified god, Mercury may have undergone a mood change and taken on the opposite temperament. Mars, Jupiter and Saturn were all associated with the god Horus, son of Osiris and therefore Set's enemy, but just why Mercury was associated with Set remains a mystery.

The Babylonians associated Mercury with their god of wisdom and the patron of scribes, Nabu, and called it Sihtu, which translates as "jumping" – a reasonable description of the planet's behavior. The Greek name Stilbon ("twinkling star") also seems inspired by the planet's appearance as it shimmers in the unsteady air near the horizon. The Greeks associated the planet with the swift-footed messenger of the gods, Hermes, who watched over travelers and brought good fortune in commerce; our name for the planet comes from his Roman counterpart. To the late Chinese it was Shuixing, the Water Star. The Anglo-Saxons named the day ruled by Mercury after their chief deity, Woden, from which we derive our name for Wednesday.

Mercury, the smallest planet, is about one-third the size of Earth and only slightly larger than our Moon; the planet's vital statistics are listed in Table 3.2. Running on the solar system's innermost lane, Mercury completes an orbit around the Sun in just 88 days on a path that is more inclined and elliptical than any other planet's. This ellipticity causes Mercury's distance to the Sun to vary by 66% between its closest point, called perihelion, and its farthest extreme. Radar studies in 1964 determined that Mercury spins once on its axis every 59 days, which is exactly two-thirds of its orbital period. This means the planet makes three complete rotations for every two orbits around the Sun. This coupling of rotation and orbital periods, together with a strongly elliptical orbit, means that the same side of Mercury faces the Sun at every other perihelion passage. As a result, two locations on opposite sides of Mercury are more Sun-baked than any others and are referred to as Mercury's "hot poles." At perihelion, where the Sun appears three times the size it does from Earth, the hot poles can reach 845°F (452°C) – not quite as hot as Venus despite its proximity to the Sun, but still more than adequate for melting zinc. Just as Earth's tidal effects on the Moon have locked its rotation period to its orbit, so the Sun's tides have locked Mercury into a 3:2 coupling of its orbit and spin.

One feature of Mercury's orbit provided a mystery that only Einstein's theory of relativity could explain. A planet's orbit isn't a fixed path in space because gravitational nudges from all the other planets continually affect it. A consequence of these nudges is that a planet's entire orbit actually makes a very slow rotation around the Sun, called precession. Put another way, the point in a planet's orbit where it lies closest to the Sun (perihelion) gradually advances around the Sun. Newton's laws explained the perihelion advance of all of the planets except Mercury. In 1859, the French mathematician Urbain LeVerrier used data from Mercury transits to show that the actual rate of the planet's orbital precession disagreed with what he calculated from Newton's theory by a small but definitive 43 arcseconds per century. Suggested explanations included the possibility of another planet or asteroid belt orbiting interior to Mercury that was providing unaccounted-for gravitational nudges, but they never panned out. Mercury's extra orbital slippage remained a mystery for decades.

Fast-forward to November 1915, as Albert Einstein completed the final equations of his general theory of relativity and began to work out their astronomical consequences. When he finished calculations related to the perihelion motion of Mercury, Einstein said that he became so excited his heart palpitated: His theory exactly accounted for Mercury's perihelion advance. According to relativity, what we experience as the force of gravity is really a curvature in space-time; the greater the mass, the more curvature it produces. Being closest to the Sun, Mercury feels the "dent" in space-time caused by the Sun's mass much more than any other planet, and it was this effect that boosted Mercury's precession. "The result of the perihelion motion of Mercury gives me great satisfaction," a cheerful Einstein wrote to fellow physicist Arnold Sommerfeld in December 1915. "How helpful to us here is astronomy's pedantic accuracy, which I often used to ridicule secretly!"

The telescope reveals little about the planet except its cycle of phases; it took a U.S. spacecraft named Mariner 10 to give us the first detailed look at Mercury. Its cameras photographed about 45% of the planet's surface during three flybys in 1974 and 1975, returning some 3,500 images. From a distance Mercury could be mistaken for our own Moon – a similarity that was expected — but a closer look revealed that even Mercury's most heavily cratered areas show fewer craters than the lunar highlands. Mercury's highlands consist of regions rich in craters intermixed with large expanses of gently rolling plains. Because the plains surround and partially cover the cratered areas, they must be younger, and scientists believe the plains were created by outpourings of lava some 4 billion years ago. Another landform, the smooth plains, resembles the maria of our Moon. Occurring near large impact basins and filling some of the largest craters, the smooth plains probably represent material that flooded onto Mercury's surface some 3.8 billion years ago, near the end of the spike in asteroid crashes known as the late heavy bombardment.

Caloris Basin, named for its location near one of the hot poles, is the largest undegraded impact structure imaged on the planet. Mariner 10 images revealed only the eastern half of the structure, but based on them scientists estimated its size to be about 800 miles (1,300 km) across, equivalent to the distance from New York to Memphis. Now, after 33 years, a NASA spacecraft called MESSENGER – short for MErcury Surface, Space ENvironment, GEochemistry and Ranging – has again reconnoitered Mercury through a series of flybys and revealed new territory (Fig. 3.9). The meat of the mission began in 2011, when MESSENGER became the first spacecraft ever to orbit the innermost planet, but scientists learned a great deal from several prior flybys. One of the first major findings is that the Caloris Basin is much bigger than previously thought: 960 miles (1,550 km) wide, or about the distance from New York to Fort Lauderdale, Fla. The basin floor is marked by a complex pattern of crisscrossing ridges and fractures and, based on current knowledge, its detailed geology appears to be unique in the solar system. Caloris is thought to have formed around 3.85 billion years ago when an object about 100 miles (162 km) wide smashed into Mercury. The powerful blow transmitted seismic waves around and through the planet; within moments these waves likely converged on the planet's surface at a location exactly opposite the impact site. The aftermath of the tremendous earthquakes that shook this spot can be seen in its uniquely jumbled and heavily fractured "weird terrain."

Ground-based radar studies of Mercury in 1991 brought the unexpected revelation that the floors of large high-latitude craters reflect radar much better than the surrounding terrain. Properties of the reflected radio waves appear similar to those bounced off of the martian polar caps and Jupiter's icy moons, which leads scientists to suggest that the planet nearest the Sun hosts polar craters containing frozen

Fig. 3.9 NASA's MESSENGER spacecraft acquired this image of Mercury shortly after its second flyby in October 2008. About 30% of the terrain in this image had never before been seen (NASA/JHUAPL/CIW)

craters near both its north and south poles may contain frozen water. With Mercury's near-zero obliquity, there can be no seasons and the Sun never shines into craters near the poles. Comparisons of maps made from radar studies and images of the surface returned by Mariner 10 show a close correspondence between radar-reflective areas and craters. In fact, radar-bright deposits fill most of the polar craters that sport uneroded rims imaged by Mariner 10. There are alternatives to ice deposits: Sulfur from meteorite impacts or from Mercury's surface could also migrate into the interiors of shadowed craters and possibly produce similar radar signals. Understanding the nature of these deposits is one of the main mission goals for MESSENGER.

Mercury's overall motion through our sky resembles that of Venus, with periods of visibility centered on the dates of its greatest elongations. Mercury averages six greatest elongations each year, but they carry the planet no more than 28° from the Sun. So although Mercury can rival even the brightest stars, it never climbs high enough into the sky to emerge from the twilight glow of dusk or dawn (Figs. 3.10 and 3.11). Elsewhere, we have discussed how the steepness of the ecliptic affects the visibility of the young Moon and Venus, and this proves even more important for Mercury. Table 3.3 summarizes Mercury's best morning and evening apparitions through 2020. A glance at the table shows that, in the Northern Hemisphere, Mercury puts on its finest shows as an evening star in the spring and as a morning star in the fall.

Locating Mercury is difficult enough that even Nicholaus Copernicus, a founding father of modern astronomy, lamented that it had always given him trouble. But there's no better way to find this speedy planet than by letting the Moon and planets point the way. The apparitions illustrated in Figs. 3.12 through 3.21 show such celestial guideposts, along with each planet's magnitude on the astronomical brightness scale. Like Venus, Mercury also replays its apparitions, but it does so with less precision and over a much longer period. The planet completes 41 synodic cycles (4,751.1 days) in a period only 3 days longer than 13 years (4,748 days). A 46-year cycle works even better, repeating within 30 h, and was employed by Babylonian astronomers.

Fig. 3.10 Pop goes Mercury. Several times a year, Mercury wanders as far from the Sun as it can get and makes a brief twilight appearance shortly before sunrise or shortly after sunset. The planet appears here as seen from Figueres, Spain, during its February 2000 apparition. Images of Mercury were taken each day when the Sun was exactly 10° below the horizon in order to emphasize the planet's day-to-day changes. All of the Mercury images were then digitally merged into the photograph with the most beautiful twilight scene (Juan Carlos Casado)

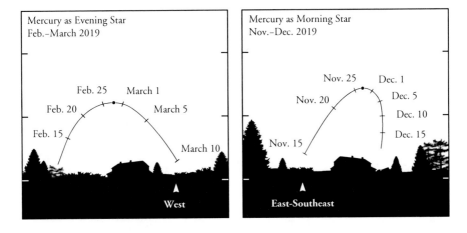

Fig. 3.11 You can get a sense of what to expect from Mercury from these diagrams, which show good evening and morning appearances in 2019 as seen about 30 min after sunset and before sunrise, respectively. *Left*: The planet is best placed for evening viewing from Feb. 19 to March 5, with greatest elongation (18.1°) occurring Feb. 27. *Right*: Mercury is visible in the predawn sky from Nov. 19 to Dec. 9; greatest elongation (20.1°) occurs on Nov. 28

Table 3.3 Mercury's best through 2020. Each entry gives the approximate range of dates for which an observer at mid-northern latitudes will find Mercury at least 10° above the horizon 30 min before sunrise or after sunset

Year	As evening star (in the west after sunset)	As morning star (in the east before sunrise)
2011	Mar. 14–30 Jul. 1–20	Dec. 28, 2010 – Jan. 16 Aug. 28–Sept. 11 Dec. 12–Jan. 2, 2012
2012	Feb. 26–Mar. 11 Jun. 11–Jul. 6	Aug. 11–25 Nov. 25–Dec. 16
2013	Feb. 12–20 May 23–Jun. 21	Jul. 26 – Aug. 8 Nov. 10 – 29
2014	Jan. 24–Feb. 6 May 9–Jun. 4	Jul. 10–22 Oct. 25–Nov. 11
2015	Jan. 9–21 Apr. 24–May 17 Dec. 27–Jan. 3, 2016	Jun. 29–Jul. 1 Oct. 10–24
2016	Apr. 7–28	Jan. 26–Feb. 7 Sept. 23–Oct. 6
2017	Mar. 22–Apr. 9 Jul. 16–23	Jan. 7–23 Sept. 7–20
2018	Mar. 7–22 Jun. 22–Jul. 14	Dec. 21, 2017–Jan. 10 Aug. 21–Sept. 4 Dec. 5–26
2019	Feb. 19–Mar. 5 Jun. 5–Jun. 30	Aug. 5 –18 Nov. 19–Dec. 9
2020	Feb. 3–16 May 18–Jun. 14	Jul. 19–Aug. 1 Nov. 3–21

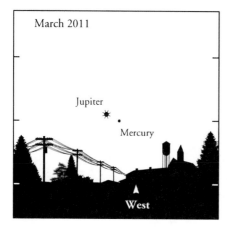

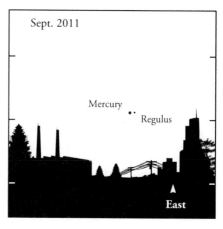

Mercury (−1.1) climbs into evening twilight in mid-March as Jupiter (−2.1) descends. Use brighter Jupiter to find Mercury; the two are closest on the 15th

On Sept. 9, in the predawn sky, Mercury (−0.9) descends past Regulus (1.4), the brightest star in the constellation Leo. Mars, as bright as but redder then Regulus, shines high above the pair

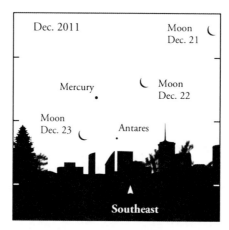

The waning crescent Moon sweeps near Mercury (−0.4) and the star Antares (0.9) on the following mornings. Saturn (0.7), slightly brighter than Antares, shines above the scene

Fig. 3.12 Mercury in 2011, looking 30 min before sunrise or after sunset. Tick marks along the sides of the horizon diagrams represent altitude intervals of 10° − about the distance covered by a fist at arm's length

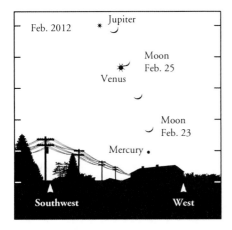

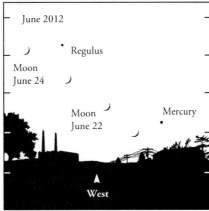

In the last week of February, the waxing Moon guides you to Mercury (–1.1), Venus (–4.2) and Jupiter (–2.2). As the sky darkens on the 26th, turn around to see Mars (–1.2), now only days from opposition, shining brilliantly very low in the east

Look for Mercury (–0.1) near the waxing Moon beginning June 21. As the sky darkens, look for Mars (0.8) in the southwest and Saturn (0.6) high in the south

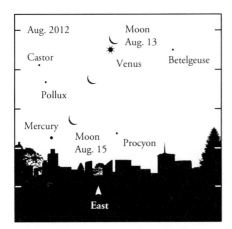

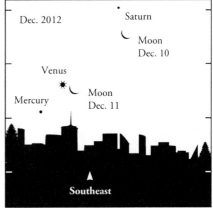

In the predawn sky in mid-August, the waning Moon guides you to Venus (–4.5) and Mercury (0.5). Many bright winter stars shine nearby

In early December, Mercury (–0.5) meets Venus (–3.9), Saturn (0.6) and the Moon in the pre-dawn sky

Fig. 3.13 Mercury in 2012, looking 30 min before sunrise or after sunset. Start viewing with enough time to get your bearings before the Sun rises too high or Mercury descends too low

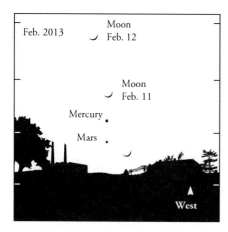

In early February evenings, the young Moon helps you find Mercury (–0.9) and Mars (1.2). Jupiter shines (–2.4) nearly overhead

The waxing crescent Moon sweeps past Venus (–3.8) and Mercury (0.3) in mid-June. Look for Saturn (0.4) high in the south-southeast

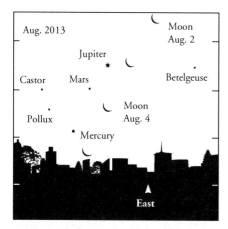

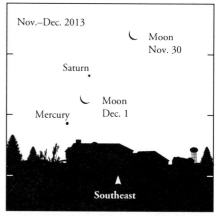

In early August, the waning Moon sweeps past Jupiter (–1.9), Mars (1.6) and Mercury (–0.4) before dawn

Look for Saturn (0.6) and Mercury (–0.6) near the Moon before sunrise as December opens. You'll find Mars (1.2) high in the south and Jupiter (–2.6) high in the west

Fig. 3.14 Mercury in 2013, looking 30 min before sunrise or after sunset

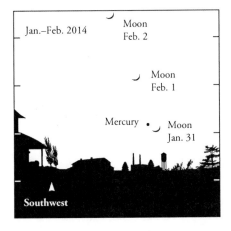

Look for Mercury (–0.6) near the waxing Moon on the evenings of Jan. 31 and Feb. 1. Jupiter (–2.6) shines high in the east

The waxing Moon lies between Jupiter (–1.9) and Mercury (1.1) on the evening of June 1. However, Mercury is much brighter and higher in mid-May, so think of this as your last chance to locate it

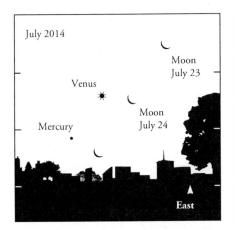

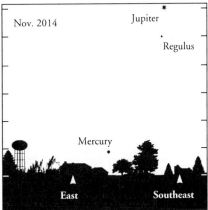

At the end of July, the waning Moon swings past Venus (–3.8) and Mercury (–0.9) in the predawn sky

Look for Mercury (–0.7) before dawn in early November. Jupiter (–2.1) shines high above it

Fig. 3.15 Mercury in 2014, looking 30 min before sunrise or after sunset

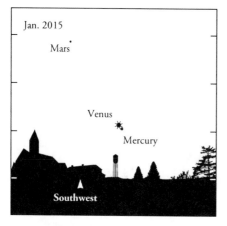

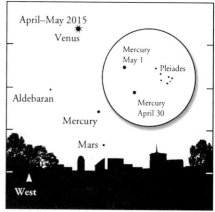

Mercury (−0.8) hovers 39 arcminutes from Venus (−3.9) on the evening of Jan. 10. As the sky darkens, look for Mars (1.1) above and to the left of the pair

Mercury (−0.4) closes April by passing near the Pleiades star cluster; use binoculars to see cluster stars as evening twilight deepens. Look for Venus (−4.2) and, higher still, Jupiter (−2.1) as well

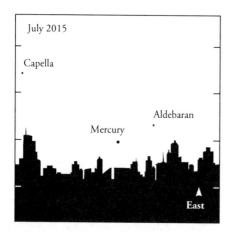

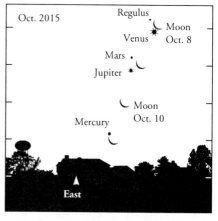

In early July mornings, look for Mercury (−0.5) near the star Aldebaran (0.9)

Early October's waning Moon guides you to Venus (−4.7), Mars (1.8) and Jupiter (−1.7) and Mercury (0.2) over successive days. Mercury brightens and the other planets grow closer as the month progresses

Fig. 3.16 Mercury in 2015, looking 30 min before sunrise or after sunset. Tick marks along the sides of the horizon diagrams represent altitude intervals of 10° – about the distance covered by your fist at arm's length

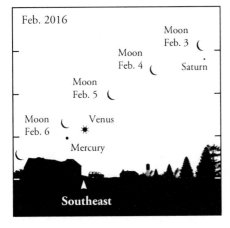

Feb. 2016

Moon
Feb. 3

Moon
Feb. 4

Saturn

Moon
Feb. 5

Moon   Venus
Feb. 6

Mercury

Southeast

On early February mornings, the waning Moon guides you to Saturn (0.6), Venus (–3.8) and Mercury (–0.1). Look south for Mars (0.7) and west for Jupiter (–2.4)

April 2016

Moon
April 9

Moon   Mercury
April 8

West

Mercury (–1.0) gleams near the young Moon on the evening of April 8. Turn to the southeast to see Jupiter (–2.4)

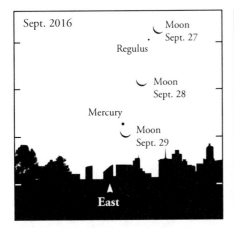

Sept. 2016

Moon
Sept. 27

Regulus

Moon
Sept. 28

Mercury

Moon
Sept. 29

East

Mercury (–0.6) stands above the waning crescent Moon on the morning of Sept. 29

Dec. 2016

Venus

Mercury

Southwest

Look low in the southwest in mid-December to locate Mercury (–0.3), with Venus (–4.3) shining brilliantly nearby

Fig. 3.17 Mercury in 2016, looking 30 min before sunrise or after sunset. Start viewing with enough time to get your bearings before the Sun rises too high or Mercury descends too low

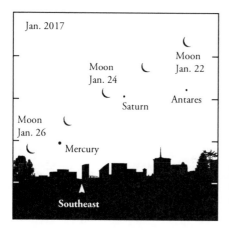

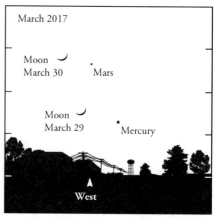

Mercury (−0.2) is joined in the predawn sky by Saturn (0.5) and the waning Moon in late January

Look for Mercury (−0.5) near the young crescent Moon during the last evenings in March. Mars (1.5) hovers faintly nearby

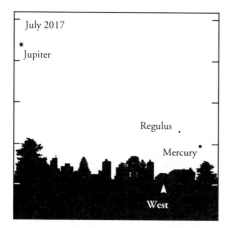

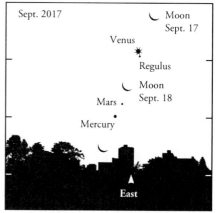

In the latter half of July, look for Mercury (0.1) low in the west in the half hour after sunset. The star Regulus (1.4) shines above it and bright Jupiter (−1.9) beams nearby

Mid-September mornings are brightened by the waning crescent Moon, brilliant Venus (−3.9) near Regulus, Mars (1.8) and Mercury (−0.3)

Fig. 3.18 Mercury in 2017, looking 30 min before sunrise or after sunset

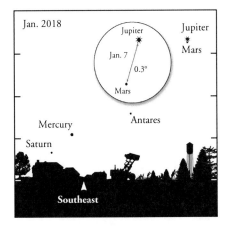

January opens with a fine morning show of shuffling planets. Bright Jupiter (–1.7) and faint Mars (1.4) pair up Jan. 7, while Mercury (–0.3) and Saturn (0.5) do so on Jan. 13

Mercury is brighter and closer to Venus (–3.9) in the first week of March, but the planetary pair is higher after sunset later in the month, when the crescent Moon joins them

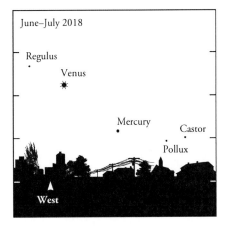

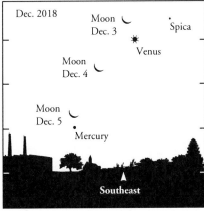

Mercury is conspicuous in the evening sky in late June, when it is brightest (–0.4), and into July. Look for it between Venus (–4.1) and the stars Pollux and Castor

The Moon sweeps past brilliant Venus (–4.9) and Mercury (0.6) before dawn in early December. Mercury brightens to –0.5 by mid-month

Fig. 3.19 Mercury in 2018, looking 30 min before sunrise or after sunset

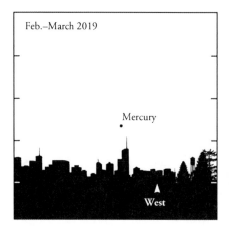

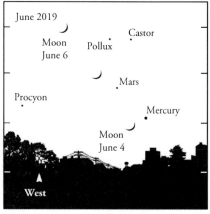

Can you locate Mercury without any other celestial landmarks after sunset in late February and early March?

The young Moon sweeps past Mercury (–0.8) and dim Mars (1.8) in the evening sky as June opens

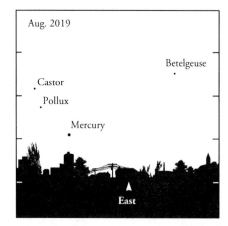

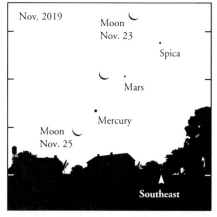

By the morning Aug. 9, Mercury (0.1) already bests the nearby stars Castor, Pollux and Betelgeuse, and it becomes more conspicuous over the following week

The waning crescent Moon swings past the star Spica, Mars (1.7) and Mercury (–0.3) in late November mornings

Fig. 3.20 Mercury in 2019, looking 30 min before sunrise or after sunset

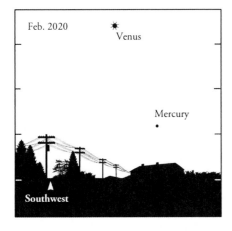

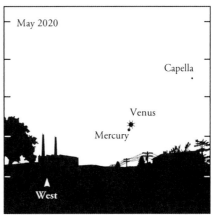

Mercury (–1.0) pops into the evening sky beneath brilliant Venus (–4.1) in early February

Mercury makes another nice evening appearance with Venus in mid-May. On May 21, catch Mercury (–0.6) just over a degree from blazing Venus (–4.3)

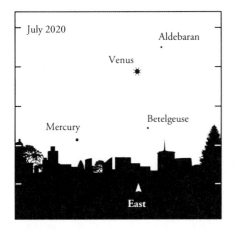

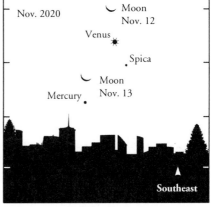

Catch Mercury and Venus (–4.6) together again before dawn on late July mornings

The Moon swings past the innermost planets Nov. 12 and 13. Venus (–3.9) is far from its brightest but still dazzling; Mercury (–0.7) far outshines the nearby star Spica

Fig. 3.21 Mercury in 2020, looking 30 min before sunrise or after sunset

Worlds in Silhouette

Since inferior conjunctions bring Mercury and Venus roughly in line between Earth and the Sun it stands to reason that, sooner or later, they will pass close enough to appear in silhouette against the Sun's disk. Such events, called transits, occur about 13 times per century for speedy Mercury and so cannot be considered rare, but only six Venus transits have been observed since the invention of the telescope, and when Venus transited the Sun in 2004, no one alive had ever seen one. In the 18th and 19th centuries, transits of Venus generated considerable excitement among astronomers and prompted the first international scientific expeditions. The reason was that transits seemed to hold the key for determining a poorly known yet essential astronomical measurement, the distance between Earth and the Sun.

Both Claudius Ptolemy and Copernicus had considered the possibility of Mercury and Venus transits, but Johannes Kepler (1571–1630) was the first to make specific predictions. Kepler noticed that Mercury would pass in front of the Sun on Nov. 7, 1631, and that, astonishingly, Venus would do the same thing a month later. On the appointed morning, French astronomer Pierre Gassendi (1592–1655) watched for the event from his Paris apartment. He used a telescope to safely project a magnified image of the Sun's disk onto a screen in a darkened room. A few hours before the time Kepler had predicted, as the Sun peeked out from occasional clouds, Gassendi saw something that looked like a very small sunspot. Could this tiny black dot – just 1/200th the apparent diameter of the Sun – really be the planet Mercury? Its slow but steady progress across the Sun's disk removed all doubt. "I found him out, and saw him where no one else had hitherto seen him," Gassendi triumphantly wrote. The following month he looked for but failed to see the Venus transit, which ended before sunrise for most of Europe.

Just 3 years before his death, Kepler published his Rudolphine Tables, which gave astronomers state-of-the-art capabilities to calculate planetary positions in the past and future. Based on Kepler's revolutionary understanding of planetary orbits, these tables elevated positional accuracy by ten times over anything that came before them. Using his tables, the German astronomer prepared an annual list of positions, called an ephemeris, for each year up through 1636. It was during the preparation of the 1631 ephemeris that Kepler noticed the transits of Mercury and Venus, but he never looked far enough down the road to notice that a second Venus transit would occur 8 years later.

That realization occurred in late October 1639, when Jeremiah Horrocks (1618–1641), a young English clergyman with a passion for mathematics and astronomy, made his own calculations based on Kepler's tables. Horrocks excitedly realized that the Venus transit would occur barely a month in the future! He "rejoiced exceedingly in the prospect of seeing Venus" and advised his friend and fellow astronomer, William Crabtree (1610–1644), to watch for it. The two planned to observe the event from their homes using telescopes to project the Sun's image, just as Gassendi had done. Horrocks monitored the Sun from sunrise on Dec. 4 until the early afternoon, when he was required to perform the Sunday services, and when he returned, he found the transit already in progress:

> I then beheld a most agreeable spectacle, the object of my sanguine wishes, a spot of unusual magnitude and of perfectly circular shape, which had already fully entered upon the Sun's disc on the left, so that the limbs of the Sun and Venus precisely coincided, forming an angle of contact.

Horrocks got back to his telescope less than half an hour before sunset, so he made what measurements he could as quickly as possible. Venus, just 1/30th of the

Sun's diameter, turned out to be some ten times smaller than some astronomers expected. Crabtree also caught the transit just before sunset and, according to Horrocks, was "rap't in contemplation ... motionless ... through excess of joy" – so overwhelmed that he nearly forgot to make measurements. Horrocks knew that the two of them were probably the only ones to have witnessed the event and, following Gassendi, he wrote up a detailed account of their observations titled *Venus Visible on the Sun*. He had planned to finish his manuscript and meet with Crabtree as 1641 opened, but on Jan. 3 he suddenly died. Twenty years later, a copy of this manuscript found its way into the hands of Polish astronomer Johannes Hevelius, who published it in 1662 as an appendix to his own observations of a Mercury transit.

The following year, Scottish mathematician James Gregory proposed a general method of using transits to measure the distance between Earth and the Sun. This number was effectively an astronomical yardstick – with it, only Kepler's laws were needed to determine the scale of the entire solar system. Efforts to measure this distance – by, for instance, observations of Mars at opposition or Mercury transits – failed to produce consistent results; values ranged over a factor of two or more. Enter Edmond Halley (1656–1742), the man who financed the publication of Newton's *Principia* – and coaxed him to write it in the first place – and who became the first to predict the return of a comet, the one that today bears his name. Halley showed in 1716 that Mercury was too close to the Sun for observations of its transits to provide effective results, but that Venus was ideal. Looking ahead, he proposed a detailed plan for observations of the 1761 Venus transit and suggested that astronomers be dispatched across the globe to observe and time the planet's passage across the Sun. Observers separated by great distances would see Venus track across the Sun in slightly different locations. Halley realized that by timing the planet's entrance onto and exit from the Sun's disk, the paths seen by each observer could be reconstructed. When combined with the latitude and longitude of the observing sites, the angle between the paths – the parallax of Venus – could be determined and from this, with a little trigonometry, astronomers could find the distance between the Earth and the Sun. Halley believed that his technique could establish the length of the astronomical yardstick to one part in 500.

More than a hundred astronomers at over 40 locations throughout Europe observed the 1761 transit and, as Halley had hoped, Britain and France mounted expeditions overseas. Astronomers were sent to Newfoundland, Sumatra, St. Helena, Siberia and the Indian Ocean. Unfortunately, the Seven Years War then raging between Britain and France proved a complication. British astronomer Charles Mason and surveyor Jeremiah Dixon were only hours into their Sumatra-bound voyage aboard *HMS Sea Horse* when the 34-gun French frigate *Le Grande* intercepted it. The two vessels blasted away at each other for an hour before withdrawing, and the badly damaged *Sea Horse* limped back to port with 11 dead and more than 3 dozen injured. Understandably, Mason and Dixon now had second thoughts about a long sea voyage and reported to the Royal Society of London that they "will not proceed thither, let the Consequence be what it will." The Royal Society replied with a strong rebuke that outlined exactly what the consequences would be. The duo's refusal to continue could not "fail to bring an indelible Scandal upon their Character, and probably end in their utter Ruin," something the society would happily guarantee by prosecuting them "with the utmost Severity of the Law." The reluctant voyagers changed their minds again.

When they learned that their destination in Sumatra had been taken by the French, Mason and Dixon decided to observe from the Cape of Good Hope at the southern tip of Africa. Their observations, which were of the highest quality, showed what the pair could do, and 2 years later they were selected for a more geographical challenge in the American colonies. For decades, Maryland and Pennsylvania, which at the time included counties that would later become Delaware, had fought over their common border. Starting in 1763 and continuing for 5 years, Mason and Dixon used astronomical observations to rigorously establish this boundary, which they marked with limestone blocks. They concluded the project in plenty of time to join expeditions organizing for the next Venus transit, but in the U.S. the team's claim to fame stems from their work defining what is now known as the Mason-Dixon Line.

For French astronomer Guillaume Le Gentil de la Galaisière, the 1761 transit expedition proved to be nothing less than a tragicomic odyssey. His ship had just arrived in India when news reached the captain that the town of Pondicherry, near the appointed observing site, had been taken by British forces. The ship immediately headed back to Mauritius, but when transit day arrived it was still at sea. Le Gentil watched the event but was unable to make any useful observations because his pendulum-driven clocks, crucial for the timings, were useless on the heaving sea. With so little to show for so long a voyage, Le Gentil decided to remain in the area until the next transit. For much of the next 8 years he roamed the Indian Ocean and explored and mapped islands from Madagascar to the Philippines. He set up for the 1769 transit at his original site of Pondicherry, which had since returned to French control. But on transit day, clouds rolled in and covered the Sun. In his journal, he wrote:

> That is the fate which often awaits astronomers. I had gone more than a thousand leagues [30,000 miles or 48,000 km]; it seemed that I had crossed such a great expanse of seas, exiling myself from my native land, only to be the spectator of a fatal cloud…. I was more than two weeks in a singular dejection and almost did not have the courage to take up my pen to continue my journal.

Le Gentil fell seriously ill and remained so for months. When he recovered enough to travel, he could bear his self-imposed exile no longer; it was time to go home. But for him, even that would not be easy. On his first homeward-bound attempt, in November 1770, his ship was nearly wrecked in a violent storm and barely managed to sail back to its originating port. He finally returned to France in October 1771 "after eleven years, six months, and thirteen days of absence." But no one had heard from Le Gentil for so long that he was presumed dead. In a final insult, his heirs and creditors had divided up his assets among them, so Le Gentil had to lawyer up and sue to reclaim his property. In the end, though, he married happily and wrote acclaimed memoirs of his extensive, if astronomically lackluster, travels.

Even with the multitude of observations, some acquired from remote parts of the globe, astronomers could determine the astronomical yardstick only to within 20%. Inadequate knowledge of the position of the actual observing sites, particularly their longitude, contributed to the low precision. More important was an unforeseen problem that foiled efforts to precisely time the moment Venus passed completely onto the Sun. At what should have been the moment of final contact, when astronomers expected to see a thread of light separating the planet's silhouette from the Sun's limb, observers noted that Venus looked like a droplet with a black tail connecting it to the Sun's edge. This "black drop" could last from seconds to a full minute and seriously degraded the timing observations. Although the black drop has sometimes been explained as being due to Venus' thick atmosphere, its roots are much more mundane.

Images made by a telescope can become smeared by a variety of effects, including unsteady air, flaws in the optics and the diffraction and scattering of light within the telescope. Even with a perfect optical system, the view through a telescope is subject to atmospheric blurring and internal scattering, and these effects tend to be more prominent in telescopes with smaller apertures, such as those used in the eighteenth century expeditions. The black drop was less noticeable in subsequent transits and was rarely reported in 2004. This is a result of the steady improvement in both the size and quality of telescope optics, which addresses two significant sources of image smearing. As Louisiana State University astronomer Bradley Schaefer has noted, the smearing explanation for the black drop was first put forward by J.J. Lalande in 1770. So, the black drop has become less pronounced because telescopes have gotten bigger and better. Astronomers observing the 1999 Mercury and 2004 Venus transits with a satellite telescope have demonstrated that an additional solar factor also plays a role: The Sun's disk is darker at its edges than at its center. We see the brightest, hottest gases at the center of the disk and slightly cooler, dimmer layers on the Sun's limb. This falloff of light at the Sun's edge distorts the smeared planet image toward the Sun's limb. Only when both effects were removed from satellite images did Mercury or Venus show perfectly circular silhouettes.

Combining the results of the 1761 and 1769 transits further narrowed the margin of error, bringing values of the Earth-Sun distance to within 4% of its true value. But even if the Venus transits could not meet what clearly had been overly optimistic expectations, they had provided astronomers with a much better measure of the scale of the solar system. The transit of 1874 was well observed around the world, but by 1882 most of the scientific enthusiasm for transits had dissipated as other techniques – photographic observations of Mars and asteroids, for instance – promised to whittle down remaining errors. Now we can measure the distances to Venus, Mars, asteroids and other solar system objects by pinging them with radar, sending a pulse of energy from a radio telescope and timing the wait for the returned signal. The average distance between the Earth and the Sun, formally called the Astronomical Unit and abbreviated AU, is now known to within a few meters, a precision far exceeding Halley's promise.

As seen in Table 3.4, Mercury transits the Sun in 2016 and 2019, but the planet is so small that observing these events requires a telescope. (Times for these transits are provided in Appendix A.) The orbit of Mercury does not lie in the same plane as Earth's, so usually Mercury passes slightly above or below the Sun at each inferior conjunction. For a transit to occur, Mercury must come to inferior conjunction when it lies near one of two points where its orbit intersects the plane of Earth's orbit, points called the orbital nodes. So for Mercury to transit, the planet must come

Table 3.4 Transits of Mercury and Venus

Mercury	Venus
Nov. 6, 1993	Jun. 6, 1761
Nov. 15, 1999	Jun. 3, 1769
May 7, 2003	Dec. 9, 1874
Nov. 8, 2006	Dec. 6, 1882
May 9, 2016	Jun. 8, 2004
Nov. 11, 2019	Jun. 6, 2012
Nov. 13, 2032	Dec. 11, 2117
Nov. 7, 2039	Dec. 8, 2125

Fig. 3.22 Venus appears as the dark circular dot on the Sun's disk. The image was taken as the Sun rose over the Canary Islands during the 2004 transit (Royal Swedish Academy of Sciences)

to inferior conjunction within 3 days of May 8 or within 5 days of Nov. 10. The "transit windows" are not equally sized in part because Mercury's orbit is very elongated. Also, perihelion occurs near the November node and because Mercury is moving faster then, we have a better chance of catching it near the node. As a result, the number of November transits is almost double the number of May transits.

All very well, but the rarity and ease of observing a Venus transit makes it a truly special event: Until 2004, no living person had seen a transit of Venus (Fig. 3.22). When it happens again in 2012, skywatchers all over the world will watch Venus slip across the solar disk. With a safe means of viewing the Sun, such as one of those described in Box 4.1 on page 98, Venus will be plainly visible even without a telescope – a perfectly circular dark spot about an arcminute across. The mechanics for Venus transits are similar to Mercury's. Venus must come to inferior conjunction when it lies near one of the nodes of its slightly tilted orbit. This works out to be within 2 days of Jun. 7 and Dec. 9. Because the orbit of Venus is nearly circular, these transit opportunities are nearly the same size and about equally likely to produce a transit. Venus transits occur in pairs separated by 8 years – the wonderful Venus cycle at work – so the transit of June 2004 is followed by one in June 2012. After each pair of June transits, there is a gap of 105.5 years before a pair of December transits occurs, and these are followed by a span of 121.5 years for the next pair of June transits. This current pattern of paired transits isn't fixed: It began in 1518, and centuries from now the pairs will again dissolve into a single transit at each node.

The last opportunity for those alive today to watch Venus transit the Sun occurs over 6 h on Jun. 5 and 6, 2012. Astronomers discuss the times of events like this in terms of an imaginary observer located at Earth's center; this is because the exact track of the transit – and thus its predicted start and end times – depends on each observer's geographic location. Geocentric times for the principal transit events are shown in Figs. 3.23 and 3.24. For the sake of simplicity, astronomers also give these contact times in Universal Time or UT (essentially Greenwich Mean Time; see Appendix A for more on this). Some of the Web resources listed in Table 3.5 provide detailed circumstances for locations world-

Table 3.5 Mercury and Venus on the Web

Active missions
MESSENGER
messenger.jhuapl.edu

Venus Express
venus.esa.int

Past missions
Venera: The Soviet Exploration of Venus
www.mentallandscape.com/V_Venus.htm

Magellan Mission to Venus
www2.jpl.nasa.gov/magellan/

Venus transits
2004 and 2012 transits of Venus
eclipse.gsfc.nasa.gov/transit/venus0412.html

Smithsonian Institution: Chasing Venus
www.sil.si.edu/exhibitions/chasing-venus

TransitofVenus.org
www.transitofvenus.org

The Transit of Venus (with local predictions)
www.transitofvenus.nl

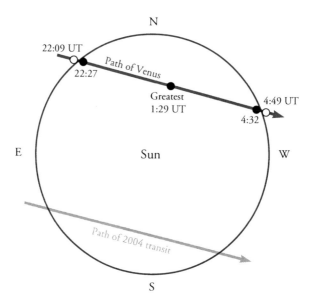

Fig. 3.23 General circumstances for the 2012 Venus transit, with times for significant events given in Universal Time (see Appendix A for more about this). During the transit, the Sun measures 31.5 arcminutes across while Venus spans slightly less than 1 arcminute. This makes Venus just visible to the unaided eye as a small black dot when seen through a safe solar filter. For comparison, Venus' path across the Sun during the 2004 transit is also shown

wide, but for most viewers it's sufficient to know that the transit contact times for any spot on the globe do not differ from the geocentric times by more than 7 min. All of North America will see Venus move onto the Sun before sunset. For major cities along the U.S. East Coast, the planet's ingress onto the Sun begins less than 90 min before sunset, but locations farther west will see improved viewing conditions. From Los Angeles, Venus first notches the Sun's disk about 4 h before sunset. From Hawaii, the Sun remains above the horizon throughout the transit, which is also true for large parts of Asia and the Pacific, including New Zealand and most of Australia. From Tokyo, the transit begins on Jun. 6 a couple of hours after sunrise. And in Europe, toward the end of the region of visibility, observers in Paris watching the Jun. 6 sunrise will see Venus in silhouette for about an hour before the planet exits the Sun's disk and the transit concludes.

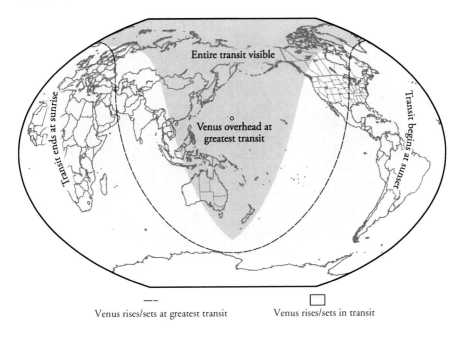

Venus rises/sets at greatest transit   Venus rises/sets in transit

Fig. 3.24 World visibility of the 2012 transit. The shaded portions of the map indicate what regions on the globe will be turned toward the Sun while the transit is in progress. Thanks to the International Date Line (not shown) the Western Hemisphere (*right*) sees the transit before sunset on June 5, and the Eastern Hemisphere (*left*) sees it after sunrise on June 6. The darker shading shows regions that will see the entire transit; the lighter shading shows areas where the Sun rises (*left*) or sets (*right*) during the transit. The dashed curve shows where the Sun will rise/set at the moment of greatest transit, when Venus lies closest to the Sun's center. All of North America will see the transit begin before sunset on June 5. For large parts of Asia and the Pacific, including Hawaii, Japan, Alaska, New Zealand and most of Australia, the Sun is above the horizon for the entire event

Once considered an important tool for astronomical progress, transits of Venus have since become novel reminders of the scale, motions, and predictability of the solar system that provide opportunities to reminisce about a heroic age in the history of astronomy. Following the transit of 1882, William Harkness of the U.S. Naval Observatory placed the event in perspective. "When the last [18th century] transit occurred the intellectual world was awakening from the slumber of ages," he wrote, "and that wondrous scientific activity that has led to our present advanced knowledge was just beginning. What will be the state of science when the next transit season arrives God only knows." As the silhouette of Venus leaves the Sun's disk in 2012, it's inevitable that we'll momentarily pause to wonder what the world will be like in 2117. Now, the scientific importance of transits lies far beyond the solar system. Astronomers have cataloged more than a hundred exoplanets by tracking the slight dip in a star's brightness caused by a planet passing over its disk; hundreds more have been found by other techniques. So far, none of these planets truly resemble Earth. But when Venus next crosses the Sun, how many earthlike worlds will astronomers know about, and will any of them show evidence of life?

References

Anon (1897) That experimental star. Boston Evening Transcript, April 15, 4
Anon (1996) NASA's Aviation Safety and Reporting System Database, report number 333382. http://asrs.arc.nasa.gov/search/database.html. Accessed 10 Sept. 2010
Anon (1999) The parallax effect. ASRS Callback, No. 246, Dec. 1999 http://asrs.arc.nasa.gov/docs/cb/cb_246.pdf. Accessed 10 Sept. 2010
Bauer BS, Dearborn DSP (1995) Astronomy and empire in the ancient Andes. Univ. of Texas Press, Austin
Bindschadler DL (1995) Magellan: A new view of Venus' geology and geophysics. http://www.agu.org/journals/rg/rg9504S/95RG00281/node14.html. Accessed 10 Sept. 2010
Bray RJ (1980) Australia and the Transit of Venus. Proc. Astron. Soc. Aust. 4:114–120
Crelintsen J (2006) Einstein's jury: The race to test relativity. Princeton Univ. Press, Princeton, New Jersey
Fernie JD (1997) Transits, travels and tribulations II. Am. Sci. 85: 418–421
Fernie JD (1998) Transits, travels and tribulations III. Am. Sci. 86: 123–126
Kollerstrom N (2004) William Crabtree's Venus transit observation. In: Kurtz DW (ed.) Transits of Venus: New views of the solar system and galaxy. Proc. IAU Colloq No. 196. doi:10.1017/S1743921305001249
Lalande JJ (1803) Bibliographie astronomique. Imprimerie de la République, Paris
Maor E (2000) June 8, 2004: Venus in transit. Princeton Univ. Press, Princeton, New Jersey
Marov MY (2004) Mikhail Lomonosov and the discovery of the atmosphere of Venus during the 1761 transit. In: Kurtz DW (ed.) Transits of Venus: New views of the solar system and galaxy. Proc. IAU Colloq. No. 196. doi:10.1017/S1743921305001390
Meeus J (1995)Astronomical tables of the Sun, Moon, and planets. Willmann-Bell, Richmond, Virginia
Mollel TM (1990) The orphan boy: A Maasai story. Clarion Books, New York
Norris R (2009) Searching for the astronomy of aboriginal Australians. In: Jonas Vaiskunas (ed.) Astronomy & cosmology in folk traditions and cultural heritage. Klaipéda Univ. Press, Klaipéda, Lithuania
Parker RA (1974) Ancient Egyptian astronomy. Phil. Trans. R. Soc. Lond. A. 276:51–65
Pasachoff JM, Schneider G, Golub G (2004) The black-drop effect explained. In: Kurtz DW (ed.) Transits of Venus: New views of the solar system and galaxy. Proc. IAU Colloq. No. 196. doi:10.1017/S1743921305001420
Prockter LM, Ernst CM, Devevi BW et al (2010) Evidence for young volcanism on Mercury from the third MESSENGER flyby. Sci. 329: 668–671. doi: 10.1126/science.1188186
Rhodes R (1986) The making of the atomic bomb. Simon and Schuster, New York
Schaefer BE (2001) The transit of Venus and the notorious black drop effect. J. Hist. Astron. 32: 325–336
Schele L, Friedel D (1990) A forest of kings: The untold story of the ancient Maya. Quill, New York
Smrekar SE, Stofan ER, Mueller N et al (2010) Recent hotspot volcanism on Venus from VIRTIS emissivity data. doi: 10.1126/science.1186785

Solomon SC, McNutt RL Jr, Watters TR et al (2008) Return to Mercury: A global perspective on MESSENGER's first Mercury flyby. Sci. 321: 59–62. doi: 10.1126/science.1159706

Warner B (1996) Traditional astronomical knowledge in Africa. In: Walker C (ed.) Astronomy before the telescope. St. Martin's Press, New York

Wolkenstein D, Kramer SN (1983) Inanna, queen of heaven and earth: Her stories and hymns from Sumer. Harper and Row, New York

Chapter 4

Eclipses of the Sun and Moon

Ask eclipse expert Fred Espenak why it's important to see a total solar eclipse sometime in your life, and he'll tell you in no uncertain terms: "It's simply the most spectacular astronomical event that you can witness firsthand with the naked eye, bar none," he said. "There's nothing else that compares to it, and it's next to impossible to try to explain what the experience is like to somebody who hasn't seen one." Having observed more than 20 of them, he ought to know. Although officially retired from NASA's Goddard Space Flight Center in Greenbelt, MD, Espenak still maintains an office and continues to update his NASA web site, publish detailed bulletins for upcoming eclipses and, of course, travel to re-experience the unique, fleeting moments of totality.

While a total eclipse of the Sun is hands-down the premier event, it's one of several types of solar cover-ups, and the Moon also undergoes its own set of disappearing acts. The Sun travels through the sky along the ecliptic and the Moon and planets travel roughly the same path, their deviations resulting from the way each object's orbit is tilted. Because these objects move at different speeds and follow similar paths around the sky, it stands to reason that, sooner or later, some of them must arrive at the same sky position at the same time. When this happens, the nearer object appears to move in front of the more distant one, briefly obscuring it from our view. Astronomers call these events "occultations." The Moon frequently occults planets and stars near the ecliptic, but the most sensational occultations occur when the Moon passes in front of the Sun, blocking its light and creating a brief eerie darkness in the middle of the day – a total eclipse of the Sun. The striking nature of total solar eclipses explains why we call the Sun's path around the sky the ecliptic instead of the "occultic."

Something similar occurs when the full Moon passes directly into Earth's shadow, where it becomes cut off from the Sun's light and turns dark. In this case,

F. Reddy, *Celestial Delights: The Best Astronomical Events through 2020*,
Patrick Moore's Practical Astronomy Series, DOI 10.1007/978-1-4614-0610-5_4,
© Springer Science+Business Media, LLC 2012

our planet is occulting the Sun as seen from the Moon. While a total lunar eclipse is less spectacular than its solar equivalent, it lasts longer and can be seen from just about anywhere on Earth where the Moon is above the horizon. The Moon doesn't disappear at totality but instead displays a striking orange color. We'll discuss more about lunar eclipses later in this chapter.

The Sun's importance to life on Earth was recognized everywhere, which made its sudden, brief disappearance during a total solar eclipse all the more frightening. In mythology, eclipses are most commonly described as attacks on the Sun and Moon either by heavenly monsters – dragons in China, snakes in Indonesia, the demon Rahu in India, and jaguars or wolves in the Americas – or, somewhat more accurately, by one against the other in battle. Some Inuit groups believed that the eclipsed body had simply left its place in the sky to check on earthly affairs. Rio Grande Pueblos feared that when the Sun Father's gleaming shield faded, he was displeased and moved away from the Earth. In areas of the South Pacific, eclipses were viewed more romantically as the lovemaking of the Sun and Moon. In many regions people were called upon to assist the eclipsed object by frightening off the forces responsible for hiding it. Often, this was done by making loud noises, but the Ojibwa of the upper Midwest went further, shooting flaming arrows into the sky.

The midday darkness of a totality holds unexpected power even today. For several minutes, the warm, brilliant Sun is transformed into a dark hole in the sky. The temperature drops, animals and insects react as if night has come, and for many people there's an emotional reaction that transcends the astronomical facts, in essence a deep sense that something has gone awry with the universe. "I've had that reaction at every single eclipse," Espenak said. "It's a very visceral in-the-gut thing that you don't anticipate. It really seems like the end of the world." More than any other event in nature, the moment of totality briefly places us at the crossroads of folklore and science, giving a little bit of insight into what ancient people who had no idea an eclipse was about to happen might have thought and felt. Some describe the moment of totality as a moving spiritual experience, and many travel the world to greet the Moon's shadow.

That the Moon covers the Sun at all results from two of the most remarkable coincidences in nature. First, as seen from the Earth, both the Sun and the Moon appear to be about the same size. The Sun is the nearest star, a self-luminous ball of gas 864,989 miles (1.39 million km) wide that glows as a result of the tremendous energy released by nuclear reactions occurring deep within it. The Moon is a rocky body 400 times smaller. Yet their distances from Earth differ by the same factor, so both objects appear about half a degree across to us. Earth's orbit around the Sun and the Moon's orbit around Earth are elliptical, and the combination of these slightly out-of-round paths can make the Sun or the Moon appear a few percent larger than the other at certain times. Obviously, if the Moon appears smaller than the Sun then a total eclipse is simply impossible. All other factors being equal, if the Moon were less than 10% smaller than it is, one of nature's grandest spectacles could never occur.

The second coincidence involves changes in the Moon's orbit over time. Gravitational interactions between Earth and the Moon cause it to recede from us by a small amount each year. This increasing distance gradually shrinks the apparent angular size of the Moon's disk. By a happy accident, human civilization arose during the period when total solar eclipses could still occur. One day, some hundreds of millions of years hence, the Moon will be too far away to completely cover the Sun, forever ending the phenomenon of a total solar eclipse.

Before venturing into the detail, we first need to get our bearings. A solar eclipse occurs when the new Moon passes in front of the Sun and the Moon's shadow is cast onto Earth, where observers see the Moon covering a portion of the Sun's disk. The Moon's shadow has two components: a broad, diffuse penumbra where the Sun's light is only partially cut off, and a much darker and smaller umbra where no part of the Sun can be seen. The Moon's precise path across the Sun dictates how these shadow components fall onto Earth's surface, and this in turn determines the type of solar eclipse that takes place and the locations that will witness it.

Sometimes the Moon obscures only a portion of the Sun. These *partial* eclipses – the most common type – are visible over much of the globe. Sometimes the Moon passes completely over the Sun, but its angular size happens to be too small to cover the disk. This is called an *annular* eclipse. From the right locations, the Moon appears centered on the Sun's disk, covering everything except a thin ring, or annulus, that surrounds the silhouetted Moon. However, this scene is visible only from a narrow strip of Earth's surface; seen from anywhere else, the Moon never completely passes onto the Sun and the event just looks like a partial eclipse. This is also true for *total* eclipses, where the Moon completely blots out the Sun but only along a comparatively narrow track. Under a rare combination of circumstances, an eclipse may be annular along one part of its track and total at another – a *hybrid* eclipse. With these basics out of the way, we'll now take a look at each type of solar eclipse in a little more detail.

A partial solar eclipse is one where only the Moon's broad penumbral shadow intersects Earth – the umbra misses us completely – and as a result, partial solar eclipses can be seen over a large region. As many as five solar eclipses can occur in any one year, but for this to occur four of these eclipses must be partial. The last time this happened was 1935; the next will be 2206. According to Espenak, this situation is quite rare, occurring only 56 times in the ten-millennia span between 3000 B.C. and A.D. 7000. As few as two solar eclipses may occur in any given year, and while both of them may be partial, different combinations are actually more common. Over the long haul, about 35% of all solar eclipses are partial, and because a portion of the Sun always remains visible, enjoying them requires some form of eye protection (see "Safe Sun-viewing" on page 98).

Sometimes the Moon's umbra points straight toward Earth but the Moon fails to cover up the Sun. Because the Moon's orbit is elliptical, its distance to us – and therefore its apparent diameter – can change from its mean value by more than 6%. Earth's orbit around the Sun is also an ellipse, which means that the Sun's distance and apparent diameter in the sky vary by some 3% from its mean value. Taken together, these differences can make the Sun's disk appear up to 11% bigger than the Moon. That makes totality impossible and results in an annular eclipse, where observers in the right locales will see a slender ring of uneclipsed Sun surrounding the Moon's dark silhouette. The path of ideal viewing locations may stretch some 10,000 miles (16,090 km) across Earth, but it's terribly narrow – typically just a few hundred miles wide. Those outside of this track won't see the annulus because from their location the Moon will never move completely onto the Sun's disk. In fact, it's helpful to regard an annular eclipse as little more than an unusually symmetrical partial eclipse – a portion of the Sun's disk always remains visible, so eye-safe viewing methods are required at all times. And while annular eclipses are actually a little more common than total eclipses, they exhibit none of the most interesting phenomena associated with totality.

Hybrid eclipses are true oddballs. They mark a transition between conditions favorable for an annular eclipse and those favoring totality. The eclipse is annular at both ends of the central visibility track. But near the middle of this path, Earth's curvature brings the surface close enough to the Moon that its disk appears big enough to cover the Sun. Observers there see a total eclipse, although the period of totality is typically quite brief, usually less than 90 s. The next hybrid eclipse, on Nov. 3, 2013, brings totality to almost the entire track, with annular eclipses only at the start and end. There are only seven hybrid eclipses this century and they make up less than 5% of solar eclipses.

That leaves us with the main event, totality, which accounts for about 27% of all solar eclipses (Fig. 4.1). For observers in the U.S., the decade holds one eclipse that

Fig. 4.1 Afternoon darkness. On August 1, 2008, the Moon's shadow passed over Novosibirsk, Russia, delivering a total eclipse lasting 2 minutes 18 seconds. A tripod-mounted camera was programmed to snap one photograph every 5 min throughout the eclipse. A solar filter protected the camera but was removed at totality to capture the Sun's corona and the foreground scene. Venus appears above the tree at left and Mercury shines above the arc of solar images. (Ben Cooper)

could hardly be better planned for maximum viewing opportunities. In Aug. 2017, the Moon's shadow will come ashore near Lincoln Beach, OR, sweep through the nation's center – bringing totality to Lincoln, NE, and Kansas City, MO – and pass into the Atlantic Ocean near Charleston, S.C. Although the Sun will be darkened for little more than a minute on the populous coasts – and just 2 min 40 s at the point of greatest eclipse, near Hopkinsville, KY – this is the first total solar eclipse visible from the continental U.S. since 1979. (See Table 4.1 for a list of all eclipses through 2020).

Sooner or later, every place on Earth experiences a total solar eclipse, but if you decided to forego travel, your wait time would be about 375 years, on average. After 2017, residents of southern Illinois and eastern Missouri will see totality again in 2024, so the average clearly doesn't tell the whole story. As in real estate, the secret for seeing totality is location, location, location.

Lacking the good fortune of being in exactly the right place at exactly the right time, the only way to see a total solar eclipse is by planning a trip to intercept the Moon's shadow. Over 100,000 eclipse watchers journeyed to Hawaii and Mexico to witness the great total solar eclipse of Jul. 11, 1991; hotels were booked years in advance, so be warned and plan ahead accordingly. In Aug. 1999, nearly a million people reportedly descended on Cornwall, England, to see the Moon's shadow come ashore for the last totality of the millennium. Meanwhile, tourists with a champagne-and-caviar bent flew through the umbra aboard the supersonic Concorde, which moved fast enough to extend the duration of totality for several additional minutes. Millions more watched along a track that brought totality across the most populous regions of Europe and the Middle East. Jordan and Syria declared national holidays, Iran issued 600,000 Sun-safe eclipse glasses and Tehran Radio announced that prayer during the eclipse was mandatory for Muslims. In 2001, when totality swept through the African nation of Zambia, more than 21,000 tourists filled Lusaka hotels to capacity – and authorities were prepared to set up tents to handle any overflow.

The rarity of these events and the fact that none of the phenomena accompanying totality can be seen during other types of eclipses explains part of the excitement. There is another aspect of totality, something that is difficult to express – a primal reaction more felt than thought and sensed differently by each observer. It elevates totality, at least in the human mind, to something grander than the predictable clockwork of celestial mechanics; see the box "A writer in the umbra" for one evocative first-hand account. But while eclipse travelers greet the daytime darkness with whoops of delight, for some locals the Sun's brief disappearance may not be regarded so pleasantly no matter how well astronomers and the media prepare them. Before the 2001 Zambia eclipse, Espenak and his group visited a village about a half mile from their observing site. "We handed out some solar filters and explained the best that we could about what was going to happen," he recalled, "and I'm sure they had heard from other people because it was well publicized." But as he watched the Moon's shadow approaching in the last seconds before totality, when the light started dropping dramatically, he heard sounds coming from the village that could only be described as screams of terror. "It just sent a shiver down your backbone to hear people reacting like that to it," he said.

Table 4.1 Solar and lunar eclipses through 2020

Year	Date (UT)	Eclipsed object	Eclipse type	Greatest eclipse (UT)	Global visibility and map page
2011	Jan. 4	Sun	Partial	08:51	Europe, Africa, and central Asia
	Jun. 1	Sun	Partial	21:16	Eastern Asia, northern North America, Iceland
	Jun. 15	Moon	Total	20:13	South America, Europe, Africa, Asia, Australia; see p. 125
	Jul. 1	Sun	Partial	8:38	Southern Indian Ocean
	Nov. 25	Sun	Partial	6:20	Southern Africa, Antarctica, Tasmania, New Zealand
	Dec. 10	Moon	Total	14:32	Europe, eastern Africa, Asia, Australia, Pacific, North America; see p. 126
2012	May 20	Sun	Annular	23:53	Asia, Pacific, North America. Annular in western U.S.; see p. 127
	Jun. 4	Moon	Partial	11:03	Asia, Australia, Pacific, Americas; see p.128
	Nov. 13	Sun	Total	22:12	Australia, New Zealand, South Pacific, southern South America; see p. 129
	Nov. 28	Moon	Penumbral	14:33	Europe, eastern Africa, Asia, Australia, Pacific, North America
2013	Apr. 25	Moon	Partial	20:07	Europe, Africa, Asia, Australia; see p. 130
	May 10	Sun	Annular	0:25	Australia, New Zealand, central Pacific; see p. 131
	May 25	Moon	Penumbral	04:10	Americas, Africa
	Oct. 18	Moon	Penumbral	23:50	Americas, Europe, Africa, Asia
	Nov. 3	Sun	Hybrid	12:46	Eastern Americas, southern Europe, Africa; see p. 132
2014	Apr. 15	Moon	Total	7:46	Australia, Pacific, Americas; see p. 133
	Apr. 29	Sun	Annular	6:03	Southern Indian Ocean, Australia, Antarctica; annular in Antarctica only
	Oct. 8	Moon	Total	10:55	Asia, Australia, Pacific, Americas; see p. 134
	Oct. 23	Sun	Partial	21:44	Northern Pacific, North America
2015	Mar. 20	Sun	Total	09:46	Iceland, Europe, northern Africa/Asia; see p. 135
	Apr. 4	Moon	Total	12:00	Asia, Australia, Pacific, Americas; see p. 136
	Sept. 13	Sun	Partial	6:54	Southern Africa and Indian Ocean, Antarctica
	Sept. 28	Moon	Total	2:47	Eastern Pacific, Americas, Europe, Africa, western Asia; see p. 137

(continued)

Table 4.1 (continued)

Year	Date (UT)	Eclipsed object	Eclipse type	Greatest eclipse (UT)	Global visibility and map page
2016	Mar. 9	Sun	Total	1:57	Eastern Asia, Australia, Pacific; see p. 138
	Mar. 23	Moon	Penumbral	11:47	Asia, Australia, Pacific, western Americas
	Sept. 1	Sun	Annular	9:07	Africa, Indian Ocean; see p. 139
	Sept. 16	Moon	Penumbral	18:54	Europe, Africa, Asia, Australia, western Pacific
2017	Feb. 11	Moon	Penumbral	0:44	Americas, Europe, Africa, Asia
	Feb. 26	Sun	Annular	14:53	Southern South America, Atlantic, Africa, Antarctica; see p. 140
	Aug. 7	Moon	Partial	18:20	Europe, Africa, Asia, Australia; see p. 141
	Aug. 21	Sun	Total	18:25	North America, northern South America. Total across the continental U.S.; see p. 142
2018	Jan. 31	Moon	Total	13:30	Asia, Australia, Pacific, western North America; see p. 143
	Feb. 15	Sun	Partial	20:51	Antarctica, southern South America
	Jul. 13	Sun	Partial	3:01	Southern Australia
	Jul. 27	Moon	Total	20:22	South America, Europe, Africa, Asia, Australia; see p. 144
	Aug. 11	Sun	Partial	9:46	Northern Europe, northeastern Asia
2019	Jan. 6	Sun	Partial	1:41	Northeastern Asia, north Pacific
	Jan. 21	Moon	Total	5:12	Central Pacific, Americas, Europe, Africa; see p. 145
	Jul. 2	Sun	Total	19:23	Southern Pacific, South America; see p. 146
	Jul. 16	Moon	Partial	21:31	South America, Europe, Africa, Asia, Australia; see p. 147
	Dec. 26	Sun	Annular	5:18	Asia, Australia; see p. 148
2020	Jan. 10	Moon	Penumbral	19:10	Europe, Africa, Asia, Australia
	Jun. 5	Moon	Penumbral	19:25	Europe, Africa, Asia, Australia
	Jun. 21	Sun	Annular	6:40	Africa, southeastern Europe, Asia; see p. 149
	Jul. 5	Moon	Penumbral	4:30	Americas, southwestern Europe, Africa
	Nov. 30	Moon	Penumbral	9:43	Asia, Australia, Pacific, North and South America
	Dec. 14	Sun	Total	16:13	Pacific, southern South America, Antarctica; see p. 150

Box 4.1 Safe Sun-viewing

The only time when it's totally safe to view the eclipsed Sun is during the total phase, when the Moon completely covers the Sun's disk. All other portions of a solar eclipse require either indirect observing, such as projecting images of the Sun onto paper or a wall, or eye protection. To ensure a safe eclipse, use only these astronomer-approved viewing methods.

Sun-safe filters. Widely available and inexpensive filters include No. 14 welder's glass and various products that use plastic film coated with aluminum. Some commonly encountered names: Eclipsers, Eclipse Glasses, Eclipse Shades® and Solar Skreen®. They're often sold as souvenirs on eclipse trips; see the resources in Table 4.2 for vendors. Use filters designed for naked-eye viewing *only* as intended. Don't use them to view magnified images through binoculars or a telescope (other filters are designed for this type of use). Handle filters carefully and inspect them for scratches or punctures, which could allow unfiltered sunlight into the eye, and replace them if they're damaged.

Pinhole cameras. Sunlight passing through a tiny aperture, like a pinhole, naturally forms a focused image on a distant screen. The width of the Sun's image is about 109 times smaller than the distance between the pinhole and the projection screen, so the longer this distance, the bigger the image. To make a pinhole camera, find a long cardboard box and cut a small square in one end. Tape a single piece of aluminum foil over the square. Carefully pierce the foil with a single hole from a needle or pushpin. Tape a sheet of white paper on the inside surface of the box on the opposite side of the pinhole; this will serve as the projection screen. Hold the box so the pinhole is aimed toward the Sun and watch the screen as you move it around; *do not* look through the pinhole. The Sun's image will show up as a bright spot on the paper.

Binocular projection. This method produces a bigger, brighter solar image than a pinhole device by projecting a magnified image of the Sun onto a sheet of paper or white wall. Mount the binoculars on a tripod. Cut a hole in a large piece of cardboard big enough to slide over one of the tubes; the other won't be needed. The cardboard will shade out direct sunlight, giving the projected image better contrast. *Do not* look through the binoculars to view the Sun, and be sure never to leave them unattended so no one else will. Tilt and sweep the binoculars until the Sun appears on the screen, then focus. Because heat build-up can soften the cement used to attach some of the internal parts of the binoculars, view for only a couple of minutes at a time. Then move the binoculars off of the Sun for a few minutes so they can cool.

Box 4.2   A writer in the umbra

James Fenimore Cooper is often considered America's first great novelist. While Cooper was living in Paris, a friend asked him to write up his recollections of the Jun. 16, 1806, total solar eclipse. The Moon's shadow swept across the United States from Baja California to Massachusetts, and the 17-year-old watched it from his family home in Cooperstown, New York. "The Eclipse," an undated essay written around 1836 and excerpted below, was found by his daughter Susan and published many years after his death in anticipation of another eclipse that tracked through the eastern U.S. The play of light and shadow clearly made an impression on the young Cooper, who wrote that his "recollections of the great event, and the incidents of the day, are as vivid as if they had occurred but yesterday."

> The birds, which a quarter of an hour earlier had been fluttering about in great agitation, seemed now convinced that night was at hand. Swallows were dimly seen dropping into the chimneys, the martins returned to their little boxes, the pigeons flew home to their dove-cots, and through the open door of a small barn we saw the fowls going to roost....
>
> Suddenly one of my brothers shouted aloud, "The Moon!" Quicker than thought, my eye turned eastward again, and there floated the Moon, distinctly apparent, to a degree that was almost fearful. The spherical form, the character, the dignity, the substance of the planet, were clearly revealed as I have never beheld them before, or since. It looked grand, dark, majestic, and mighty, as it thus proved its power to rob us entirely of the Sun's rays... This was no interposition of vapor, no deceptive play of shadow; but a vast mass of obvious matter had interposed between the Sun above us and the earth on which we stood... Darkness like that of early night now fell upon the village....
>
> At 12 minutes past 11, the Moon stood revealed in its greatest distinctness – a vast black orb, so nearly obscuring the Sun that the face of the great luminary was entirely and absolutely darkened, though a corona of rays of light appeared beyond.... That movement of the Moon, that sublime voyage of the worlds, often recurs to my imagination, and even at this distant day, as distinctly, as majestically, and nearly as fearfully, as it was then beheld....
>
> Thus far the sensation created by this majestic spectacle had been one of humiliation and awe. It seemed as if the great Father of the Universe had visibly, and almost palpably, veiled his face in wrath. But, appalling as the withdrawal of light had been, most glorious, most sublime, was its restoration! The corona of light above the Moon became suddenly brighter, the heavens beyond were illuminated, the stars retired, and light began to play along the ridges of the distant mountains... I can liken this sudden, joyous return of light, after the eclipse, to nothing of the kind that is familiarly known. It was certainly nearest to the change produced by the swift passage of the shadow of a very dark cloud, but it was the effect of this instantaneous transition, multiplied more than a thousand fold....
>
> I shall only say that I have passed a varied and eventful life, that it has been my fortune to see earth, heavens, ocean, and man in most of their aspects; but never have I beheld any spectacle which so plainly manifested the majesty of the Creator, or so forcibly taught the lesson of humility to man as a total eclipse of the Sun.

Shadow Play

It's been said that the only time the Moon has a day in the Sun is during a solar eclipse. In all types, the moment when the Moon first encroaches on the Sun's disk is known as first contact. Over the course of about an hour the dark silhouette slides over the Sun, giving it the appearance – viewed through eye-safe filters or projected onto a screen – of a bright cookie slowly being eaten. The eclipse ends when the limb of the Moon detaches from the Sun, a moment known as fourth contact (only total and annular eclipses exhibit second and third contacts, as we'll see below). The Moon covers the Sun to its maximum degree and then, just as slowly, slides off the Sun's face. For a partial eclipse this expanding and receding cookie-bite represents the extent of the visible phenomena. The specific circumstances of each eclipse and of the observer's location on Earth determine whether or not a partial eclipse is visible and if so, how deeply the Moon appears to penetrate. Remarkably, most people won't notice a change in the level of sunlight even when the Moon covers 70% of the Sun.

For an annular eclipse, the moment of greatest interest comes at second contact, when the body of the Moon passes completely onto the Sun. From this point on and for a period of up to twelve and a half minutes, a bright ring of sunlight surrounds the silhouetted Moon. At third contact, the limbs of both disks come together and break the ring, ending the period of annularity.

For total eclipses, the time between second contact and third contact marks the duration of totality. Because the Moon appears wider than the Sun, the definitions of these moments are slightly different than for annular eclipses. Second contact is the moment when the Moon just touches the eastern edge of the Sun; third contact is the moment when the Moon begins to reveal the Sun's western limb. The path of totality spans thousands of miles but is typically just a few hundred miles across, and the shadow sweeps eastward at about 2,100 mph (3,380 km/h). That's about twice the speed Earth rotates at the equator in roughly the same direction as the Moon's shadow travels. These factors conspire to lengthen the duration of totality by slowing down the umbra's apparent speed as it passes observers on the ground. The theoretical maximum for totality is just seven and a half minutes, and all of the longest total eclipses occur close to Jul. 3, the mean date of Earth's greatest distance from the Sun (that is, when the solar disk appears smallest).

Totality is the main event, but it takes an hour for the Moon's disk to fully cover the Sun. As the total phase of the eclipse nears, skywatchers must suddenly come to terms with a rush of new sensations and different phenomena, so it's helpful to know what to look for in advance. What follows, in countdown style, is a typical sequence of events for a total solar eclipse, with tips on what to observe as the Moon's umbra approaches and departs.

T minus 60 min and counting: First contact, when the western edge of the Sun meets the limb of the Moon. The Sun's disk develops a progressively larger "bite" as the dark Moon glides into view. Its slow progress can be monitored directly

Fig. 4.2 Hong Kong was south of the umbra's track for the Jul. 22, 2009, total eclipse, but an east-facing grid of skyscraper windows projected the partial eclipse onto the city. Photographer Alfred Lee captured these multiple reflected images of the crescent Sun from the 27th floor of Two Pacific Place (Alfred Lee)

through protective devices, such as eclipse glasses and properly shielded telescopes, or indirectly by projecting the Sun's image onto a screen (Fig. 4.2). When looking at a magnified image of the Sun, keep an eye out for large sunspot groups that will be gradually occulted by the encroaching Moon (Fig. 4.3).

T minus 20 min: By this time, most people are at least vaguely aware of an unusual character in the light of the landscape around them. Some may report that everything has a peculiar yellowish cast, or that the rich blue color of the clear sky has become duller, flatter. The temperature has fallen noticeably, too. In a humid climate, the temperature may drop enough to trigger cloud formation in what was previously a clear sky – a source of unexpected drama that weather-conscious eclipse viewers could live without! Look for multiple overlapping images of the crescent Sun forming in the shade of foliage. The small gaps between intersecting branches and leaves act as a multitude of pinhole cameras.

T minus 15 min: Now the pace quickens. The crescent Sun is dwindling – it's now about 80% covered – the light is failing noticeably, and the sky and landscape are shaded a steely blue-gray. Animals and plants respond to the dropping light levels

Fig. 4.3 Enormous sunspot groups covered the Sun on Oct. 28, 2003. During a solar eclipse, sunspots provide landmarks for monitoring the Moon's progress. The Sun's visible surface, called the photosphere, has a temperature of about 9,900°F (5,500°C). Sunspots, which contain intense magnetic fields, are cooler regions and therefore appear dark. Their numbers, recorded since 1750, rise and fall over a roughly 11-year cycle (NASA/ESA/SOHO)

as if night were falling. Insects seem to react most dramatically – bees return to their hives, butterflies nestle in the grass, and nocturnal insects emerge. Viewing an eclipse from the South Pacific islands of Tonga in 1911, William J. S. Lockyer described the burst of insect noise at totality as "most impressive, and will remain in my memory as a marked feature of the occasion."

Recalling the 2001 eclipse in Zambia, Fred Espenak noticed a loud chorus of crickets filling the air with background sound from about 10 min before totality until the Sun returned. Many bird species go to roost, cows come home, bats may even appear. From observing locations with an unobstructed view of the western horizon, a dusky shading reminiscent of a gathering storm becomes visible. This is

Fig. 4.4 Seconds before (*left*) and after (*right*) totality, the eclipsed Sun resembles a diamond ring as the last bit of sunlight shines through deep lunar valleys and the faint corona outlines the dark Moon. At totality (*center*), the corona's pearly glow fully emerges (Ben Cooper)

the oncoming shadow of the Moon, already bringing totality to sites located a few hundred miles to the west.

T minus 5 min: A few minutes before (and also after) totality, when the exposed portion of the Sun forms a thinning crescent, narrow bands of light and shadow can be seen racing across the ground. First described in 1706, these so-called shadow bands occur because wind-blown pockets of warmer and cooler air refract light from the narrow crescent slightly differently. The effect resembles the ripples of light that travel across the bottom of a swimming pool.

T minus 1 min: Now the crescent Sun's remaining light can be blocked with a finger or solar filter to reveal a ghostly halo around the Moon's dark circle. This is the Sun's corona, which becomes steadily more obvious.

T minus 15 s: In seconds the Moon will obscure all of the Sun, but the limb of the Moon is not perfectly smooth. Mountains obscure the sliver of Sun sooner than do valleys, and the narrowing crescent breaks up into several brilliant globs known as Baily's beads. They're named for Francis Baily (1774–1844), the English astronomer who called attention to them during the annular eclipse of 1836 as "a row of lucid points, like a string of bright beads." They can be seen for a few seconds along the Moon's limb at second and third contacts during both annular and total eclipses. The last bead to fade seems so dazzling against the darkened sky and black Moon that it conjures up the image of a gleaming jewel set in a silvery halo – a "diamond ring" (Fig. 4.4).

Totality! Second contact begins when the diamond ring winks out. It's now safe to view the Sun without eye protection through cameras, binoculars, and telescopes. The temperature cools several degrees more than it already has – over the course of the eclipse it may fall by more than 20°F (11 °C) – and the landscape grows about as dark as during evening twilight. This isn't exactly "night at noon," but it's dark enough to reveal the brightest planets and, at least according to some observers, the brightest stars. According to Espenak, seeing stars during a total eclipse is quite a challenge, and he think that many reports of "stars coming out" at totality are really referring to planets that wouldn't otherwise have been visible during the day because they were simply too close to the Sun.

The glow of dusk appears all along the horizon. The Moon's umbral shadow covers only a small region, and sunlight beyond its limits paints the sky near the horizon with twilight colors.

The big attraction, of course, is the Sun. For a few seconds the eastern rim of the Sun shows a bright pink or rosy coloring. This is the chromosphere, the lowest and coolest layer of the Sun's atmosphere, and above it, huge arcs of hot gas called "prominences" lie suspended by intense magnetic fields (see Figs. 4.5 and 4.6). The prominences extend into space like tongues of reddish flame. Those on the eastern rim are quickly swallowed up by the Moon's progress, but watch the western side to see if more are uncovered. Large prominences in the north and south may remain visible throughout the Moon's passage.

Fig. 4.5 A composite of 25 images reveals prominences along the Sun's limb and fascinating structure in the corona. We're seeing the Sun's eastern limb just after second contact during the Jan. 8, 2008, total eclipse as seen from Bor Udzuur, Mongolia. Exposures ranging from 1/4,000 of a second to 8 s captured the full range of illumination, and computer processing was used to bring out subtle details – even including surface features on the silhouetted Moon (Miloslav Druckmüller, Martin Dietzel, Peter Aniol and Vojtech Rušin)

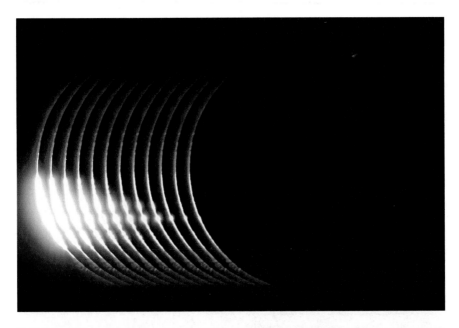

Fig. 4.6 Fringed in pink. The last portion of the solar surface winks out in this sequence of images taken 5 s before totality on Aug. 1, 2008. Mountains along the Moon's edge reach the far side of the Sun's disk before lunar valleys, so the narrow crescent Sun breaks into a series of dwindling blobs known as Baily's beads. As the final bead fades, a thin pink line emerges. This is the chromosphere, which glows in the color characteristic of ionized hydrogen gas. Flame-like prominences, also in pink, extend far from the Sun's surface. These are dense gas structures supported by powerful magnetic fields (Tunç Tezel)

With the Sun's brilliant light completely obscured, the pearly white corona at last becomes fully visible (Fig. 4.7). The corona is the Sun's extremely hot, tenuous outer atmosphere, about a million times fainter and more than 200 times hotter than the solar surface. The corona is about as bright as a full Moon and it sports visible streamers that extend several times the Sun's diameter. The corona also changes its overall shape from eclipse to eclipse, showing the least amount of symmetry in times of low solar activity, when the equatorial region is marked by long streamers stretching into space. At times of high solar activity, the polar corona expands, which makes the overall glow more circular. The corona's outermost portion expands into interplanetary space and creates the solar wind, a thin interplanetary outflow that streams the Sun's gases and magnetic field throughout the solar system at an average speed of 745,000 mph (1.2 million km/h). Solar storms can expel large amounts of additional gas at much faster speeds, creating gusts and gales in the solar wind that scientists now refer to as "space weather." We'll discuss more about this in Chap. 9.

Now the phenomena noted in our countdown can be seen in reverse order. As totality draws to a close, look for the pink glow of prominences and the chromosphere

Fig. 4.7 Near the end of the Jul. 11, 2010, solar eclipse, the Moon's umbra swept across southern Argentina and treated observers to totality near sunset. The lights of El Calafate in Patagonia have switched on in the premature darkness and glint off the waters of Largo Argentina. The Moon, ringed by the Sun's corona, hovers above the rugged horizon formed by the Andes mountains. At left, the sky glows brighter outside of the umbra's track (Janne Pyykkö)

on the Sun's western edge. This is followed by a second diamond ring, which marks third contact and the need for a return to eye-safe viewing methods. Next comes Baily's beads, and then an uninterrupted sliver of sunlight.

There is a genuine emotional release with the end of totality, especially for those who have traveled far and wide to see it, but the eclipse isn't over yet. For a few seconds after totality, from observing sites with a clear view of the eastern horizon, the Moon's retreating shadow may be visible racing across the landscape. Maria Mitchell (1818–1889) of Vassar College described this moment from the eclipse she observed from Denver in 1878:

> Happily, someone broke through all rules of order, and shouted out, "The shadow! The shadow!" And looking toward the southeast we saw the black band of shadow moving from us, 160 miles over the plain, and toward the Indian Territory. It was not the flitting of the closer shadow over the hill and dale: it was a picture which the Sun threw at our feet of the dignified March of the Moon in its orbit.

Spacecraft now allow us to view that picture from a vantage point Mitchell could only imagine. Today many of the unique observing opportunities eclipses provide are available to astronomers by other means. Satellites carrying instruments above the atmosphere can observe the Sun in greater detail and over a broader range of its emitted radiation than is possible from Earth's surface. Space-based observatories study the faint outer atmosphere of the Sun with instruments called "coronagraphs" that use an opaque disk to occult the Sun. But they also block the innermost corona, which is bright enough to scatter light within the device. For example, the Large Angle Spectrometric Coronagraph (LASCO) aboard the Solar and Heliospheric Observatory (SOHO) spacecraft places more than twice the Sun's diameter in artificial eclipse. Ground-based instruments using the Moon as an occulting disk can see much closer to the Sun than their space-borne counterparts, and eclipse observations still provide important insight into the innermost regions of the lower corona and chromosphere. It's here that some of the most interesting and puzzling science takes place, where coronal streamers transform into the outward-flowing solar wind and the corona itself becomes mysteriously heated to hundreds of times the temperature of the visible solar surface. Moreover, eclipse trips cost hundreds of times less than a space-borne experiment and can be planned in months rather than years, enabling researchers to use the latest technology available. Although now a scientific niche, eclipse observations remain a valuable complement to other methods of studying the Sun, and professional astronomers will continue to chase the Moon's shadow for the foreseeable future.

Tales of the Shadowlands

The visual impact of solar eclipses has occasionally shaped human events. The most dramatic example is recorded by Herodotus, the first Greek historian, who wrote around 430 B.C. Two of the major military powers in Asia Minor, the Lydians and the Medes, had been fighting one another with about equal success for 5 years. In the sixth year, as they pounded away at each other in another battle, day was suddenly changed into night. This was taken as a serious portent; the fighting ceased and peace talks began, ultimately leading to a treaty sealed by marriage. Working with modern eclipse calculations and the dates of kings referred to in the story, astronomers generally agree that the eclipse mentioned occurred on May 28, 585 B.C. The Greek historian Herodotus records that the year of the eclipse was predicted by the philosopher Thales. Astronomers know that by this time a framework for predicting lunar eclipses was becoming established in Mesopotamia, but even centuries later these sky watchers couldn't predict where on Earth a solar eclipse would be total. Thales likely was aware of at least one type of eclipse cycle that indicated the event's possibility, but not where the eclipse would have been seen as total.

Knowledge of an upcoming eclipse could be a useful tool for persuading those who did not know their cause. During his fourth voyage to the West Indies, Christopher Columbus and crew found their remaining two vessels so infested with shipworms and damaged by rough weather that the ships were no longer seaworthy. They beached the caravels on the north coast of Jamaica, and while awaiting the arrival of a relief ship, the crew traded with the natives for food and other supplies. But after 6 months and a mutiny of half the crew that led to raids on some of the villages, the Europeans had worn out their welcome. Columbus was out of trade-worthy goods, and the natives decided to cease helping the guests who wouldn't leave. From an almanac, Columbus learned that a total lunar eclipse would take place on Feb. 29, 1504, and with this in mind he hatched a desperate plan. Three days before the event, Columbus met with local chieftains, warning that their withholding of food had angered the Christian god. Three days hence, he said, they would receive a heavenly sign of his displeasure – the Moon would all but disappear! How absurd this must have seemed to them given the circumstances, with a leader of a band of shipwrecked, starving and barely controlled men claiming that a powerful god would come to his aid by, of all things, messing with the Moon. But as the full Moon rose over the island on the appointed evening, half of it was missing – the eclipse was already in progress. At totality, the reddened orb had precisely the effect Colombus had hoped for and the natives agreed to keep him supplied until a relief ship arrived.

Tecumseh and his brother Tenskwatawa, two influential members of the Shawnee tribe of the Ohio River valley in the early 1800s, used a similar trick. They recognized that the only way to stop the increasing westward advance of white settlers was for the various tribes of the area to put aside their differences and unite against their common threat. Tenskwatawa, known to white settlers as The Prophet, was a religious leader who preached a rejection of all white customs and a return to traditional cultural values. As his fame and influence grew, Tenskwatawa eventually came to the attention of William Henry Harrrison, the governor of Indiana Territory and a future U.S. president. Harrison encouraged neighboring tribes to be skeptical of The Prophet's claims.

> Demand from him some proofs, at least of his being the messenger of the Deity…. If he is really a prophet, ask of him to cause the Sun to stand still – the Moon to alter its course – the rivers to cease to flow – or the dead to rise from their graves. If he does these things, you may then believe that he has been sent by God.

But this plea backfired by essentially giving the brothers a blueprint for demonstrating a cosmic connection. Somehow, likely through access to a farmer's almanac or contact with astronomically inclined settlers, Tecumseh heard about an upcoming solar eclipse. He shared this information with Tenskwatawa, who announced that on Jun. 16, 1806, the Great Spirit would hide the Sun from the world. At Greenville, Ohio, he was there with a thousand followers when the Moon's umbra arrived, and he demonstrated his ties to the Great Spirit by "restoring" the darkened Sun with appropriate showmanship and timing. As word of this miracle spread, so did The Prophet's fame. Open conflict finally came on Nov. 7, 1811, after Harrison led troops to the confederation's capital, Prophetstown, located by the Tippecanoe River near present-day Lafayette, Ind. Despite instructions by Tecumseh to avoid hostilities until he returned from a trip to rally more support, Tenskwatawa convinced his warriors to ambush the encampment in the morning.

The attack initiated a 2-h battle that was decisively won by Harrison; the loss so heavily discredited The Prophet that his settlement was essentially abandoned when troops entered it the following day. Decades later, Harrison was elected the ninth U.S. president with a memorable campaign slogan that referenced this battle: "Tippecanoe and Tyler too."

These historical incidents provided inspiration to novelists. The hero of Mark Twain's *A Connecticut Yankee in King Arthur's Court* uses knowledge of a total solar eclipse to bargain his way out of being burned at the stake. H. Rider Haggard's protagonist in *King Solomon's Mines* similarly interprets a lunar eclipse to his own advantage. In both cases, though, these eclipses are fictional.

While solar eclipses continue to play an important role in our current understanding of the Sun, the scientific bonanza from eclipses came between the years 1840 and 1920 thanks to the emergence of two important technologies: photography and spectroscopy. Photography allowed recording of an event for later analysis, while spectroscopy let astronomers determine the Sun's composition by looking for emission lines in its spectrum. This works because each element emits light in certain characteristic colors. In 1863, William Huggins (1824–1910) showed that the Sun and stars are primarily composed of hydrogen gas. In 1868, French scientist Jules Janssen (1824–1887) and British astronomer J. Norman Lockyer (1836–1920) called attention to a yellow line in the corona's spectrum; it didn't seem to belong to any element then known. They daringly proposed that it was an element not yet discovered, and Lockyer even went so far as to name it: helium, after the Greek Sun god Helios. Validation came in 1895, when helium's signature was identified on Earth in a uranium-bearing mineral.

One particular eclipse from the middle of this period offers an interesting glimpse of American science and celebrity at the time. On Jul. 29, 1878, the shadow of the Moon raced southeast across the United States, passing through the Montana and Wyoming Territories, the new state of Colorado, Oklahoma Territory, Texas, and Louisiana. The path of the umbra passed right across the Rocky Mountains. The prospect of clear, dry high-altitude observing sites attracted some of the world's most distinguished astronomers. The new transcontinental railroad made it much easier to move personnel and equipment than in previous eclipses, and the Pennsylvania Railroad even gave astronomers a special discount on tickets. Simon Newcomb (1835–1909) of the U.S. Naval Observatory went in hopes of spotting a new planet astronomers had predicted. Lockyer joined a party led by Henry Draper (1837–1882) of New York University. Samuel Langley (1834–1906), director of the Allegheny Observatory, led a stalwart expedition to the summit of Pike's Peak in Colorado.

But the best-known observer of the 1878 eclipse was, by his own admission, neither astronomer nor scientist. He was a thin, graying, 31-year-old New Jersey inventor named Thomas Alva Edison (1847–1931). He had already patented more than 150 inventions, including important improvements to both the telegraph and the telephone. In late December 1877, Edison had demonstrated his most original invention, one that catapulted him to a level of fame he had yet to experience.

The device was called the phonograph, and it was nothing less than the first machine capable of recording and reproducing sound. Newspapers and magazines raved about it all winter, and when the National Academy of Sciences asked for a demonstration during its April meeting in Washington, D.C., so many people crammed into the hall to hear Edison's presentation that the doors had to be taken off their hinges. During this meeting, Edison mentioned that he had designed a super-sensitive gadget for measuring heat he called the tasimeter, an off-shoot of his telephone improvements. George Barker, a professor at the University of Pennsylvania, was heading west with the Draper expedition and wanted to use the device to measure the heat of the corona. To make sure he got one, he invited Edison along.

In 1878, Wyoming was still a dozen years from statehood. Draper had selected the little town of Rawlins, a railroad switching point that consisted of a dozen or so unpainted buildings, a hotel, and about 800 people, for his observing location. Thanks to the eclipse, the town could boast among its visitors some of the leading astronomers of the day – and the world's most famous inventor. Newspapers along the Rockies enjoyed the presence of the scientists, or "wise men from the east" as one called them. Edison, though, was a bona-fide celebrity the press could not resist. Mostly self-educated, Edison frequently disdained academics and theoretical science when dealing with reporters, whom he was always happy to oblige, while astronomers could only wince at headlines like "Professor Edison attended by a party of scientists."

Edison shared his hotel room with Edwin Fox, a reporter for the *New York Herald*. The night before the eclipse, the two men were jolted awake by thunderous knocking. "Upon opening the door," recalled Edison, "a tall, handsome man with flowing hair, dressed in western style, entered the room." This, Edison was to learn, was Texas Jack. Hands on his gunbelt, Jack quickly took in the room with blood-shot eyes and asked "Which one of you is Edison?"

The inventor gulped hard and introduced himself. Jack explained that he was the best pistol shot in the West and wanted to meet the great inventor of the phonograph he had read about in the newspapers. Then he pulled his Colt revolver from its holster, aimed it out the window at the freight depot across the street, and fired, setting a weather vane into a wild spin. "The shot awakened all the people, and they rushed in to see who was killed," wrote Edison. "It was only after I told him I was tired and would see him in the morning that he left." Fame, the inventor was discovering, had its drawbacks.

Tourists flocked to towns all along the eclipse path. Weather had been a problem for some of the sites but, in the words of one Wyoming paper, the sky was "as slick and clean as a Cheyenne free-lunch table." Astronomers busied themselves as the Moon slowly covered the Sun and darkness fell. An eerie wind arose, kicking up whorls of dust, and then the twilight of totality arrived.

A dozen miles west of Rawlins, Simon Newcomb's group searched for a new planet. As noted in Chap. 3, a quirk of Mercury's orbit continued to frustrate astronomers: Although Newton's theory of gravity successfully accounted for most of the planet's motions, it consistently underestimated the advance of its perihelion point, which crept along the planet's orbit by almost 1 min of arc per century. Many astronomers – including Urbain Le Verrier, who had predicted the existence of Neptune in 1846 from discrepancies in the orbit of Uranus – believed that Mercury's mysterious perihelion advance was being caused by a gravitational influence of an unknown planet that circled even closer to the Sun. It was even given a name: Vulcan, after the Roman god of fire and metalworking.

Newcomb planned to telegraph the position of any planet his team spotted to astronomers in Texas, who would try to confirm its presence when totality came their way. During totality, James Craig Watson (1838–1880) of the University of Michigan thought he saw an object south of the Sun. But at the critical moment, Newcomb was busy with his own work and neglected to send the information – a failure for which he publicly apologized. Watson nevertheless announced his discovery to the press. Proving Vulcan's existence to a skeptical astronomical community remained his foremost scientific concern in the last years of his life. While the hypothetical planet doesn't exist, its name proved to be inspiration for the fictional homeworld of *Star Trek's* Mr. Spock. Yet something really was affecting Mercury's orbit, and we'll get to that shortly.

At Rawlins, Edison set up a borrowed telescope in the yard of the railroad superintendent. "I had my apparatus in a small yard enclosed by a board fence 6 ft high, at one end there was a house for hens," he recalled years later. "I noticed that they all went to roost just before totality. At the same time a slight wind arose, and at the moment of totality the atmosphere was filled with thistle-down and other light articles." According to a more embarrassing version of the story, Edison had set up near the entrance to the hen house, and as he adjusted his equipment in the minutes leading up to totality, he suddenly was besieged by the roosting birds. But with only 1 min of totality remaining, the inventor did manage to place his pocket-sized infrared detector at the end of the telescope. He connected the battery, and the needle of a galvanometer quickly traveled its full range.

Edison announced to reporters that he had detected the heat of the corona, but the scientists weren't convinced and ignored his claim. Edison's name appears nowhere in the official eclipse report produced by the U.S. Naval Observatory, and even the inventor himself quickly dismissed his effort, admitting that his device was just too sensitive; it was also unstable, poorly calibrated, and largely untested for its astronomical role. Edison quickly lost interest in the gadget, never even bothering to patent it. Other infrared detectors at the time, called thermopiles, were generally not sensitive enough to detect coronal heat, to the frustration of solar astronomers. With most thermopiles there was no response, with the tasimeter, the needle pegged. We know now that the corona is far hotter than anyone expected at the time, and this probably added to skepticism about Edison's claims. In the 1970s, John Eddy of the High Altitude Observatory in Boulder, CO, took another look at the inventor's data and concluded that Edison's detector had done what he'd claimed.

Meanwhile, the problem with Mercury's orbit remained all too obvious, and with zero evidence for the hypothetical planet Vulcan, astronomers were perplexed. The mystery would be solved in an entirely unexpected way, through an extreme makeover of physics led by an obscure patent clerk named Albert Einstein (1879–1955).

In 1905, Einstein rose to scientific prominence with four groundbreaking papers, including a first take on his famous relativity theory. It took another decade for him to complete a general form of the theory, which describes how space distorts in the presence of a strong gravitational field. Near the end of 1915, he enthusiastically described his efforts in a letter to another physicist:

> I have lived through the most exciting and the most exacting period of my life; and it would be true to say that it has also been the most fruitful.... The wonderful thing that happened was that not only did Newton's theory result from it as a first approximation, but also the perihelion motions of Mercury, as a second approximation.

Einstein's theory seemed to explain Mercury's behavior as a natural consequence, but he was aware of additional phenomena that could be used to test it. In particular, he believed that the warping of space near a massive object like the Sun would bend the path of any light passing close to it. So light from a star near the limb of the Sun, where the effect is strongest, would be deflected by a tiny yet detectable amount. If measured carefully enough, this displacement would make it appear that the star had moved from its normal position. An early analysis showed that this type of deflection could exist under Newton's laws, but under Einstein's completed relativity theory its value was two times larger. So accurate measurements of stars located near the Sun would either favor Newton or Einstein, and the only way to accomplish this at the time was during a total eclipse.

The first effort to carry out the observations came in 1914, and an expedition of German astronomers headed to Crimea on the Russian Black Sea for the August eclipse. But science was outpaced by political events: On Jun. 28, Archduke Franz Ferdinand of Austria-Hungary was assassinated in Sarajevo. As August opened, the busy German astronomers learned that their homeland had declared war on Russia – part of the opening act of World War I – and the expedition suddenly found itself in the wrong place at the wrong time. The team was promptly arrested, but after a few weeks the scientists were returned to Germany as part of a prisoner exchange.

As World War I drew to a close, British astronomers began to turn their attentions to Einstein's theory and draw up plans to test it. The eclipse of May 29, 1919, was considered ideal for the purpose, with the eclipsed Sun nestled among the bright stars of the constellation Taurus. Two teams, one at Sobral in northern Brazil and the other at Principe Island off the west coast of Africa, observed the eclipse and photographed the darkened Sun and nearby stars. By comparing the star positions during the eclipse with those on photographic plates exposed when the Sun was in a different part of the sky, astronomers hoped to be able to measure their positional displacement with enough accuracy to rule on whether Newton or Einstein was correct.

The English physicist Arthur Eddington (1882–1944), an early supporter of Einstein's theories, led the expedition to Principe. After a month of preparations, eclipse day began with heavy rain, but by totality the sky was only partly cloudy. Eddington wrote in his diary: "I did not see the eclipse, being too busy changing plates, except for one glance to make sure it had begun and another halfway through to see how much cloud there was." Clouds interfered with the star images on many plates, but he was optimistic that some of the last six photographs would be suitable. "One plate that I measured gave a result agreeing with Einstein," he wrote, and he later referred to this moment as the greatest of his life. The 1919 measurements confirmed that the Sun bent light rays by roughly the right amount – less than predicted in Principe, more than predicted in Brazil – and the dramatic eclipse results were made public in November. It was clear that the predictions from relativity had surpassed Newton's, and the international press immediately made Einstein a household name.

Observations from a 1922 eclipse showed even better agreement with relativity, but the errors associated with all of these measurements remained large. As new theories of gravity challenged Einstein's, these comparatively large errors made it difficult for astronomers to choose between them. By the 1970s, radio astronomy offered opportunities to test the deflection of radio waves – like light, a form of electromagnetic radiation – without the need of an eclipse. The positions of bright and distant radio sources known as quasars can be monitored as the Sun approaches and occults them. Such studies reduced measurement errors to about 1% and validated relativity over its challengers. More recently, emphatic confirmation came from the European Space Agency's astrometric satellite Hipparcos, which between 1989 and 1993 measured over 100,000 star positions with unprecedented accuracy. Whereas previous observations of light deflection had been confined to objects seen within a degree or two of the Sun's limb, Hipparcos detected the bending of light-rays as far as 90° from the Sun. According to Hipparcos, Einstein's prediction is correct to within one part in a thousand.

Nevertheless, scientists continue to test relativity even today because they know that at some level Einstein's famous theory must fail. This is because relativity's take on gravity, which is considered a fundamental force in the universe, is inherently incompatible with quantum mechanics, the theoretical framework that rules the other fundamental forces of nature (electromagnetism and the so-called weak and strong forces at work in the nucleus of an atom). Now physicists are eagerly searching for a single overarching theory – a "theory of everything" – that could supplant both relativity and quantum mechanics, effectively unifying them.

Eclipse Cycles

Because the Moon courses through the ecliptic once a month, you might at first imagine that lunar and solar eclipses occur with the same frequency. And it's true, we would see a total solar eclipse with every new Moon and a total lunar eclipse with every full Moon *if* the Moon followed exactly the same path as the Sun. But it doesn't. Instead, the Moon's path runs about 5° above and below the ecliptic, so most new Moons don't block out the Sun and most full Moons never hit Earth's shadow. An eclipse can occur only when a new or full Moon lies near one of the Moon's nodes – one of two points where its orbit intersects the ecliptic – and the Sun, the Moon, and the nodes are almost aligned with Earth.

For example, the Sun can be eclipsed anytime it lies less than 18.5° from one of the Moon's nodes. Because the Sun slides along the ecliptic by about a degree each day, we know that it stays within the "solar eclipse window" for about 37 days. (Similarly, the faster-moving Moon remains in a "lunar eclipse window" for a shorter time, about 24 days.) So with a new Moon happening every synodic month (about 29.53 days), this means we're guaranteed at least one solar eclipse every time one of the nodes lines up with the Sun. On the face of it, you'd think these alignments would recur every 6 months, but it's really about 173.31 days. The reason for the shortfall is that the nodes themselves move, gradually sliding eastward along the ecliptic by about half a degree each cycle.

Various cycles merge to form an important interval called the saros, which was first recognized in Mesopotamia sometime before 500 B.C. There, astronomer-priests kept detailed records of eclipses and other sky events because they were regarded as portents from the heavens, memos to the king on his conduct of earthly affairs. Once a sufficiently long record of observations was available, anyone looking for a pattern could find one, and what the Mesopotamian priests discovered was that very similar eclipses were separated by an interval corresponding to about 18 years, 11 days and 8 h. Astronomers believe that the cycle was first established for total lunar eclipses because these events can be seen over a greater geographical range than total solar eclipses.

In any case, the saros comes about because three different lunar cycles mesh together almost exactly. We know that solar and lunar eclipses depend on certain lunar phases, which repeat every synodic month. The time elapsed over 223 synodic months is 6,585 days and just under 8 h, so this forms one necessary interval. For an eclipse to occur, the Moon must be located at one of its nodes, and it returns to the same node after 242 orbits – elapsed time, 6,585 days and about eight and a half hours. Because the Moon's orbit isn't precisely circular, it makes a closest approach to Earth, called perigee, every month, and after 239 successive perigees, the elapsed time is 6,585 days and 13 h. The saros interval divides into each of these fundamental periods with an error of less than 1 part in 28,000 – equivalent to less than 4 s out of a single day.

So about 18 years, 11 days and 8 h after one eclipse, the Moon returns to the same phase, lies nearly the same distance from us – and therefore has the same apparent size – and is again aligned with the Sun, Earth, and one of the nodes of its orbit. Eclipses separated by a saros interval aren't exactly replays, but they share similar characteristics and occur near the same date. For instance, during the eclipse of Aug. 21, 2017, the arc of the Moon's shadow across Earth will be almost identical to the one it made during the eclipse of Aug. 11, 1999. But that fraction of a day in the saros period affects where on Earth the shadow falls, and thus where totality will be seen. Because each succeeding eclipse in a saros occurs about a third of a day later, Earth has rotated eastward by an additional 120° before the eclipse begins. In 1999, totality ran through Europe, but in 2017 the Moon's shadow swings across the United States.

An interesting consequence of the extra spin is that every third eclipse in a saros returns to more or less the same geographic region. Totality for the June

1806 solar eclipse – the one "predicted" by Tenskwatawa and observed by the young James Fenimore Cooper – swept across the United States from California to New England, crossed the Atlantic and ended in western Africa. The next eclipse in this series, occurring exactly one saros interval later, in 1824, brought totality to China, Korea, Japan and the northern Pacific. But three saros intervals after the 1806 eclipse, in 1860, totality returned to North America, its path passing farther north into Canada and just grazing the western United States. This triple-saros interval, which equals about 54.09 years, is useful for predicting when similar eclipse circumstances will return to roughly the same region of the globe. The related eclipses separated by one saros cycle make up a saros series, a family of similar eclipses running into the past and the future. While a single saros family cannot run forever because the various periods don't mesh precisely, a typical saros endures for 1,300 years and hosts more than 70 eclipses. When one series ends, a new one begins. We see several eclipses a year because multiple saros families run concurrently; in fact, 40 or so are in progress at any given time.

According to astronomical nomenclature, the eclipses seen by Cooper and Edison belong to Saros 124. To understand the rise and fall of a typical saros, we'll take a brief look at this family's other eclipses. Every saros family starts with a modest partial eclipse visible only from high northern latitudes (for an odd-numbered saros) or deep southern latitudes (even-numbered saros). So begins the story of Saros 124, with an insignificant partial eclipse – less than 2% of the Sun was covered – visible, had anyone been there to see it, from Antarctica on Mar. 6, 1049. As time passed, the slight flaws in the eclipse cycle slowly improved the alignments in the series. The next eight partial eclipses were progressively deeper until, in the summer of 1211, the first of 43 total eclipses darkened the Sun along a track across the Southern Ocean. Improvement continued until, 30 eclipses later, Saros 124 produced its best. The duration of totality on May 3, 1734, was the longest of the series, reaching 5 min and 46 s along a path from the east coast of South America to India.

Then began a rapid decline. For the 1806 eclipse, the longest duration of totality fell just short of 5 min. Four eclipses later, when Texas Jack and Edison were viewing in Wyoming, the greatest duration of totality barely exceeded 3 min. A triple-saros interval later, in 1932, brought an eclipse that darkened the Sun for less than 2 min on a path that ran through Canada and the northeastern United States (Fig. 4.8). The fading astronomical alignments produced the family's last total eclipse in 1968; the shadow curved through China and Russia, and totality lasted less than 40 s. October 1986 brought the only hybrid eclipse of the series, annular everywhere except for a spot off the coast of Greenland, where the Sun blinked dark for just a tenth of a second. The next Saros 124 eclipse, in October 2004, marked the beginning of the end, ushering in a final phase of dwindling partial eclipses that spans three centuries. The family expires in May 2347 – after having produced a total of 73 solar eclipses in 1,298 years – with a small partial eclipse visible from northern Canada and Alaska.

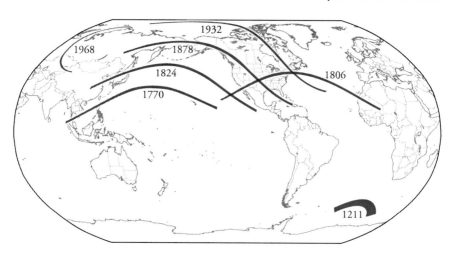

Fig. 4.8 This map plots the totality tracks of a single eclipse family, Saros 124. During its first total solar eclipse, in 1211, the Moon's umbra traced a path across the Southern Ocean near Antarctica. Eclipse paths moved to more northerly latitudes, with the last, in 1968, running through Russia. In between, a total eclipse occurred every saros interval, each track falling about 120° farther west from the previous one. So while the 1806 eclipse tracked through the U.S., the 1824 eclipse swept over the Pacific. After three consecutive eclipses, the umbra returns to roughly the same part of the globe, as shown by the stacked tracks of eclipses from 1770 to 1932

Bad Moons Rising

Just as Earth occasionally passes through the Moon's shadow, so the Moon sometimes passes through the shadow cast by Earth. Like solar eclipses, lunar eclipses require a precise geometry between Earth, Sun, and the nodes of the lunar orbit, and their periodicity is similarly ruled by the saros cycle. At the Moon's distance, the apparent size of Earth's umbra is more than two-and-a-half times the apparent size of the Moon's disk, which means that the lunar version of totality is measured in hours rather than minutes. Better yet, the shadowed-immersed Moon can be seen by the entire night side of Earth. While lunar eclipses are not nearly as impressive as their solar counterparts, they have an eerie beauty all their own that more than rewards the small effort required to see them.

Stonehenge, the famous megalithic monument near Salisbury in southern England, has been associated with lunar eclipses since the 1960s. Stonehenge was built in four stages between 2800 B.C. and 1500 B.C., beginning as little more than a circular embankment and evolving into the iconic stone circle we see today. Increased computer power allowed investigators to test the monument's many possible alignments against objects in the sky at various stages of its construction. Such research led two astronomers, Gerald Hawkins and Fred Hoyle, to propose that Stonehenge incorporated an impressive amount of astronomical knowledge – so much, in fact, that it could have been used to predict when lunar eclipses would occur or, in a weaker form, to predict the "danger times" when eclipses were possible.

Today, few astronomers and even fewer archaeologists agree with these imaginative assertions. While it may have been possible to use Stonehenge as some sort of eclipse warning system, there is no evidence it was used in such a fashion. There are solar and possibly lunar alignments built into Stonehenge, but they lack the precision often claimed for them: Sightlines to the Sun or Moon are too short – much shorter than at other megalithic monuments. A given observation, such as the summer solstice sunrise, would not shift appreciably within a week on either side of the solstice. At other megalithic sites, horizon landmarks could have served as long and accurate sightlines, but the horizon seen at Stonehenge is largely flat and featureless. Stonehenge remains an enigma, built by an ancient people for a purpose we have not yet completely fathomed. What we do know is that it functioned as a ritual center, not as an astronomical observatory in any modern sense.

Without knowing its cause, any eclipse is disturbing because it appears to be a dramatic violation of the natural order. The Bible contains many allusions to events that sound suspiciously like eclipses, usually with a stock description along the lines of "the Sun shall be turned into darkness, and the Moon into blood"; all occur in connection to events leading up to the Day of Judgment. Islamic tradition is more specific about this association, holding that both a solar and a lunar eclipse during the holy month of Ramadan will occur before Judgment Day.

This may have played a role in the reaction of Turkish soldiers to the total lunar eclipse of Jul. 4, 1917, which occurred during Ramadan. Thomas Edward Lawrence, better known as Lawrence of Arabia, served as a British adviser during World War I and helped organize local tribes into a guerrilla force operating against troops of the Ottoman Empire. Lawrence planned to take Aqaba, a strategic Red Sea port in what is now Jordan, by attacking it from its most poorly defended side. This meant approaching it through the forbidding desert landscape known as Wadi Rum at the height of summer. Lawrence made the crossing with a force of 50 Bedouin troops, augmented them with local rebels, and headed for the first of two well-fortified outposts en route to the port. "By my diary," wrote Lawrence, "there was an eclipse. Duly it came, and the Arabs forced the post without loss, while the superstitious soldiers were firing rifles and clanging copper pots to rescue the threatened satellite." The second outpost fell a few days later, and Aqaba surrendered without a fight.

The Moon passes through Earth's shadow moving west to east, its exact path varying greatly from one eclipse to another. Earth's shadow has the same basic structure as the Moon's – a broad faint penumbra surrounding a darker, more compact umbra. Earth's penumbra is so wide that the entire Moon can pass through it without ever crossing into the umbra. While these penumbral eclipses make up about 37% of all lunar eclipses, the darkening of the Moon is so subtle that casual observers rarely notice it.

When only a portion of the Moon's disk dips into the umbra, the eclipse is said to be partial. As the full Moon contacts and enters the umbra, a zone of darkness appears to envelop its disk. The darkness reaches its maximum extent at mid-eclipse, and then it gradually withdraws as the Moon passes out of the shadow.

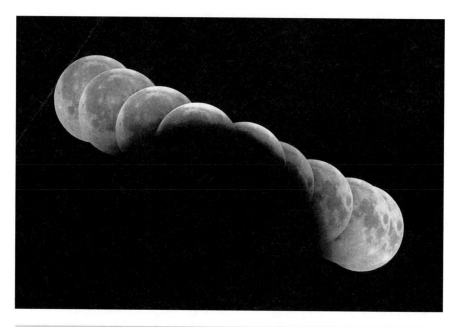

Fig. 4.9 Earth's circular umbra is revealed in this nine-image composite of the Aug. 2008 partial lunar eclipse (Anthony Ayiomamitis)

During partial eclipses – and the partial phases of total eclipses – Earth's shadow looks almost black in contrast with the part of the Moon still in sunlight (Fig. 4.9). Partial lunar eclipses make up about 35% of the total.

The rest, about 28%, are total, with the Moon passing completely into Earth's umbra. A total lunar eclipse may last as long as 6 h, counting from the Moon's first to its final contact with the penumbra, the outermost shadow zone where the Moon still receives some sunlight. The most noticeable darkening, during the Moon's passage through the umbra, may take place for more than half of this time. Typically, the Moon spends about an hour traversing the penumbra. Expert observers will notice a subtle dusky shading on the leading edge of the Moon a little more than 10 min after it contacts the penumbra. Most observers, however, won't detect a change on the Moon's face for another half hour, or about 10 min before contact with the umbra. When the Moon begins its passage through the umbra, a dark circular shadow slowly creeps across its face for the next hour. The Moon lies completely in shadow, faintly glowing with an eerie color ranging from brownish to brick red. The Moon may remain in this state for nearly 2 h before its leading edge exits the umbra, beginning the hour-long emergence. The Moon's last contact with the umbra marks the end of the noticeable part of the eclipse, but the Moon will traverse the penumbra for another hour before finally clearing Earth's shadow.

The most striking aspect of a total lunar eclipse is the dramatic color change the Moon's silvery disk undergoes once it fully enters the umbra. The hue ranges from a bright coppery orange or red, to brick-colored to even brownish (see Fig. 4.10). Because Earth has an atmosphere, some scattered sunlight always reaches the

Fig. 4.10 These simulated images show the Moon as it will appear at mid-eclipse for upcoming total lunar eclipses. The images were computed using a model that tracks how sunlight refracts, disperses and scatters in Earth's atmosphere. Each image shows the Moon at mid-totality as seen from the point on Earth where it appears directly overhead, when its light is passing through the least amount of atmosphere. As a result, these images should be regarded as showing the minimum coloration of totality for each eclipse. The Moon will show deeper coloring if viewed near moonrise or moonset and if any large volcanic eruptions have injected gases into the stratosphere during the previous year. Compare these images to your own observations (Theodore C. Yapo)

totally eclipsed Moon. The atmosphere scatters the shorter wavelengths in sunlight, such as blue and violet, much more than longer wavelengths, like red and orange. This is why the clear sky is blue, but it's also why the rising and setting Sun appears reddish. Near the horizon, sunlight is taking its longest path through the atmosphere and travels through hundreds of miles more of it than at noon. With the Sun's shorter wavelength light scattered away even more efficiently, the disk takes the color of the least scattered light and looks reddish. Likewise, any sunlight making it into Earth's umbra similarly is enriched in reddish light, so what illuminates the totally eclipsed Moon is a sunset glow that encircles our planet.

The condition of the atmosphere along Earth's edge also contributes to the brightness and hue of the totally eclipsed Moon. For example, large numbers of tall thunderclouds during a summer eclipse may block some light from entering the umbra and produce a darker Moon at totality. The most significant impact, though, comes from sulfuric acid droplets that form in the stratosphere following exceptional volcanic eruptions. These particles strongly scatter blue light and they can persist for years. The darkest lunar totality in half a century occurred in Dec. 1963, when the Moon all but disappeared thanks to an eruption by Indonesia's Mount Agung 9 months earlier. In fact, this was the darkest eclipse since 1816, a year so unusually cool it's often referred to as the "year without a summer." Both the unusual cooling and the dark eclipses had the same cause – sulfuric acid particles formed by gases belched out by the Indonesian volcano Mount Tambora in April 1815 – one of the largest eruptions in recorded history. Likewise, the 1982 eruption of El Chichón in Mexico and the 1991 eruption of Mount Pinatubo in the Philippines also resulted in especially dark eclipses.

Imagine for a moment what observers on the Moon would see during totality. For them, it would be an eclipse of the Sun by Earth. Because it's four times larger, our planet blots out the Sun far more completely than the Moon ever can. With no light-scattering atmosphere on the Moon, planets and constellations would easily be visible during totality, and the gray lunar landscape would be bathed in reddish light. The circle of Earth's black disk, rimmed by the ruddy glow of a continuous sunset, would hang serenely among the stars, silhouetted against the ghostly backdrop of the Sun's outer corona.

The first question to arise about any eclipse is simply "Will I be able to see it?" The following pages provide maps illustrating solar and lunar eclipses through 2020 in chronological order Figs. 4.11 to 4.36. For the solar eclipse maps, a saddle-shaped region encloses the part of world touched by the Moon's penumbra, and a shaded path centered within this region marks the central path of the eclipse – the track of the Moon's umbra from which totality or annularity may be viewed. The region's eastern and western boundaries indicate where maximum eclipse – the moment where the Moon covers the Sun to the greatest degree – occurs at sunrise or at sunset. Everywhere within the saddle-shaped region will experience at least a partial eclipse, with the eclipse becoming deeper at locations closer to the central path. The northern and southern boundaries indicate the northern and southern limits of eclipse visibility; from locations beyond these lines, the Moon misses the Sun's disk. Dashed lines indicate the hour of Universal Time (UT) when maximum eclipse occurs at different parts of the world; small maps of selected regions show populated areas near the

Total Lunar Eclipse, June 15, 2011

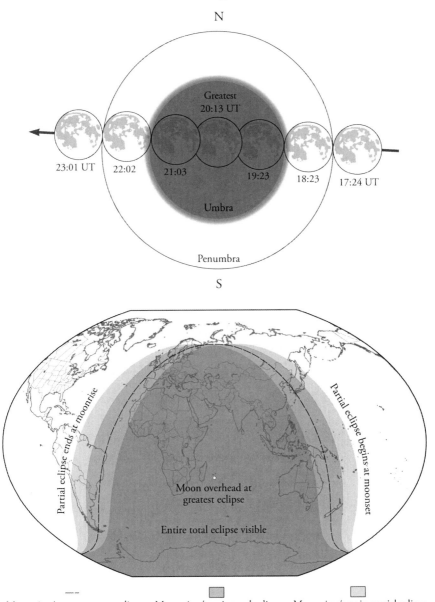

Fig. 4.11 At mid-eclipse, the Moon is nearly directly overhead for observers in Reunion and Mauritius, which means that all of totality will be visible throughout the Indian Ocean, most of Africa, the Middle East, central Asia and western Australia. The moon sets in totality as seen from Eastern Asia, eastern Australia, and New Zealand; likewise, it rises in totality throughout Europe – only northern Scotland and northern Scandinavia miss out – and eastern South America. The eclipse is not visible from North America. It's the 34th lunar eclipse of Saros 130

Total Lunar Eclipse, Dec. 10, 2011

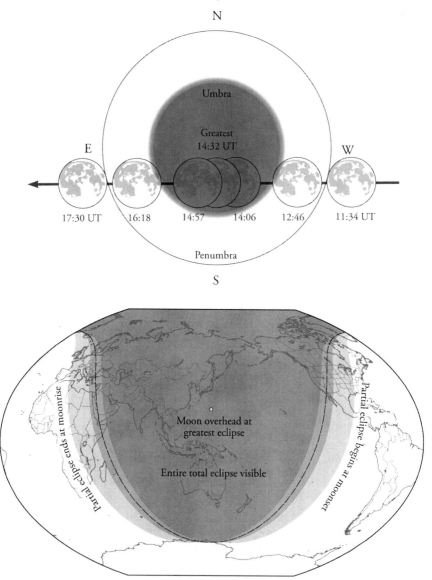

Moon rises/sets at greatest eclipse   Moon rises/sets in total eclipse   Moon rises/sets in partial eclipse

Fig. 4.12 For North Americans, the Moon sets in totality along a line from Montana to California, with areas farther north and west treated to more of the eclipse before moonset. All of totality can be seen along most of the U.S. West Coast, Alaska and Hawaii, all of Asia, Australia and New Zealand and much of the Pacific Ocean. The Moon rises in totality along a line from Iceland through eastern Europe, Saudi Arabia and Somalia. At mid-eclipse, the Moon is near the zenith as seen from Guam and the Northern Mariana Islands. This is the 23rd lunar eclipse of Saros 135

Annular Solar Eclipse, May 20–21, 2012

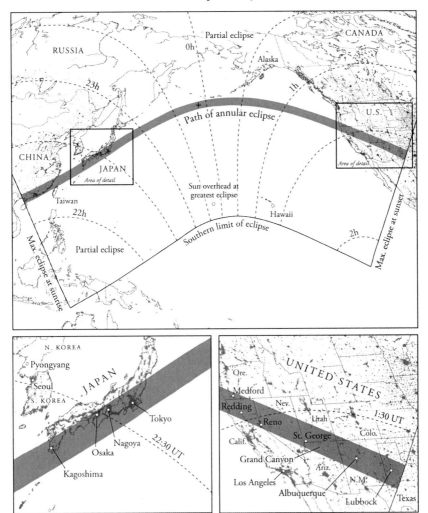

Fig. 4.13 The morning Sun turns into a ring for several minutes along the populous southern coast of China, including Macau, Hong Kong and Guangzhou, with maximum eclipse occurring around 22:09 UT on May 20; maximum eclipse occurs at Taipei several minutes later. The Moon's shadow sweeps across southeastern Japan, arcs over the Pacific, crosses the International Date Line, and terminates along the U.S. West Coast on the afternoon of May 21, bringing a ringed sunset to New Mexico and Texas. The duration of annularity lasts 5 min 46 s at the point of greatest eclipse (+), and this is the 58th solar eclipse of Saros 128. The morning annular eclipse maxes out at 22:22 UT for Kagoshima; 22:29 UT for Osaka; 22:31 UT, for Nagoya; and at 22:35 UT for Tokyo. The Moon covers 81% of the Sun as seen from Seoul and 77% from Pyongyang at about 22:31 UT. Gray patches indicate other city locations. The afternoon of May 21 brings a ringed Sun to locations from Oregon to Texas. The eclipse is maximum for Medford, OR, at 6:26 P.M. PDT and peaks at Reno, NV, several minutes later. It reaches maximum at 7:34 P.M. MDT from St. George, Utah; at 7:36 P.M. MDT from Albuquerque; and from Lubbock, Texas, where the Sun is nearly setting, at 8:36 P.M. CDT

Partial Lunar Eclipse, June 4, 2012

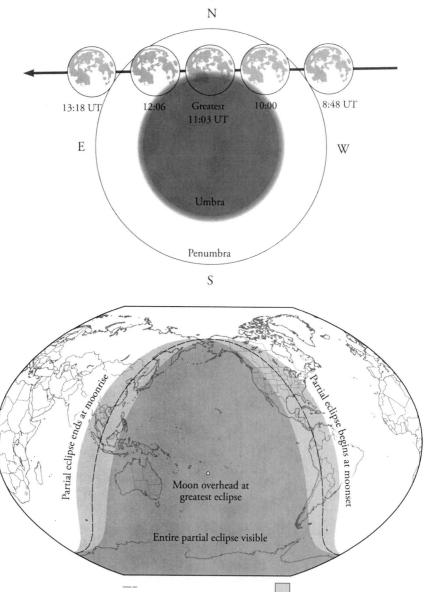

Fig. 4.14 The Moon sets partially eclipsed for North Americans west of a line from North Carolina to Michigan. Most of the western half of Canada and most of Alaska, as well as the Caribbean islands and western South America, also see the setting Moon in partial eclipse. At mid-eclipse, the Moon is nearly overhead as seen from the South Pacific Islands of Tonga and American Samoa, so Indonesia, Australia, New Zealand and Antarctica witness the entire event. The last stages are visible from Russia to Malaysia as the Moon rises. This is the 25th lunar eclipse of Saros 140. Look for Antares beneath the Moon

Total Solar Eclipse, Nov. 13, 2012

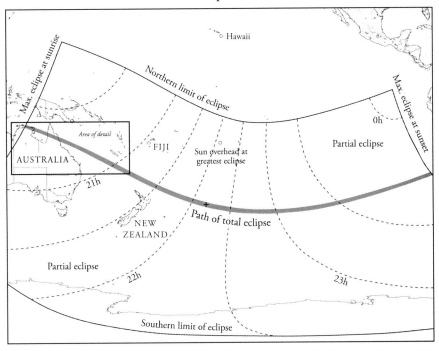

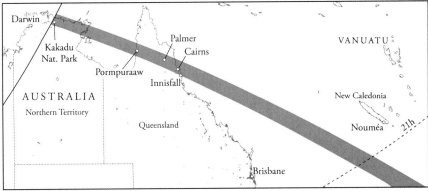

Fig. 4.15 The Moon's shadow seems to be trying its hardest to avoid dry land. It sweeps across the northern fringes of Australia and heads into the Coral Sea and the South Pacific. Maximum partial eclipse at Auckland, New Zealand, (21:28 UT) shows the Sun 87% obscured. Fiji sees a 64% partial eclipse at 21:04 UT. This is the 45th solar eclipse of Saros 133. The duration of totality at the point of greatest eclipse (+) is 4 min 2 s. Maximum totality occurs at sunrise (20:36 UT) on the northwestern edge of Kakadu National Park in Australia's Northern Territory. The shadow glides across the Gulf of Carpentaria to Pormpuraaw, where the Sun is 9° above the horizon at maximum eclipse (20:37 UT). Palmer sees mid-totality about 90 s later; Innisfall, at 20:40 UT. For Brisbane, outside the umbral track, eclipse maximum (20:54 UT) obscures 83% of the Sun's disk. At Nouméa, the partial eclipse reaches 91% at 20:57 UT. Gray patches indicate other city locations

Partial Lunar Eclipse, April 25, 2013

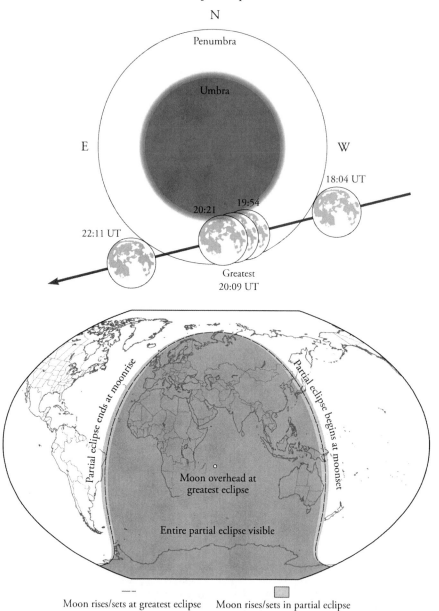

Fig. 4.16 The Moon just skirts the umbra in this marginal partial eclipse, which is not visible to North Americans. The Moon is nearly overhead as seen from Madagascar, so all of the eclipse can be seen throughout the Indian Ocean, across most of Antarctica and Asia, and all of Australia, Africa and Europe. Easternmost Brazil catches the partial eclipse as the Moon rises. This is the 65th lunar eclipse of Saros 112

Annular Solar Eclipse, May 9–10, 2013

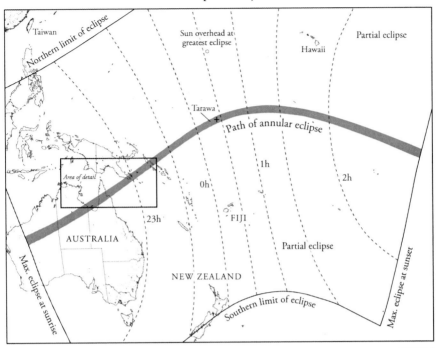

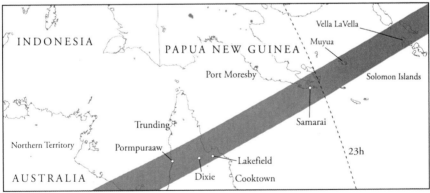

Fig. 4.17 The eclipse delivers a ring of Sun to Australia, Papua New Guinea, the Solomon Islands and Kiribati's Tarawa atoll. The island lies near the point of greatest eclipse (+), where annularity lasts 6 min and 3 s. The ringed Sun is just above the horizon as seen from Newman, Western Australia, at 22:32 UT on May 9. From Tarawa, mid-eclipse occurs at 0:15 UT on May 10, with annularity lasting nearly 6 min. At about the same time, the Mili atoll of the Marshall Islands sees an 84% partial eclipse. From Hawaii, the Sun is less than 50% obscured (1:53 UT). This is the 31st solar eclipse of Saros 138. The eclipse's tracks over many Australian towns that saw last year's total eclipse, including Pormpuraaw (22:44 UT, May 9). Maximum eclipse comes to Dixie 2 min later and to Lakefield 90 s after that. The shadow grazes Papua New Guinea, with Samarai seeing the ring of Sun at 22:59 UT, but from Port Morseby the eclipse is partial (89% at 22:55 UT). The shadow proceeds across the Solomon Sea, bringing the annulus to Muyua (23:04 UT) and Vella LaVella (23:13 UT) islands

Hybrid Solar Eclipse, Nov. 3, 2013

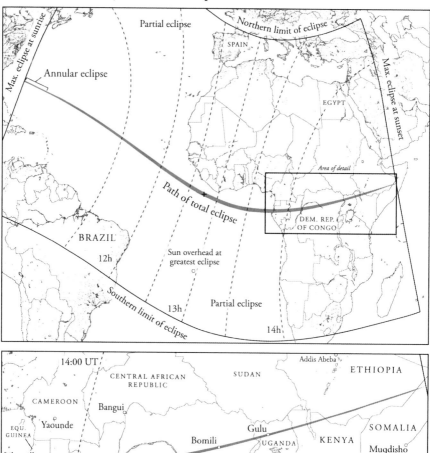

Fig. 4.18 This is the 58th solar eclipse of Saros 128 and the only hybrid eclipse of the decade. The eclipse is total everywhere along the umbra's track except for a short span off the U.S. East Coast, where the Moon is just too far away to completely cover the Sun; there, an annular eclipse results. The duration of totality at the point of greatest eclipse (+) is only 1 min 40 s. Look for Saturn (0.5) just 3° east of the Sun, with Venus (−4.6), now 2 days past greatest elongation, shining brilliantly 45° farther away. The Moon's umbra makes African landfall at Gabon's Wonga Wonga Reserve, just north of Port-Gentil. Totality there lasts just over a minute and reaches maximum at 13:51 UT. The coast hosts the best viewing because totality dwindles in length as the shadow moves eastward. At Oyabi, totality lasts only 37 s and maxes out at 14:01 UT. Only the northern part of Mbandanka will see totality (14:08 UT), but there, on the shadow's fringe, it lasts less than 11 s. At Gulu, near the shadow tracks's center, the eclipse lasts just 20 s, with maximum at 14:23 UT. Gray patches indicate other city locations

Total Lunar Eclipse, April 15, 2014

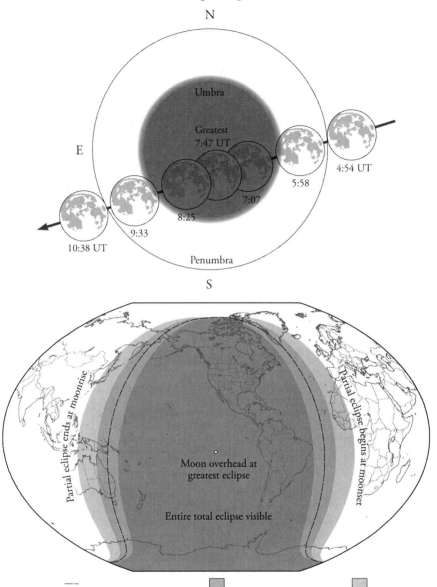

Moon rises/sets at greatest eclipse   Moon rises/sets in total eclipse   Moon rises/sets in partial eclipse

Fig. 4.19 This is the first of four consecutive total lunar eclipses, a sequence known as a tetrad. The Moon, located over the South Pacific at mid-eclipse, is well placed to bring early morning totality throughout the Americas, including the Caribbean, Alaska and Hawaii. Observers from Brazil to New Zealand and from Canada to the South Pole will see totality in its entirety. The Moon rises in total eclipse for the eastern third of Australia, and it sets in totality as seen from western Africa, the Azores and the Cape Verde Islands. This is the 56th lunar eclipse of Saros 122. Look for bright Mars (–1.4) nearby and the star Spica closer to the Moon

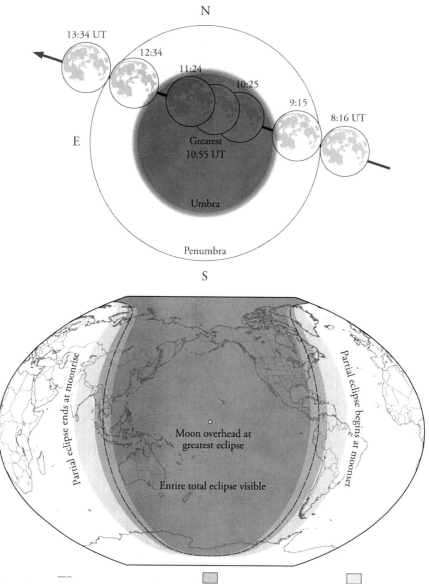

Fig. 4.20 At mid-eclipse, the Moon is almost directly overhead as seen from Kiribati in the South Pacific. All of North America sees at least some portion of totality before moonset, with states west of Maryland and New York able to witness the entire event. Also favored are Australia, New Zealand, Indonesia, Japan, and eastern Asia. This is the 42nd lunar eclipse of Saros 127

Total Solar Eclipse, March 20, 2015

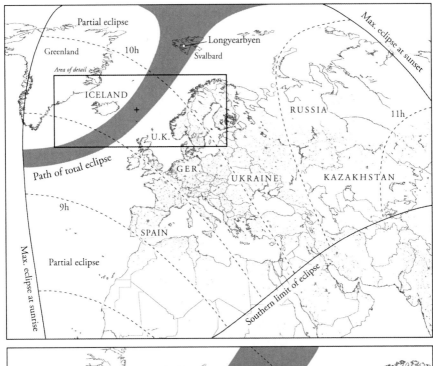

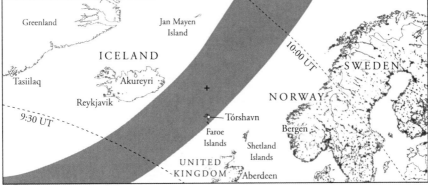

Fig. 4.21 The curious shape of totality's path in this eclipse is an artifact of the map projection; the track terminates at the North Pole. The umbra carves a swath that barely touches land – only Denmark's Faroe Islands and Norway's Svalbard. At the point of greatest eclipse (+) the Sun will be blotted out for 2 min 47 s. From Longyearbyen, totality lasts just under 2.5 min and peaks at 10:12 UT. Mercury (–0.4), nearly a month past its greatest western elongation, shines 18° west of the Sun; Mars (1.3 and 22° east) and Venus (–4.0, 34° east) also shine nearby. From Tórshavn, the capital and largest town on the Faroe Islands, totality lasts just under 2 min and peaks at 9:41 UT. A deep partial eclipse comes to Tasiilaq (94% at 9:35 UT), Reykjavík (97%, 9:37 UT), Aberdeen (94%, 9:38 UT), the Shetland Islands (97% at 9:43 UT) and Bergen (94%, 9:49 UT). This is the 61st eclipse of Saros 120

Total Lunar Eclipse, April 4, 2015

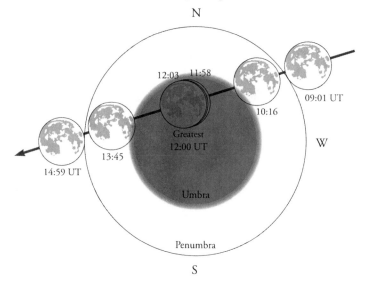

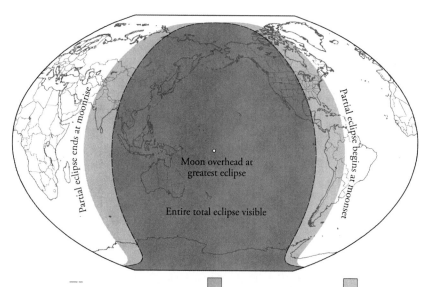

Moon rises/sets at greatest eclipse   Moon rises/sets in total eclipse   Moon rises/sets in partial eclipse

Fig. 4.22 Totality for this eclipse falls into the "blink and you'll miss it" category. The Moon is again over the South Pacific, and it appears near the zenith at mid-eclipse as seen from Tuvalu. For the eastern half of North America, only the partial stages can be seen before the Moon sets, but west of a line from Texas to North Dakota all of totality will be seen. Western Mexico and Canada, all of Alaska, eastern Russia and China, Japan, Indonesia, Malaysia, Australia, and New Zealand are similarly favored. Totality is over before the Moon comes up for a swath of the globe that includes India; these regions will see only the final partial phase of the eclipse, which is the 30th of Saros 132

Total Lunar Eclipse, Sept. 28, 2015

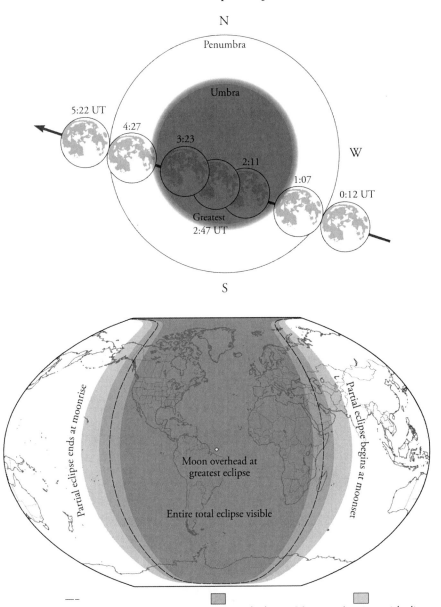

Fig. 23 The final total eclipse of the 2014–2015 tetrad is an excellent one for the Americas. All of the contiguous U.S., most of Canada, and all of Central and South America will witness totality in its entirety. It's a good eclipse for all of Europe and most of Africa too, and totality is under way as the Moon sets from Madagascar through the Middle East. The Moon is nearly overhead at mid-eclipse as seen from the northeastern Brazil. This is the 28th lunar eclipse of Saros 137

Total Solar Eclipse, March 9, 2016

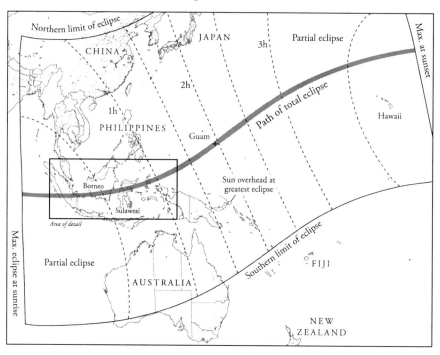

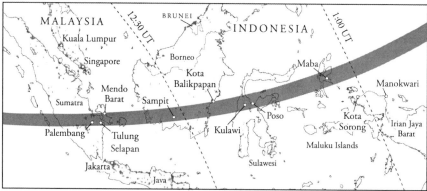

Fig. 4.24 At greatest eclipse (+), which falls between Guam and the westernmost islands of Micronesia, the Moon covers the Sun for 4 min 9 s centered on 1:57 UT. Before that, the umbral track crosses Sumatra, Borneo, Sulawesi, and the Maluku Islands. This is the 52nd solar eclipse of Saros 130. When the Sun winks out, look for Mercury (−0.6) 13° west of the it and Venus (−3.8) 10° farther away. The umbra peaks at the Mentawai Islands off Sumatra's west coast, where totality lasts 1 min 46 s, at 0:19 UT; it reaches maximum at Palembang almost 3 min later. Mid-totality arrives in Sampit at almost 0:29 UT, producing a nearly 2-min-long totality. At Kulawi totality has lengthened to 2.5 min (max. at 0:38 UT), and at Mbaba (0:54 UT) 3 min 20 s. A deep partial eclipse comes to Kuala Lumpur (83%, 0:23 UT), Singapore (89%, 0:24 UT), Jakarta (90%, 0:21 UT) and Kota Sorong (94%, 0:59 UT). From Honolulu, Hawaii, the Moon obscures nearly 70% of the Sun at 3:36 UT

Annular Solar Eclipse, Sept. 1, 2016

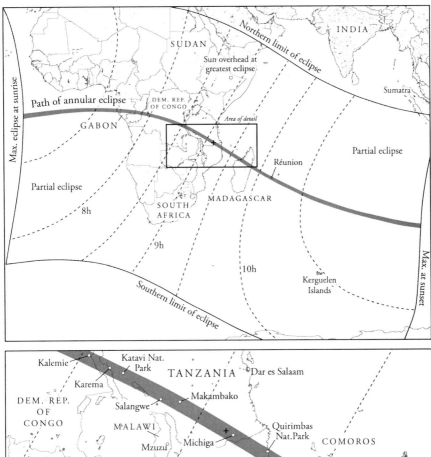

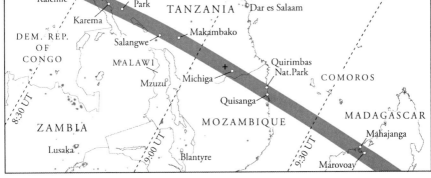

Fig. 4.25 The Moon's shadow returns to central Africa, bringing a ring of Sun to Gabon starting at 7:39 UT. The annulus moves on to Congo, Tanzania, Mozambique, Madagascar and Réunion. At greatest eclipse (+, 9:07 UT) in southern Tanzania, the Sun is nearly 99% obscured for 3 min 5 s. By the time the shadow reaches southwestern Réunion, at 10:09 UT, the Moon's coverage of the Sun has dropped to 97%. This is the 39th eclipse of Saros 135. The eclipse reaches 98% at Kalemie (8:32 UT), then the shadow crosses Lake Tanganyika and enters Tanzania. At Karema, near the center of the track, maximum eclipse (98%, 8:37 UT) lasts just under 3 min. The shadow then traverses Katavi National Park, the Rungwa Game Reserve and Lake Rukwa; the eclipse is 98% at Salangwe (8:49 UT) and Makambako (8:54 UT). The shadow treks across the Selous Game Reserve, with greatest eclipse (+) occurring near the park's southwestern fringe. Annularity enters Mozambique at the Niassa Game Reserve (9:08 UT), leaves the continent near Quisanga (max. at 9:18 UT), crosses the Mozambique Channel, and peaks at Mahajanga and Marovoay near 9:41 UT

Annular Solar Eclipse, Feb. 26, 2017

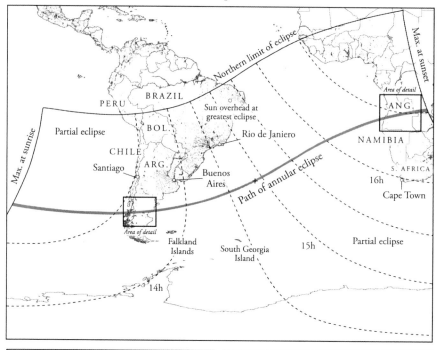

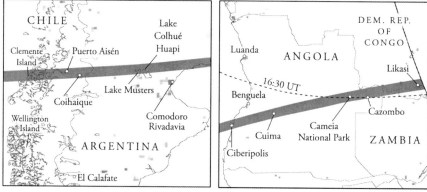

Fig. 4.26 This eclipse brings a solar ring to Chile and Argentina near its start and to Angola, Zambia and the Democratic Republic of Congo near its end. At the point of greatest eclipse (+, 14:53 UT), the Sun is more than 99% obscured for just 44 s. A partial eclipse is seen at Santiago (65%, 13:32 UT), the Falkland Islands (up to 80 percent, 13:52 UT), Buenos Aires (73%, 13:53 UT) and Cape Town (51% at 15:59 UT). This is the 29th solar eclipse of Saros 140. Annularity first comes to western Chile, peaking at Clemente Island at 13:34 UT. It reaches maximum at Puerto Aisén about 90 s later and at Coihaique just after 13:36 UT. Passing into Argentine Patagonia, the eclipse maxes out at Lakes Musters and Colhué Huapi around 13:39 UT, where the solar ring lasts just over a minute. Residents of Comodoro Rivadavia see a partial eclipse only (97%, 13:42 UT). In Africa, the eclipse reaches maximum at Ciberipolis (16:26 UT) and proceeds eastward to Cuima (16:28 UT) and Cameia National Park (16:30 UT) – all near the center of the track, where 99% of the Sun is obscured and annularity lasts a bit longer than 1 min. At Likasi, maximum eclipse (16:31 UT) occurs nearly at sunset

Partial Lunar Eclipse, Aug. 7, 2017

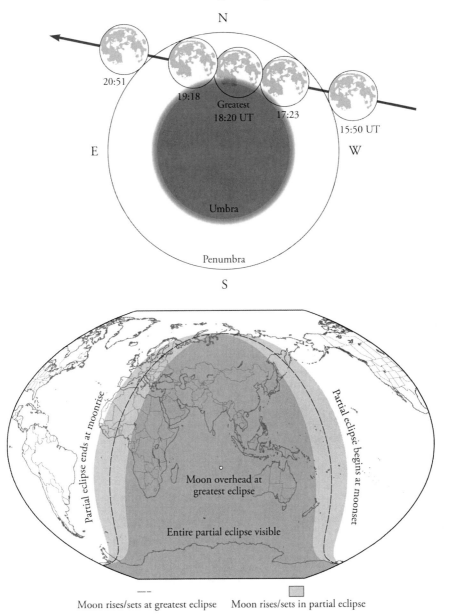

Fig. 4.27 The Moon stands directly above the Indian Ocean at mid-eclipse, bringing a partial lunar eclipse to eastern Africa, the Middle East, most of Asia, Japan, Indonesia, Malaysia, Australia and Antarctica. For Fiji, Vanuatu and the eastern half of New Zealand's North Island, the Moon sets before it reaches its greatest immersion in the umbra. The later stages of the eclipse will be seen after moonrise across much of Europe and western Africa. The eclipse is not visible from North America. This is the 62nd lunar eclipse of Saros 119

Total Solar Eclipse, Aug. 21, 2017

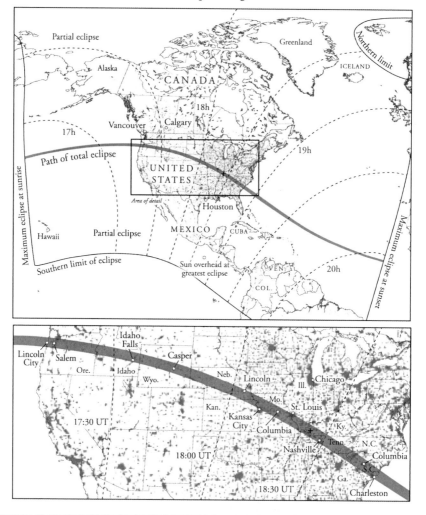

Fig. 4.28 This is the eclipse Americans have been waiting decades for: The Moon's umbra swings across the continental U.S. from Oregon to South Carolina. The Sun goes dark for 2 min 40 s at the point of greatest eclipse (+, near Hopkinsville, KY, beginning at 1:25 P.M. CDT). A deep partial eclipse comes to Vancouver (89%, 10:21 A.M. PDT), Calgary (82%, 11:33 A.M. MDT), Houston (73%, 1:17 P.M. CDT), Chicago (89%, 1:20 P.M.) and Atlanta (97%, 2:34 P.M. EDT). This is the 22nd eclipse of Saros 145. As totality arrives, look for Venus (–3.9) shining 23° west of the Sun. A nearly 2-min-long totality comes ashore at Lincoln City (10:16 PDT), reaching Salem a bit over a minute later. Proceeding eastward, the shadow marches through Idaho (max. at Idaho Falls, 11:34 A.M. MDT), Wyoming (Casper, 11:44 A.M. MDT), and Nebraska (Lincoln, 1:03 P.M. CDT). Kansas City is on the fringe; the northern part of the city experiences a brief 18-s totality shortly after 1:08 P.M. Columbia fares better 4 min later, when a 2 min 38 s totality commences. The umbra passes just south of downtown St. Louis at 1:18 P.M. and will produce a brief darkening there. Nashville experiences totality at 1:27 P.M. CDT. Columbia sees the Sun go dark at 2:41 P.M. EDT and Charleston follows 5 min later

Total Lunar Eclipse, Jan. 31, 2018

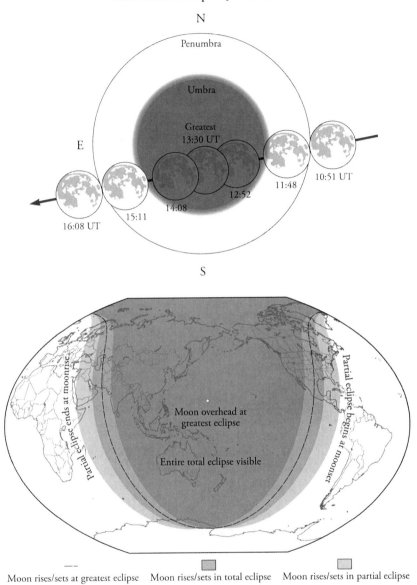

Fig. 4.29 Lunar totality comes to much of North America, with regions farther west treated to longer views of the eclipse before the Moon sets. Hawaii, Alaska, and continental U.S. states west of a line from New Mexico to North Dakota will see the entire total potion of the eclipse. At mid-eclipse, the Moon is nearly overhead as seen from the Pacific islands of Micronesia, and New Zealand, Australia, Indonesia, Malaysia, nearly all of India, China, Japan, and eastern Russia also witness the entire total portion of the eclipse. The Moon sets during totality along the central third of the U.S. and rises during totality from Iran, central Asia and western Russia. This is the 49th lunar eclipse of Saros 124

Total Lunar Eclipse, July 27, 2018

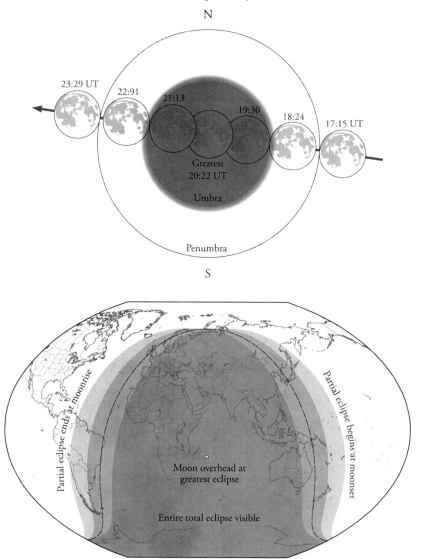

Fig. 4.30 With the Moon in the umbra for 1 h and 43 min, this eclipse has the decade's longest stretch of lunar totality. All of it is visible to observers in Antarctica, all but extreme eastern Africa, central Europe, eastern Russia, the Middle East, India, Malaysia, parts of Indonesia, and the western two-thirds of Australia. The Moon sets in totality from the South Island of New Zealand, southern Japan, Scandanavia, the United Kingdom, northeastern France, Portugal, the eastern half of Spain, the Canary and Cape Verde Islands and most of Western Sahara. At mid-totality, the Moon is nearly overhead as seen from Madagascar. This is the 38th lunar eclipse of Saros 129. Brilliant Mars (−2.8), just hours past its opposition, lurks nearby

Total Lunar Eclipse, Jan. 21, 2019

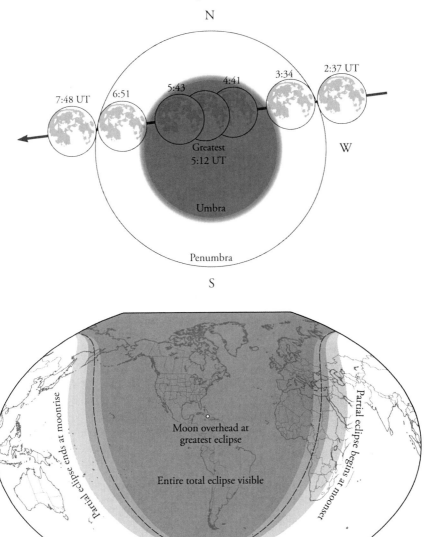

Fig. 4.31 The Moon is directly overhead from Cuba during mid-totality, which makes this a perfectly placed eclipse for North and South America. From Europe to Hawaii, all of the eclipse's total portion is visible. The Moon sets during totality in a band that runs from Namibia to Egypt, the Black Sea and Russia, while northeastern Russia and Pacific islands see the Moon rise while totality is under way. This is the 27th lunar eclipse of Saros 134

Total Solar Eclipse, July 2, 2019

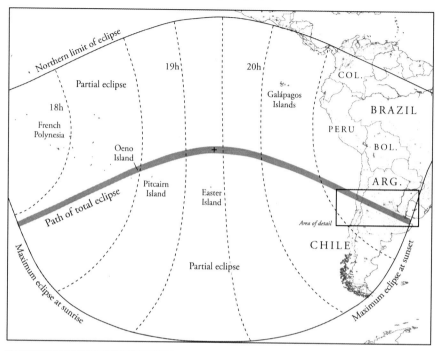

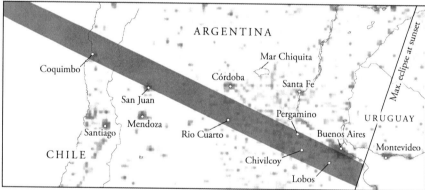

Fig. 4.32 The Moon's shadow courses over the South Pacific, avoiding a significant land mass until the end of the eclipse; only Oeno Island, about 90 miles (145 km) northwest of Pitcairn, will witness totality. At the point of greatest eclipse (+, 19:20 UT), which lies about 680 miles (1,094 km) north of Easter Island, the Sun darkens for 4 min 33 s. The length of totality dwindles to more than half by the time the shadow makes landfall in Chile. This is the 58th solar eclipse of Saros 127. During totality, look for Venus (−3.9) about 12° west of the Sun. When totality commences at Coquimbo (20:38 UT), the afternoon Sun stands just 14° above the horizon. San Juan, near the edge of the shadow track, sees a 39-s-long totality at 20:40 UT. Eclipse maximum reaches Rio Cuarto about 90 s later with totality lasting just under 2 min. At Pergamino, the Sun is just 3° above the horizon as totality begins (20:43 UT). The track passes south of Buenos Aires, but from Lobos the eclipse peaks at 20:44 UT with the Sun 1.3° above the horizon. A deep partial eclipse comes to Santiago (93%, 20:37 UT), Mendoza (96%, 20:39 UT), Córdoba (98%, 20:44 UT) and Santa Fe (95%, 20:46 UT)

Partial Lunar Eclipse, July 16, 2019

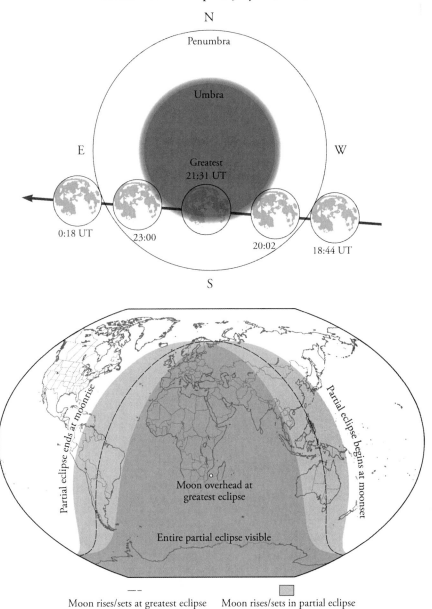

Fig. 4.33 Our final partial lunar eclipse of the decade brings a shadowed Moon to Africa, much of Europe, the Middle East, Central Asia, India, and parts of Malaysia and Australia. These regions witness the Moon's full umbral passage. Most of South America, the United Kingdom, most of Scandanavia will see the later stages of the eclipse after moonrise, while Mongolia, most of China, southernmost Japan, and the South Island of New Zealand will catch the early stages before the Moon sets. This is the 22nd lunar eclipse of Saros 139

Annular Solar Eclipse, Dec. 26, 2019

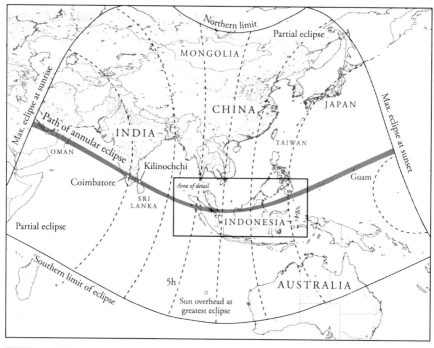

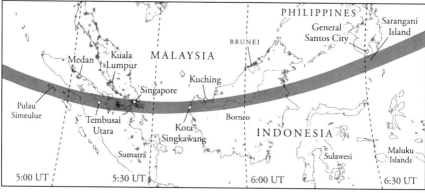

Fig. 4.34 A ring of Sun greets early risers on the eastern shore of Saudi Arabia, southern Qatar and the United Arab Emirates and Oman. There, maximum eclipse (97%, 3:38 UT) occurs along Oman's eastern shore with the Sun less than 12° high. The shadow sweeps across the Arabian Sea to India, peaks at Coimbatore (3:59 UT) and Kilinochchi (4:06 UT), and rushes to Sumatra and the point of greatest eclipse (+, 5:18 UT), where the annular Sun lasts 3 min 35 s. It then sweeps over Singapore and Borneo, clips the southernmost Philippines at Mindinao, and brings annularity to Guam at 6:54 UT. This is the 46th eclipse of Saros 132. The annulus reaches Pulau Simeulue at 4:51 UT and Tembusai Utara at 5:09 UT. Greatest eclipse (+) occurs just off the Sumatran coast east of Sungai Apit. At Singapore, the solar ring appears at 5:23 UT and lasts 2 min 10 s. The annular Sun comes to Kota Singkawang at (5:41 UT), southern Kuching (5:47 UT) and grazes the southern tip of Mindanao, with Sarangani Island afforded the best view at 6:29 UT

Annular Solar Eclipse, June 21, 2020

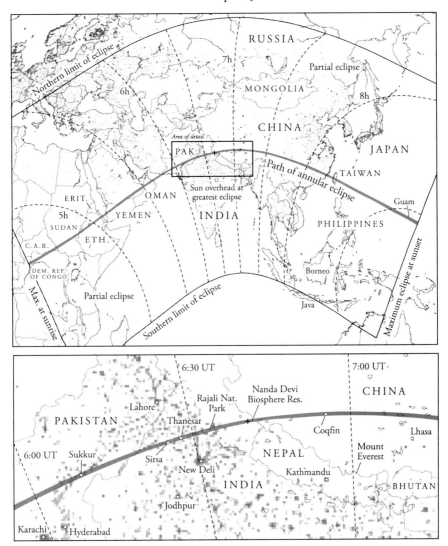

Fig. 4.35 In this annular eclipse, the greatest duration occurs near the sunrise and sunset points of the path and not at the point of greatest eclipse (+, 38 s at 6:40 UT), which actually defines when the axis of the Moon's shadow passes closest to Earth's center. From the extreme western Republic of Congo to the Red Sea, annularity lasts about a minute in the track's center, and lengthens to 52 s by the time it passes Taiwan. Annularity lasts up to 1 min 15 s at the track center directly south of Guam. This is the 36th solar eclipse of Saros 137. The ringed Sun sweeps over Sukkur (6:07 UT), Sirsa (6:26 UT), Thanesar (6:32 UT), the northern portion of Rajali National Park 2 min later, and greatest eclipse (+) occurs in the Nanda Devi Biosphere Reserve. The annulus moves on to Coqfin (6:56 UT), after which it begins an arc through southern China and Taiwan

Total Solar Eclipse, Dec. 14, 2020

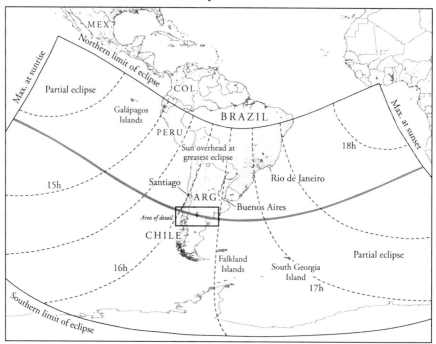

Fig. 4.36 The decade's final solar eclipse again brings totality to Chile and Argentina. At the point of greatest eclipse (+, 16:13 UT), totality lasts just under 2 min 10 s. A deep partial eclipse comes to Santiago (83%, 16:01 UT), Buenos Aires (79%, 16:32 UT), Montevideo (79%, 16:40 UT), the Falkland Islands (72%, 16:40 UT), South Georgia Island (63%, 16:59 UT) and Rio de Janeiro (42%, 17:14 UT). This is the 23rd solar eclipse of Saros 142. During totality, look for Mercury (−1.1) 3° and Venus (−3.9) 24° west of the Sun. Jupiter (−2.0) and Saturn (0.6), now less than a degree apart, shine 36° east of the Sun. The Moon's umbra sweeps over Mocha Island beginning at 15:59 UT and proceeds to the Chilean coast. Totality peaks at Toltén at just before 16:02 UT, at Pitrufquén about 45 s later, and Loncoche follows at just after 16:03 UT. Crossing into Argentina, the shadow passes over Junín de los Andes (16:06 UT), Piedra del Águila (16:08 UT), Sierra Colorada (16:13 UT), Ministro Ramos Mexia a minute later, and San Antonio Oest at 16:19 UT. The eclipse is partial in Viedma (99.9%, 16:24 UT) but its southern suburbs see totality. A deep partial eclipse comes to Los Angeles (96%, 16:02 UT), Puerto Montt (94%, 16:05 UT), Neuquén (97%, 16:12 UT) and to Bahía Blanca (93%, 16:25 UT)

Table 4.2 Eclipses on the Web

Eclipse information

NASA's Eclipse Web Site
eclipse.gsfc.nasa.gov

Eclipses Online
www.eclipse.org.uk

Eclipse filters

American Paper Optics, LLC
www.3dglassesonline.com

Rainbow Symphony, Inc.
www.rainbowsymphonystore.com

Science Stuff, Inc.®
www.sciencestuff.com

Eclipse photography

Astronomy Picture of the Day
apod.nasa.gov

Mr. Eclipse
www.mreclipse.com

Ben Cooper's Launch Photography
www.launchphotography.com

Miloslav Druckmüller's Eclipse Photography
www.zam.fme.vutbr.cz/~druck/Eclipse

Other eclipse information

Solar eclipse: Stories from the path of totality
www.exploratorium.edu/eclipse

Shadow and Substance: Animated diagrams
shadowandsubstance.com

central track. For much more detailed eclipse predictions, refer to the resources listed in Table 4.2, particularly Fred Espenak's NASA Eclipse Web Site, which publishes lists of local circumstances for major cities and also provides predictions interactively via Google Maps.

Lunar eclipse maps are much easier to understand. First, only partial and total lunar eclipses are illustrated, and each eclipse is shown with a pair of drawings. At the top of the page, the Moon's track through Earth's penumbra and umbra is shown to scale along with the UT time for each of the shadow contacts. Below this is a map of the world illustrating where on Earth the eclipse's partial and total phases will be visible. A dashed curve indicates where the Moon is either rising or setting at the moment of mid-eclipse, and because the eclipse occurs at full Moon, the Moon is rising around sunset and setting at about sunrise.

Before each eclipse, either solar or lunar, review the expected phenomena and try to look for them during the event. Also, keep an eye out for nearby planets and, during lunar eclipses, bright stars and constellations that may become visible in the dimmed light.

Box 4.3 Danjon's lunar eclipse scale

The coloration and brightness of a totally eclipsed Moon varies considerably thanks to the scattering of sunlight by dust, clouds and volcanic particles in Earth's atmosphere. French astronomer André Danjon (1890–1967) devised a five-point scale for evaluating the visual appearance and brightness of the eclipsed Moon. Things to look for: How visible are the familiar lunar features at mid-eclipse? Are there color or brightness variations within the umbra? How sharp is the umbra's edge? Record your impressions, then try to rate the eclipse on Danjon's scale below.

1. Very dark eclipse; the Moon is almost invisible, especially at midtotality.
2. Dark eclipse, gray or brownish in color; difficult to distinguish features.
3. Deep red or rust-colored eclipse; dark central shadow, while umbra's outer edge is relatively bright.
4. Brick-red eclipse. Umbral shadow usually has a bright or yellow rim.
5. Bright copper-red or orange eclipse. Umbral shadow has a bluish and very bright rim.

References

Brunier S, Luminet J (2000) Glorious eclipses: Their past, present and future. Storm Dunlop, (trans). Cambridge University Press, New York

Clark RW (1971) Einstein: The life and times. World Publishing Co., New York

Conot R (1979) Thomas A. Edison: A streak of luck. Da Capo Press, New York

Cooper J F (1869) The eclipse. Putnam's Mon. Mag. 21 (n.s. 4), Sept., 352–359. http://etext.lib.virginia.edu/toc/modeng/public/CooEcli.html. Accessed 18 Nov. 2010

Crelinsten J (2006) Einstein's Jury: The race to test relativity. Princeton University Press, Princeton, New Jersey

Dyer FL, Martin TC (1910) Edison, his life and inventions, vol. 1. Harper & Brothers Publishers, New York http://etext.lib.virginia.edu/toc/modeng/public/Dye1Edi.html. Accessed 18 Nov. 2010

Eddy JA (1973) The great eclipse of 1878. Sky Telesc. June, 340–346

Espenak F Five millennium catalog of lunar eclipses. http://sunearth.gsfc.nasa.gov/eclipse/LEcat/LEcatalog.html. Accessed 18 Nov. 2010

Espenak F Six millennium catalog of solar eclipses. http://sunearth.gsfc.nasa.gov/eclipse/SEcat/SEcatalog.html. Accessed 18 Nov. 2010

Espenak F Solar eclipses of saros 124. http://sunearth.gsfc.nasa.gov/eclipse/SEsaros/SEsaros124.html. Accessed 18 Nov. 2010

Espenak F (1987) Fifty year canon of solar eclipses, 1986–2035. NASA Reference Publication 1178 Revised, July 1987. Sky Publishing Corp., Cambridge, Massachusetts

Espenak F (1989) Fifty year canon of lunar eclipses, 1986–2035. NASA Reference Publication 1216, March 1989. Sky Publishing Corp., Cambridge, Massachusetts

Espenak F (2009) Interview with author, 25 June 2009

Fernie JD (2000) Eclipse vicissitudes: Thomas Edison and the chickens. Am. Sci. 88: 120

French AP (ed.) (1979) Einstein: A centenary volume. Harvard University Press, Cambridge, Massachusetts

Harrington PS (1997) Eclipse! The what, where, when, why & how guide to watching solar and lunar eclipses. John Wiley & Sons, Inc., New York

Lawrence TE (1997) The seven pillars of wisdom. Wordsworth Editions Limited, Hertfordshire

Littman M, Wilcox K, Espenak F (1999) Totality: eclipses of the Sun. Oxford University Press, New York

Meeus J (1982) The frequency of total and annular solar eclipses for a given place. J. Br. Astron. Assoc. 92: 124–126

Kendall PM (ed.) (1896) Maria Mitchell: Life, letters and journals. Lee and Shepard, Boston

Peterson S (1999) Mania in the path of an eclipse. Christ. Sci. Monit. 11 Aug

Schaefer BE (1994) Solar eclipses that changed the world. Sky Telesc. May, 36–39

Schaefer BE (1992) Lunar eclipses that changed the world. Sky Telesc. Dec., 639–642

Stephenson RF (1997) Historical eclipses and Earth's rotation. Cambridge University Press, Cambridge, England

Yapo TC, Cutler B (2009) Rendering lunar eclipses. Proc. Graphics Interface, May, 63–69. http://www.cs.rpi.edu/research/groups/graphics/eclipse_gi09. Accessed Nov. 15, 2010

Zirker JB (1984) Total eclipses of the Sun. Van Nostrand Reinhold Co. Inc., New York

Chapter 5

Mars: The Red Wanderer

At the close of the 19th century, Mars was imagined to be the most likely abode of extraterrestrial life. The available science hinted that conditions on its surface were similar to those on Earth. Observations of unusual linear surface features by some of the leading astronomers of the day ultimately led a few of them to promote the idea that life, and even intelligent beings, lived there. Even among scientists, the belief that Mars should have some form of life proved hard to shake; as late as the 1950s some astronomers felt that color changes detected on the martian surface were best explained by the seasonal growth of vegetation. While early scientific speculation about a martian civilization was quickly discredited, the notion settled into popular culture and inspired writers from Ray Bradbury to H. G. Wells. These stories stoked public imagination about the possibilities of contact with extraterrestrial civilizations. By the mid-1970s microscopic life was the most advanced organism anyone seriously expected to find on Mars. Biological experiments sent to the surface detected *something*, but probably not life. As we'll see, the scientific story doesn't end there.

Long before it was considered a home to aliens, Mars already had established itself in the human imagination. Its reddish coloring, which contrasts beautifully with the deep blue of a twilight sky, is unique among the planets. Mars also undergoes exceptional changes in brightness within a single apparition – and its peak brightness varies from one appearance to the next. Near the time when Mars shines best, a point called opposition, the Red Planet takes a seemingly chaotic whirl through the starry sky. Every planet makes one of these so-called retrograde loops, but the one Mars takes is by far the most obvious and dramatic. The planet's distinctive color, remarkable brightness variations and bizarre sky motion combine to make Mars the most outstanding of the celestial "wanderers." So in both a cultural and an astronomical sense, Mars is the archetypal planet. Table 5.1 provides its physical and orbital characteristics and Fig. 5.1 shows how the planet compares to Earth and Moon.

F. Reddy, *Celestial Delights: The Best Astronomical Events through 2020*,
Patrick Moore's Practical Astronomy Series, DOI 10.1007/978-1-4614-0610-5_5,
© Springer Science+Business Media, LLC 2012

Table 5.1 Facts about Mars

Mars	
Diameter	4,222 miles
The total surface area of Mars is about 97% of the total land area on Earth.	6,794 km
	53.3% of Earth's
Surface temperature	
Maximum (30° south, summer solstice)	86°F (30°C)
Minimum (poles, winter solstice)	−200°F (−130°C)
Surface atmospheric pressure (average)	7.5 millibars
	0.74% of Earth's
Atmospheric composition	95.3% carbon dioxide (CO_2)
	2.7% nitrogen (N_2)
	1.6% argon (Ar)
	0.13% oxygen (O_2)
	0.07% carbon monoxide (CO)
	0.03% water vapor (H_2O)
Moons	2
Largest	Phobos ("Fear")
	Deimos ("Terror")
Rotation period	24.62 h
Obliquity	25.19°
Sidereal period (time for one full orbit around the Sun)	686.98 days or 1.88 years
Synodic period (time between successive conjunctions with the Sun; longest of any planet)	779.94 days or 2.14 years
Average distance from Sun	141.6 million miles
Light takes 13 min to travel this far.	227.9 million km
	1.524 Astronomical Units (AU)
Orbit inclined to Earth's	1.85°

Mars seems anything but a wanderer when it first appears in the morning sky. The Red Planet is singularly unimpressive, a faint tawny "star" that only gradually increases its lead on the Sun, lingering for weeks above the eastern horizon before sunrise. But a closer look reveals more interesting behavior. The background stars progress noticeably westward each week, but Mars resists the flow, traveling eastward through the constellations. By the time Mars is rising around 9 P.M., it's barely moving through the background stars at all and nearly matches their normal day-to-day progression. When the Red Planet shines brightly in the south at midnight, it's now moving westward each day – and doing so much faster than the stars. And when Mars appears in the west after sunset, it seems to stubbornly linger in the afterglow as the background stars slip toward the Sun.

In Mesopotamia and the classical world, Mars was associated with gods of war: Nergal of the Babylonians, Ares of the Greeks, and of course, the Roman Mars.

Fig. 5.1 Earth, Mars and the Moon compared (Montage by the author; photos by NASA and NOAO)

Rome was believed to have a special relationship with Mars, for he was said to have fathered the mythical twins Romulus and Remus, who founded the city. His name was given to the first month of the Roman calendar, March, but the Anglo-Saxon war god Tiw is the one who gives his name to Tuesday. It's often said that the Red Planet's association with war stems from its reddish appearance, suggestive of bloodshed, but to the eye Mars is more peachy or orange than crimson. If color is an imperfect explanation, perhaps the planet's idiosyncratic departure from the order of the heavens justifies its link with war, which is itself a radical deviation from life's ordinary rhythm.

Chinese astronomers called the planet Yinghuo (Dazzling or Sparkling Deluder) and Huoxing (Fire Star), a name that nicely evokes its ember-like hue. To them Mars was a portent of a variety of troubles and astrological texts were quick to point to its color, as in this omen from the T'ang Dynasty (A.D. 618–907): "When Sparkling Deluder enters the Southern Dipper [eastern Sagittarius] and its color is like blood, there will be a drought." The Chinese thought Venus to be a better indicator of coming war, while Mars was more of a force of nature with judicial functions: "Sparkling Deluder is the Master of the Proprieties; when the proprieties are misdone, then the punishment issues from it." Other names that T'ang astronomers used for Mars – Star of Punishment, Holder to the Law – suggest just the opposite of the chaos of war.

In ancient Egypt, Mars was a manifestation of Horus, the son of Isis and Osiris. Murdered by his brother Set, Osiris nevertheless was able to conceive a son through special rites performed by Isis. Horus was the rightful heir to his father's throne, but after all Set had done he wasn't going to let Egypt slip through his fingers. Set contested Horus' rule and fought him in many battles, but eventually Horus won the war, killing Set and at last avenging his betrayed father. Each pharaoh was regarded as a manifestation of Horus, and the god was worshipped in different locales under numerous monikers. Mars, Jupiter and Saturn were all linked to him, but the names "Horus of the Horizon" and "Horus the Red" refer to Mars. Another epithet – "he who travels backwards" – seems inspired by the Red Planet's retrograde motion.

Among the Skidi Pawnee of central Nebraska, Morning Star was a powerful god who helped his younger brother, the Sun, to have heat and light. Morning Star and the Sun together ruled the stars of the eastern sky, who were male, while Evening Star and her sister, the Moon, ruled the female stars of the western sky. One day, Morning Star sought to marry Evening Star, so he and the Sun set out to her village in the west. But the Moon had other ideas: She barred their way by opening cracks in the earth that threatened to swallow them up. Morning Star struck his war club on the ground, and the earth again became firm. So began a contest between Morning Star and the Moon, who used her powers to attempt to block his progress. The Moon materialized beds of sharp flints and cactus, and animal attacks from magical snakes, buffalo and, finally, bears. With one blow from his war club, Morning Star made the obstacles disappear or killed the attacking animals. When the Moon's powers were exhausted, Morning Star and the Sun entered the western star village. Morning Star wed Evening Star, and the Sun wed the Moon. Through their unions, the stars had a girl and the Sun and Moon had a boy – the first humans.

To honor Morning Star, the Skidi Pawnee performed a ceremony that occasionally required the sacrifice of a young girl from an enemy tribe. Usually, the need for the ceremony arose when Morning Star appeared in a warrior's dream and demanded it, but it might be performed for other reasons. After consulting with priests who determined the validity of the dream, the warrior would lead a raiding party against an enemy tribe, locate and capture an appropriate girl, and immediately dedicate her by saying "Opirikuts," a word that translates to "mighty star of fire." This certainly suggests that Mars played the role of Morning Star. When the season was right, and after days of feasts, dances and singing, the girl was tied to a scaffold a few hours before dawn. Then her captor shot an arrow through her heart, another struck her head with a war club, and the other males of the tribe fired arrows into the body. The ritual symbolized both the stellar union and the obstacles Morning Star overcame to achieve it, and may have been performed in abbreviated fashion, without the sacrifice, every year. However, the identity of Morning Star with Mars was promoted by ethnographers in the early twentieth century, who were writing decades after the last sacrifice in 1838. Astronomically, this association doesn't jibe with known performances of the sacrifice or the abbreviated ritual. "The description of its appearance would seem to indicate Mars, but the times given for its appearance suggest that Venus and even Jupiter could serve in the proper place," noted James Murie, who was both a member of the Skidi band and a translator for two ethnographers.

The Maya of Mesoamerica also maintained an interest in the Red Planet. The principle aim of Mayan astronomical observation appears to have been the discovery of commensurate relationships between the seasons, their 260-day sacred calendar and various other celestial cycles. In the case of Venus, such a cycle was relatively straightforward, but the erratic wanderings of Mars would seem to be another matter altogether. The *Dresden Codex*, one of four surviving Mayan texts, contains a table that was first identified as having something to do with Mars more than a century ago. It lists ten intervals of 78 days each, a number close to the average time it takes Mars to swing through its retrograde loop (75 days). The ten intervals together equal 780 days, very close to the planet's mean synodic period. They also equal three cycles of the sacred calendar. That's just the sort of the thing that would appeal to Maya astronomers, who discovered a novel link between the Red Planet's first visibility in the morning sky and the timing of its curious loopy motion (see box, "How the Maya made Mars make sense").

Box 5.1 How the Maya made Mars make sense

The messy motions of Mars beviled all astronomers until Kepler. No single cyclic relationship can explain, for example, the variable time spans between the planet's first visibility and the start of its retrograde motion. Today, we understand that this variability is due to Mars following an eccentric elliptical orbit.

In 2001, Harvey and Victoria Bricker of Tulane University, together with Colgate University's Anthony Aveni, called attention to a little-appreciated table in the Maya's *Dresden Codex* that they believe reveals a particularly ingenious solution to the problem of predicting Mars. It appears that Maya astronomers discovered two directly observable time cycles that together "not only accurately described the planet's motion, but also related it to other cosmic and terrestrial concerns," the team wrote.

(continued)

Box 5.1 (continued)

The longer cycle included the time Mars spent in its retrograde loop (702 days), but the shorter cycle (about 543 days) did not. Maya astronomers then linked multiples of these periods together to arrive at a simple and practical formula:

$$7 \text{ long} + 1 \text{ short} + 7 \text{ long} + 1 \text{ short} + 8 \text{ long} + 1 \text{ short}$$

This cycle contains a total of 25 time spans and contains nearly the same number of days as it takes Mars to revolve around the Sun 25 times, so the formula tracks Mars relative to the stars and the seasonal year. The scientists say that the values the Maya chose for the long and short intervals are less precise than they could be. But given the Maya penchant for cleanly nested cycles, it's likely that this imprecision was designed so that the Mars cycle would be commensurate with the Red Planet's 780-day synodic period. Both the synodic period and the long interval are even multiples of 78, and the short interval very nearly is, too. So a Maya astronomer could think of all of these cycles as built from different numbers of 78-day units.

Best of all, the last four terms of this Mars formula match the number of days in 20 synodic Venus cycles to within about 1 day. Remarkably, the Maya found a straightforward way to integrate the Red Planet's erratic motions with the seasonal year, their own sacred calendar and the actual appearances of Venus.

Mars Attacked

The Greeks were the first to attempt a theoretical framework that explained the motions of the planets. Constrained by the purely philosophical considerations of Plato (c. 427–327 B.C.), Greek cosmologists allowed the planets to move around Earth only in perfect circles and only at a constant speed. The earliest such system was devised by one of Plato's students, Eudoxus of Cnidus, in the fourth century B.C. He placed the planets on concentric spheres centered on Earth. To replicate the observed motions, the Sun and Moon each required three moving spheres, the planets needed four apiece, and a single sphere held the "fixed stars." Callippus (c. 370–300 B.C.), a student of Eudoxus, later refined the system by adding another seven spheres, and Aristotle (384–322 B.C.), another of Plato's students, made further refinements that brought the total to 55.

That these nested-sphere models were even approximately successful helped validate the conceit of uniform circular motion. But they failed to account for changes in appearance that today we know are caused by changing distances from Earth. The dramatic brightness variations of Mars and Venus, and changes in the size of the Moon made apparent by annular and total solar eclipses, were simply ignored.

Some promoted far more radical ideas. The most notable was Aristarchus of Samos (c. 310 to c. 230 B.C.), who was the first to argue that the Sun resided at the center of the universe and that Earth both revolved around the Sun and spun daily on its own axis. A Sun-centered or heliocentric system went against the grain of mainstream thought by challenging the teachings of Plato and Aristotle, who argued that Earth was immovable. By retaining circular orbits and uniform motion, one could say that the early Sun-centered models weren't radical enough, yet refinements to these models might have progressed faster had they displaced Earth-centered theories. And let's face it: The evidence of our senses tells us that the sky and the objects in it are moving, not Earth, so heliocentrism waited in the wings throughout antiquity while geocentrism held center stage.

A geocentric system that enjoyed phenomenal success was the purely practical system of epicycles. Its two main proponents, Hipparchus (fl. c. 125 B.C.) and Claudius Ptolemy (c. A.D. 90–168), focused more on computing planetary positions than explaining the physical nature of the cosmos. Although uniform circular motion remained a significant constraint, Ptolemy arranged multiple circles of different sizes and allowed them to move at different speeds. His perfected system contained 43 circles – the Sun, Moon and planets rode on circular epicycles, which were set on larger circular deferents that spun around Earth. But Earth wasn't located at the exact center of Ptolemy's system. It was slightly offset and located directly opposite the "equalizing point," or equant. Only to an observer on the equant would an epicycle appear to move with uniform speed. As top-heavy and clumsy as all this seems today, there was no arguing with the system's results. Nothing better would come along until the 17th century.

Although Ptolemy's equant gimmick provided an improved method to match the varying speeds of planetary motion, it rubbed many scholars the wrong way. One of them was a Polish cleric named Mikolaj Kopernigk, who in 1543 published *On the Revolutions of the Heavenly Spheres*, the book that resurrected the idea of a Sun-centered universe. We know him today as Nicolaus Copernicus (1473–1543), a physician, lawyer, church administrator and an astronomer. His original motivation was to refine the Ptolemaic system and, in particular, excise the equant's "gross conflict" with the ideal of uniform motion. "I often considered whether there could perhaps be found a more reasonable arrangement of circles, from which every apparent irregularity would be derived while everything in itself would move uniformly, as is required by the rule of perfect motion," he wrote in the *Little Commentary*, a manuscript he circulated to friends sometime after 1512. This was his first heliocentric writing, and it took him decades to develop the details. He lay on his deathbed when *On the Revolutions* arrived from the printer.

While it's often said that Copernicus' system was simpler, easier to use and more precise than Ptolemy's, the contest was really more of a draw. While Copernicus freed himself from the mandate of an unmoving Earth, his system retained small epicycles to achieve uniform circular motion; indeed, in total, his theory required five more circles than Ptolemy's, so by this measure was the more complex one. Predictions derived from the Copernican model were generally no better than those made within a Ptolemaic framework and, according to scholars, planetary positions were no easier to calculate. The motions of Mars remained problematic in both systems. On the other hand, Copernican theory offered a natural and compelling explanation for retrograde loops as the combined effect of both the planet's and Earth's motion. As far as most astronomers of the day were concerned, replacing Ptolemy's distasteful equants with little epicycles was Copernicus' most important achievement.

What Copernicus accomplished was opening the door to a new way of thinking about the cosmos. He created a system that connected the computation of planetary positions to a model of the solar system where each planet's distance from the Sun was directly linked to the size of its orbit, and he showed that such a system could perform just as well as the standard model that had been in use for centuries. But while Copernicus had opened the door, it was a young adherent named Johannes Kepler (1571–1630) who first stepped across the threshold.

Kepler served as an assistant to the most famous astronomer of the age, Tycho Brahe (1546–1601). Tycho rejected Copernicus' ideas because it appeared to conflict with biblical teachings. He was driven to make his own detailed observations in support of a hybrid Ptolemaic-Copernican planetary model he was developing. In his system, Mercury and Venus revolved around the Sun and everything else revolved around Earth. Among Tycho's best traits was a passion for accurate observation; his treatise on the new star of 1572 and the comet of 1577 spread his fame throughout Europe. Among his worse traits were an aristocratic sense of self-importance and a domineering nature that guaranteed he would always wear out his welcome. With an amazing level of financial support from King Frederick II of Denmark, Tycho built on the island of Ven an extraordinary pair of observatories with state-of-the-art equipment of his own design, a residence for himself and eight assistants, a paper mill, fishing ponds and more. Tycho recorded thousands of planetary and stellar positions and his catalog represented the pinnacle of naked-eye observation. But when funding steadily diminished following the king's death, Tycho and his entourage pulled up stakes and eventually landed in Prague, where he met Kepler.

Even before going to work for Tycho in 1600, Kepler realized the importance of Tycho's observations in developing any theory of planetary motion. He also understood that the motions of Mars would make or break any model of the solar system. "For Mars alone enables us to penetrate the secrets of astronomy which otherwise would remain forever hidden from us," he wrote. Kepler details his years of tortured work on the Red Planet – he calls it his "war against Mars" – in his book *The New Astronomy*, published in 1609. Tycho's observations ultimately led Kepler to two great discoveries: The planets move in ellipses, not circles, and their orbital speeds change predictably as their distances from the Sun vary. These form two of what we now call Kepler's three laws of planetary motion; he would publish a third, which mathematically related the period of an orbit to its size, 10 years later. There was no longer any question in his mind of the Sun's place in the cosmic scheme. "The Sun will melt all the Ptolemaic apparatus like butter," he triumphantly wrote. Near the end of his life, Kepler published his *Rudolphine Tables*, which were set up to allow the calculation of planetary positions at any time in the past or future.

This work, which was fully grounded in a heliocentric cosmos, set new standards for precision and placed the insights of Copernicus on a firm foundation.

Make Way for Martians

Just a year after Kepler's *The New Astronomy*, which upended theoretical understanding of the cosmos, a new tool became available to astronomers that would do the same for the observational side. Although the first telescopes were too crude to permit interesting discoveries on Mars, it wasn't long before improved instruments revealed a planet much more interesting than the bland Venus. By 1677 astronomers had discovered polar icecaps, along with bright and dark markings that were regarded as deserts and seas. These features allowed astronomers to measure the length of the martian day, which at 24 h, 37 min is remarkably similar to Earth's. The icecaps were seen to grow and shrink with the martian seasons, and the bright and dark patches also varied from year to year. Compared to other celestial sights, Mars was looking downright mundane. In *Conversations on the Plurality of Worlds*, a popular-level astronomy book written by the French author Bernard le Bovier de Fontenelle and published in 1686, the Red Planet was barely an afterthought:

> Mars has nothing curious that I know of; its days are not quite an hour longer than ours, and its years the value of two of ours. It's smaller than the Earth, it sees the Sun a little less large and bright than we see it; in sum, Mars isn't worth the trouble of stopping there.

It was the "discovery" of another type of martian feature – canals – that truly dialed up popular interest in the planet. If the premier astronomical curiosity of the eighteenth century was the black drop seen when Venus transited the Sun, then for the 19th century it was canals on Mars. They were first reported in the work of an Italian astronomer, Angelo Secchi, during the 1860s. He plotted several vague linear features on the planet and called them *canali*, which can be translated either as "channels" or "canals." The term was picked up by fellow Italian astronomer Giovanni Schiaparelli (1835–1910), who during the close opposition of 1877 found the features to be sharper, more distinct and more numerous. "They traverse the planet for long distances in regular lines, that do not at all resemble the winding courses of our streams," he later wrote.

Meanwhile other astronomers strained to see the *canali* at all. The features were visible only under the best observing conditions. Moreover, successive Mars oppositions, which recur every 26 months, had taken the planet to more distant parts of its orbit, making detection of the *canali* increasingly difficult. The orbit of Mars is much more eccentric than Earth's, so as the Red Planet progresses around its orbit, the difference between its nearest and farthest opposition distances can exceed 28 million miles (45 million km). As seen in Fig. 5.2, this makes a significant impact on the apparent size of the planet's disk as seen through a telescope. When Mars made its record close approach in August 2003 – the nearest the planet has been to

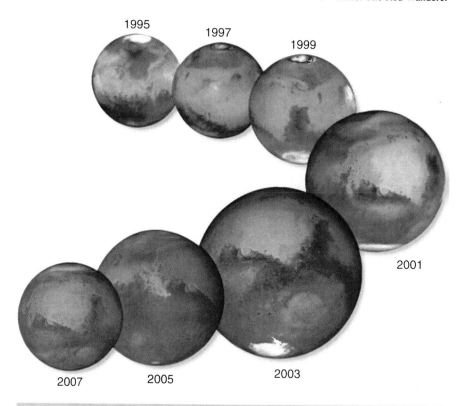

Fig. 5.2 Every 26 months, Mars is opposite the Sun, visible high in south at midnight and shining at its brightest. But because the Red Planet's orbit is more eccentric than Earth's, some oppositions are better than others. These Hubble Space Telescope images show Mars at oppositions from 1995 to 2007. The Red Planet made a record close approach to Earth in 2003, appearing more than 25 arcseconds across, but it has since moved on to more distant parts of its orbit. Mars won't appear nearly as large again until July 2018 (NASA/ESA/Z. Levay (STScI))

us, to the best anyone can determine, in more than 59,000 years – its disk was over 25 arcseconds across, or about 72 times smaller than a full Moon. That's big for Mars, but it's only about half the apparent diameter of Jupiter, which is why Mars remains an observing challenge even at its best. For the 2012 opposition, the Red Planet will be 1.8 times farther away from us and will appear correspondingly smaller, less than 14 arcseconds across – 128 times smaller than a full Moon. The situation will then steadily improve at oppositions through July 2018, when the tawny disk of Mars will appear over 24 arcseconds wide (Fig. 5.3).

Only as Mars returned to more favorable approaches, after 1886, did reports of *canali* – now translated into English as "canals," which implied an artificial construction – begin again. Not all observers saw them, and those who did weren't always seeing the same ones. The scientific question of whether the lines were real

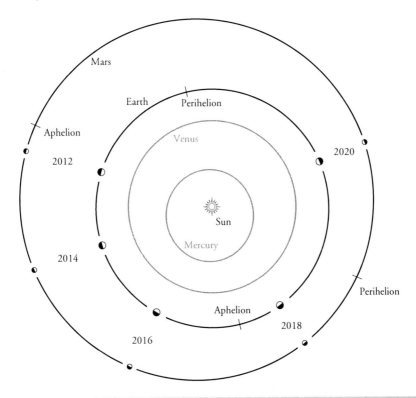

Fig. 5.3 The closest distances between Earth and Mars vary as we encounter the Red Planet in different parts of its eccentric orbit. As seen here, oppositions steadily become more favorable this decade. For the 2018 opposition, the Red Planet will be 1.8 times closer to us than in March 2012. Aphelion and perihelion are the points in the orbits of Earth and Mars where the planets, are, respectively, farthest from and closest to the Sun

features on the planet's surface or simply optical illusions would not be determined for decades. But the possibility that the canals were indeed some sort of artificial geometric network powered an intense burst of popular interest in astronomy as the favorable 1892 opposition drew closer. Even a few scientists found themselves caught up in the excitement. Camille Flammarion, a French astronomer and writer of popular science, enthusiastically declared in 1892 that "the present inhabitation of Mars by a race superior to ours is very probable."

Not everyone was buying it. The American astronomer E. E. Barnard, widely regarded as a gifted observer, concluded that the canals were illusions. "To save my soul I can't believe in the canals as Schiaparelli draws them," he wrote in a letter to astronomer Simon Newcomb in 1894. "I verily believe…that the canals as depicted by Schiaparelli are a fallacy and that they will be so proved before many oppositions are past."

This might have happened more quickly were it not for the efforts of an American astronomer who heavily promoted the canals as alien engineering. He was

Percival Lowell (1855–1916), a wealthy Bostonian who built an observatory in Flagstaff, AZ, specifically to acquire high-quality observations of Mars. "There are celestial sights more dazzling, spectacles that inspire more awe, but to the thoughtful observer who is privileged to see them well, there is nothing in the sky so profoundly impressive as the canals as Mars," he wrote in 1895. Lowell was convinced that the canals were created by an advanced race of beings to irrigate the planet's vast deserts. "To account for these phenomena, the explanation that at once suggests itself is, that a direct transference of water takes place over the face of the planet, and that the canals are so many waterways." Lowell believed that he was seeing not the irrigation channels themselves but the adjacent strips of fertilized land. He catalogued over 180 canals – more than four times the number Schiaparelli had recorded.

The canals were close to the resolution limit of telescopes at the time and many astronomers remained skeptical that they existed at all, let alone that they were the product of an advanced alien race. Others accepted the canals but interpreted them as geographical features, a novel set of rifts and valleys gouged into the planet's crust. Still others argued that the canals were small features blurred together by the turbulence of Earth's atmosphere. Today we know that the canals were illusions caused by the combined effects of telescopic resolution limits, atmospheric smearing, and wishful thinking on the part of observers. Orbiting spacecraft have mapped the planet in detail for decades, and the canals mapped by Lowell and other astronomers do not correspond to any natural features.

Lowell had the best story to tell, and the public was captivated by the idea of a martian civilization so advanced yet so desperate for water that it would turn to planetary-scale engineering. The English author H. G. Wells capitalized on the craze with his 1898 science-fiction novel *The War of the Worlds*, in which "intellects vast and cool and unsympathetic, regarded this earth with envious eyes" and unsuccessfully attempted to take our planet for themselves, a victory Wells credited to terrestrial germs rather than human weapons or ingenuity.

With an apparently intelligent civilization essentially right next door, it would be rude not to say hello, and techniques that might be used to bridge the gulf of both space and culture were discussed. "Geometry is the same for inhabitants of every world," wrote Flammarion in 1888. He suggested that signals could be sent to Mars by setting up and changing giant geometrical shapes like triangles and squares. The shapes would be dozens of miles on a side, but by altering them – say, forming a square out of right triangles, something that indicated intelligence at work – we could announce our presence to Mars. The French poet and inventor Charles Cros (1842–1888), a friend of Flammarion's, argued that signaling could be accomplished via modern technology, with powerful electric lights trained on parabolic mirrors. By flashing the lights using a kind of interplanetary Morse code, Cros said that strings of numbers could be communicated to build up a common language, and he felt that bright spots reported on Mars could well be signaling attempts. A French amateur astronomer named A. Mercier wrote a booklet titled *Communication with Mars* in which he suggested placing mirrors on the Eiffel Tower in Paris. When illuminated by the setting Sun, the mirrors could be directed toward Mars and interrupted with a movable screen, indicating an intelligent presence on Earth. Mercier even attempted to raise funds to put the scheme into practice.

Humorists had their day, too. In 1897, the French writer Tristan Bernard questioned the efficiency of communication with Mars once it was definitively established. In his story, astronomers discovered light signals from Mars and responded by covering the Sahara Desert with paper printed with the words "I beg your pardon?"

"Nothing," signaled Mars.

Astronomers deployed another massive sheet of print across the desert, asking "Why are you making signs, then?"

Mars tersely replied: "We're not talking to you. We're talking to the Saturnians."

As the favorable 1909 opposition approached, Harvard University astronomer William H. Pickering made a splash in the press by proposing the construction of a huge array of 5,000 mirrors, at an estimated cost of some $10 million, to signal the planet. A group of enthusiastic citizens in Stamford, Texas, had pledged $50,000 for a similar scheme that would use a giant version of a mercury mirror then being tested at Johns Hopkins University in Baltimore. However, Harold Jacoby, an astronomer at Columbia University in New York, reflected the majority scientific view: Better to wait for an obvious signal from the unlikely martians before pouring that kind of money into interplanetary communications – let the martians be the first to foot the bill!

Astronomers ultimately dismissed the canals as illusions, but many had no difficulty believing that the planet might be capable of plant or even animal life. Hope for animal life faded in the 1930s as it became clear that oxygen did not exist in any appreciable amount in the martian atmosphere. Plant life was still considered possible, if unlikely, by the end of World War II, and shortly thereafter a new discovery reinvigorated the search. In 1947, Gerard Kuiper (1905–1973) detected infrared emissions from carbon dioxide, a key ingredient for photosynthesis by plants, in the martian atmosphere. Comparison of the dark grayish areas on Mars with different types of plants on Earth eliminated most possibilities, but Kuiper pointed out that lichens were consistent with the observations. "The hypothesis of life," he wrote in 1955, "appears still the most satisfying explanation of the various shades of dark markings and their complex seasonal and secular changes." More than half a century after Lowell was writing about martian civilization, the idea of life on the Red Planet still proved hard to shake.

Yet one of the opening acts of the Space Age brought only disillusionment to those hoping for a habitable Mars. Mariner 4 sped past the planet in July 1965, returning 21 images that covered about 1% of the planet's surface. Only about half of the images contained any appreciable detail, but those that did revealed a pockmarked landscape of overlapping impact craters reminiscent of the Moon. Instruments measured frigid temperatures and an extremely low atmospheric pressure of about 10 millibars, less than half the value expected from telescopic studies and equivalent to the pressure at an altitude of 19 miles (31 km) in the Earth's atmosphere. It was a glimpse of the planet that surprised and disappointed. "The Mars we had found was just a big Moon with a thin atmosphere and no life," wrote Caltech planetary scientist Bruce Murray, recalling his reaction to the flyby. "There were no martians, no canals, no water, no plants, no surface characteristics that even faintly resembled Earth's."

Nothing changed with the 1969 flybys of Mariners 6 and 7, which together imaged about 20% of the surface. Carbon dioxide turned out to be the main ingredient of the atmosphere. The polar caps appeared to be frozen carbon dioxide – dry ice – and not water. With no oxygen, the thin atmosphere could not develop an ozone layer and absorb the most damaging ultraviolet radiation from the Sun, so organisms could not survive on the surface. There was no getting around the fact that Mars appeared to be hostile to any known form of life, although a few scientists, most notably Cornell University's Carl Sagan (1934–1996), publicly promoted the idea that life might exist even under these conditions.

Science by its very nature is conservative, tied to the vast collection of facts and concepts available at any given time. Flipping the problem of Mars on its head, what might a hypothetical martian scientist conclude about the possibility of life here? Earth is clearly far too warm over most of the surface, and only in the Antarctic is the climate suitable for life. Vast volumes of solvent – water – occupy most of the surface and the stuff rains or snows everywhere else on the planet. A highly reactive and flammable gas – oxygen – makes up one-fifth of the atmosphere and blocks the Sun's most energetic ultraviolet light. Viewed from the perspective of an organism able to thrive on present-day Mars, life on Earth seems improbable, too.

While the first close-up views at Mars amounted to drive-by photo shoots, in November 1971 a new probe, Mariner 9, settled into orbit for long-term study of the Red Planet. A great global dust storm raged below, obscuring everything but the south polar ice cap and four large, mysterious spots until January 1972. As the dust cleared, it became apparent that these spots were enormous, ancient volcanoes reaching up to 16 miles (25 km) above the surface. Mariner 9 also found an enormous canyon system dubbed Valles Marineris to honor the spacecraft. The canyons span over one-fifth of the planet's circumference – equal to the width of the continental U.S. – and in places plunge to a depth of 6 miles (10 km), dwarfing any similar feature on Earth (Fig. 5.4). Finally, Mars was revealed in all its glory. Planetary scientists were surprised and delighted by the unexpected diversity of landforms. Mars wasn't earthlike in the sense that Percival Lowell imagined it, but it also wasn't the battered Moon suggested by the first probes. Mars was like nothing ever seen before. Mars was unique.

As luck would have it, all earlier Mariner spacecraft had passed over heavily cratered areas that were not representative of Mars as a whole. Mariner 9 established that the apparent seasonal changes seen by Earth-based observers were caused by the windblown redistribution of surface dust, not plant life. The craft detected water vapor in the atmosphere over the south polar cap, which scientists interpreted as a sign that the polar caps were primarily made of frozen water dusted by a veneer of dry ice – now thought to be as little as 8 yards thick – when carbon-dioxide snows fell during the winter (see box, "Blue note on the Red Planet"). This meant that each polar cap held about half the volume of frozen water as the

Fig. 5.4 The enormous scar of the Valles Marineris canyon system runs across this mosaic of images returned from the U.S. Viking Orbiter in the 1970s. The feature spans about 2,500 miles (4,000 km), long enough to span the United States from coast to coast. The three dark spots along the left edge of the planet are three of its largest volcanoes, each about 16 miles (25 km) tall (NASA/USGS)

Greenland ice sheet on Earth. By far the most important discovery was a wide variety of features that could only be formed by a flowing liquid: flood plains, canyons and what looked like dried-up river beds. The atmospheric pressure on Mars is too low for water to run freely on the surface today, but could the situation have been different millions of years ago?

There are good reasons to expect dramatic changes in the martian climate. Like Earth, Mars has seasons because its rotational axis tilts toward its orbital plane. The angle of this tilt is known as obliquity. Earth experiences only minor changes in obliquity due to the stabilizing influence of the Moon, which acts like a giant flywheel to prevent large excursions. But Mars lies even closer to the disrupting gravitational tug of Jupiter, which drives some of these changes. And its two moons, Phobos and Deimos, are too small to act as planetary flywheels; they are probably asteroids that wandered too close and were captured into Mars orbit. The upshot is that the obliquity of Mars, which is currently about 25° and similar to Earth's, changes chaotically, ranging from as little as 11° to as high as 60° over the past few million years. The planet's tilt is also sensitive to geological processes that alter the distribution of mass, such as the vast lava outpourings that occurred around those giant volcanoes.

How does this affect the Red Planet's climate? Polar summers become much warmer when the planet's axis is severely tipped. And because the Red Planet's orbit is so eccentric, the pole that tilts toward the Sun when the planet is closest to it (at perihelion) experiences even warmer summers. Right now that's the south pole. But thanks to a slow change in the direction of Mars' axis, a wobble called precession, the "hot" pole alternates every 51,000 years. Finally, cyclic variations in the shape of Mars' orbit provide still another mode for martian climate change.

Taken together, these effects may have warmed Mars enough to unfreeze some of the carbon dioxide and water locked under the planet's surface as permafrost and in the ice caps. Increasing amounts of both of these atmospheric gases would give rise to increased "greenhouse effect" warming, creating a positive feedback that would liberate more gases and increase the pressure still further. If Mars was once warmer and wetter than it is today, then perhaps microbes could have developed and managed to evolve in a way that would allow them to tolerate the deteriorating climate.

The U.S. Viking mission, often regarded was one of the most ambitious planetary missions ever undertaken, followed Mariner 9. Twin spacecraft, each consisting of an orbiter and a sophisticated lander, arrived at Mars in 1976. The two orbiters provided high-resolution imaging of the planet – it remained the state of the art for two decades – while the landers returned images, weather data and chemical analyses of the soil. Iron-rich clays cover much of the planet's surface, which is what gives Mars its rusty hue. Of greater general interest, however, was a biology package the Vikings used to test for living organisms. Its three experiments worked by adding water and different nutrients to soil samples, incubating them for a time, and then looking for substances indicating that organisms consumed and metabolized the nutrients. Only one test, called the Labeled Release experiment, showed any response, but it was a dramatic one, with rapid release of gases containing the radioactive carbon used to "label" the various nutrients added to the soil sample. The official verdict was that the tests were inconclusive, and the prevailing scientific view is that the experiments failed to find any sign of martian life. However, Gilbert Levin, the principal investigator for the Labeled Release (LR) experiment, has never wavered in his belief that his experiment detected life on Mars. "Over the 30 years since the landing of Viking, more than 40 attempts have been made to explain the LR results abiologically," he wrote in 2006. "To this date, no experiment has duplicated or realistically approximated the Mars LR positive and control results except when using living microorganisms."

Results of another test that looked for organic molecules, such as methane, hydrocarbons and amino acids in soil samples, seemed to dash the last hopes for martian life. Organic compounds are always associated with life on Earth and they were expected on Mars even if life did not exist. They are common substances in interplanetary dust, meteorites and comets, all of which reach the surface of Mars at one time or another. Because the Sun's ultraviolet light eventually destroys organic material on the surface, the experiment was designed to be extremely sensitive. Yet to a level of a few parts per billion, Viking reported no organics, a finding that has been called the mission's most surprising single discovery. Because the molecules should be there even without life, scientists suggested that the martian surface contains a powerful oxidizing agent, a substance that rapidly transforms organic compounds into chemicals the Viking experiment could not detect. Moreover, the presence of such a "superoxide" might drive chemical reactions that could explain the positive biology results. On the other hand, Levin has maintained that the Viking organics detector had more limited sensitivity than is usually acknowledged.

Two decades would pass before the question of martian life again created widespread excitement.

Box 5.2 Blue note on the Red Planet

In addition to its thin carbon-dioxide atmosphere and ultraviolet-irradiated surface, Mars is just plain cold. One of the best records of the planet's global temperatures comes from the Thermal Emission Spectrometer (TES) aboard NASA's Mars Global Surveyor spacecraft.The TES instrument returned spectra for nearly three martian years, from March 1999 through August 2004.

The highest surface temperature measured, 86°F (30°C), occurred at 30° south latitude each year in afternoons near the martian summer solstice. This sounds pleasant, but the air temperature decreases so quickly with height that it falls to near freezing just a meter or so above the surface. During a barefoot stroll at the hottest spot on Mars, your feet might be comfortable but your head could be below freezing. Of course, that would be the least of your troubles.

The lowest temperatures are harder to pinpoint because they naturally provide a minimal signal to the instrument, but the TES observations are consistent with a minimum temperature of about −200°F (−130°C), which occurs near both poles at the winter solstice. When it's this cold, the carbon dioxide in the atmosphere freezes out, coating the winter polar region with a veneer of dry ice.

Mars Rocks

The detailed picture of Mars' atmospheric composition made possible by the Viking landers allowed planetary scientists to identify a small group of meteorites as rare visitors from Mars. The Shergotty, Nakhla and Chassigny (SNC) class of meteorites are named for the locations in India, Egypt and France, respectively, where the first three members of the type were found. Today, scientists know of at least 56 specimens. To be launched from the Red Planet's surface, these rocks had to reach martian escape velocity – about 3 miles (5 km) a second, so fast that the rocks were nearly destroyed in the process. The key to discovering the meteorites' true origin lay in small pockets of gas locked inside them. The composition of this gas closely matches both the distribution of elements and the specific isotopes, such as Argon-40, that the Viking landers had measured in the martian atmosphere. The long-imagined invasion from Mars actually had happened, but it was geological rather than biological.

One of these SNC meteorites renewed the scientific debate about the possibility of martian life. The softball-sized chunk of igneous rock first solidified from lava 4.1 billion years ago. About 200 million years later, water inundated the rock and altered some of its minerals. Then, about 60 million years ago, a large, nearby impact launched the rock off of the Red Planet. It circled the Sun until 13,000 years ago, when it skirted Earth's atmosphere, briefly lit up the sky, and crashed onto Antarctica. There it lay buried in the ice until 1984, when scientists recovered it from the continent's Alan Hills region. The rock was the first specimen to be found in that year's Antarctic meteorite search, so it was designated ALH 84001. It was destined to become the most famous meteorite in the world.

In August 1996, a team led by David McKay of NASA's Johnson Space Center in Houston made the surprising announcement that ALH 84001 contained three lines of evidence that suggested ancient martian microbes had lived within it. The first is the presence of mineral grains similar in size, shape and structure to those produced by bacteria on Earth. One of the minerals, magnetite, a form of iron oxide, is a natural magnet, and some bacteria secrete this substance so that they can follow Earth's magnetic field. Unlike Earth, there is no internally generated magnetic field at Mars now, but shortly after the ALH 84001 study became public, NASA's Mars Global Surveyor picked up signs of a weak remnant magnetic field imprinted on the planet's surface rocks, presumably the leftovers of a strong planetary field that faded out billions of years ago. Of course, both the magnetite and the other chemical grains the team found can also form without the aid of biology.

The next piece of evidence comes in the form of organic molecules called polycyclic aromatic hydrocarbons (PAHs), which are chemicals with structures similar to naphthalene, better known as the active ingredient in mothballs. PAHs form when living organisms die and decay, but they're also created through a variety of other processes; for example, the soot from a candle flame contains them. The team showed that the type and abundance of PAHs found in the Mars rock were different from those found in other meteorites, even other martian ones. Perhaps these substances were related to martian microbes, or perhaps they formed as a result of natural organic chemistry that might one day have led to life if conditions on Mars had remained favorable. Interestingly, the organics detector on Viking was not sensitive enough to detect these molecules if they existed in the samples tested. Definitive proof that organic chemistry was occurring on Mars would excite scientists almost as much as proof that life once – or still – existed.

Finally, the team found elliptical and rope-like features that appeared to be microscopic fossils. They were 100 times smaller than any microbes on Earth and could only be detected using a scanning electron microscope. The largest were less than 1/100 the width of a human hair, and most were about ten times smaller. Until recently, some biologists had argued that there was evidence for organisms this small on Earth, but further studies have shown that these putative "nanobes" result from chemical interactions and are not a new class of organism.

In the years since, the strongest evidence for fossil life in ALH 84001 remains the magnetite. In 2009, the team showed that the leading non-biological mechanism for producing the grains also added large amounts of chemical contaminants to them. The meteorite grains, however, are in most cases pure iron oxide, just like grains produced by terrestrial bacteria. Now, if you're a bacterium trying to make a micro-compass to negotiate a planetary magnetic field, contaminants will only get in the way and weaken your navigation system. "We believe that the biogenic hypothesis is stronger now than when we first proposed it 13 years ago," said team member Everett Gibson, a senior scientist at NASA's Johnson Space Center in Houston. The wider scientific community is a harder sell, and the team is continuing to study ALH 84001 and other martian meteorites. The episode illustrates that even when scientists are afforded the ability to examine fresh samples returned from some future Mars mission, they may not immediately recognize fossilized martians even if the robot happens to collect the right rock. The story is yet another example of how Mars keeps us guessing.

Life as we know it cannot exist without water, so a new wave of interest in exploration focused on clarifying the past and present role of water on Mars. On July 4, 1997, NASA's Mars Pathfinder literally bounced onto the Red Planet as a demonstration of low-cost methods for exploring the surface; the spacecraft was wrapped within an envelope of airbags. Investigations carried out by both the lander and a small rover, named Sojourner, over the next 83 days seemed to validate the idea that Mars was once warmer and wetter in the past. That same year, the U.S. Mars Global Surveyor spacecraft entered orbit around the planet. By the time contact was lost in late 2006, the mission had returned over 200,000 images at resolutions as small as a few yards, as well as much other important data. Scientists found city-block-sized systems of small, apparently fresh gullies that resemble terrain cut by flash-floods on Earth. Elsewhere, large floods appear to have originated from the same vents that produced extensive lava flows, providing evidence that volcanism and the release of water have occurred on Mars in the geologically recent past – possibly even the present.

Another mission revealed the global extent of martian water. NASA's 2001 Mars Odyssey spacecraft began mapping the planet's surface composition in early 2002. One of the mapping instruments was a neutron spectrometer, a device similar to the ones that detected subsurface hydrogen on the Moon by measuring changes in neutrons given off by the soil. The following year, the team released a global map that translated the hydrogen signal into mass-percentages of water. In the central complex of Valles Marineris, where the map shows an equivalent water abundance of about 6%, 1 lb (454 g) of heated soil would yield just about 1 oz (27 g) of water. It gets much better at higher latitudes, closer to the poles, where the same amount of heated soil could yield half its weight in water. According to the team, if the hydrogen Mars Odyssey detected represents a permafrost layer just a yard thick, if melted all at once it could cover the planet to a depth of 5.5 in. (14 cm). However, exactly how much of this hydrogen signal actually *is* frozen water remains unclear.

Later, Mars Odyssey located hundreds of spots on the surface that were rich in chloride-containing salts. The chemicals may have been dissolved in briefly standing pools of water, left behind when the liquid evaporated. The deposits date back to 3.9 billion years ago and earlier, about the time ALH 84001 became chemically altered by exposure to water. "The more we look at Mars, the more fascinating a place it becomes," said Jeffrey Plaut, the mission's project scientist, in March 2008. By then, three additional spacecraft were busily scrutinizing the Red Planet, with a fourth on the way.

Mars Express, an orbiter launched by the European Space Agency, reached the planet in December 2003. Among its instruments was a novel radar sounder that scientists hoped would provide a glimpse of what lies beneath the surface. By pinging radio waves at the planet, timing the returned echoes and analyzing the reflected signal, scientists used the instrument to successfully detect large quantities of ice several kilometers underground. Radar signals in the south polar region established that subsurface ice there was at least 90% frozen water, and the study allowed the best-ever determination of the volume of water in the region – about 1/800th of the total water volume in Earth's oceans. If all ice in and under the south polar region were completely melted, it could cover Mars in a liquid layer 36 ft (11 m) deep.

Weeks after the arrival of Mars Express came the U.S. Mars Exploration Rovers, named Spirit and Opportunity. Using a larger version of the airbag landing system pioneered by Mars Pathfinder, the two six-wheeled coffee-table-sized rovers bounced onto opposite sides of the planet and immediately set to work. Evidence for surface rocks altered by water quickly emerged. For the scientists, Opportunity was practically a hole in one, for it landed in a crater that included a rocky outcrop formed by minerals deposited in acidic, briny water. The mission team guided the rover to other, deeper craters in an effort to track the same deposits as deeply as possible (Fig. 5.5). On the other side of the planet, Spirit had to travel much farther to locate signs of water, but it eventually succeeded, too. Spirit became mired in soft sands in 2009 and communications were lost the following year, but at this writing Opportunity is still going strong. Between them, the rovers have logged over 25 miles (40 km) of martian travel, and the mission's scientific return has far exceeded all expectations.

In March 2006, just months before contact was lost with Mars Global Surveyor, another U.S. spacecraft began operations: Mars Reconnaisance Orbiter (MRO). MRO's imaging system includes a 20-in.-wide (0.5-m) telescope, the largest ever sent to another planet, and is revealing features as small as the width of an office desk – a level of resolution only landers had previously enjoyed. In fact, the spacecraft's imager is so good that it was able to capture another arriving U.S. spacecraft even before it set down on the planet. On May 25, 2008, as the Phoenix Mars Lander descended by parachute over the martian arctic, MRO's camera caught the spacecraft in midair.

The Phoenix Mars Lander boasts the first true soft-landing on Mars since Viking, and it was the first mission ever dedicated to study the planet's north polar region. By July, scientists announced that they had "touched and tasted" martian water excavated by a robot arm that found ice-laden soil just inches below the surface. Perhaps the most important result from the mission was the discovery of a chemical in the soil that provided a missing piece for understanding Viking results.

Fig. 5.5 In August 2008, NASA's Opportunity climbed out of Victoria crater after spending nearly a year exploring it. The six-wheeled rover followed its own tracks up the slope of the half-mile-wide (800 m) bowl. The rover's navigation camera recorded this scene as Opportunity rolled onto level ground (NASA/JPL-Caltech)

Organic molecules are the essential building blocks of life and are common in comets, meteorites and interstellar dust, but the only organics detected by Viking were substances thought to be contamination from fluids used to clean the apparatus between experiments. The Phoenix scientists announced that they had identified a highly reactive substance called perchlorate in martian soil.

Was this the mysterious "superoxidizer" suggested to explain the lack of Viking organics? In a 2010 study led by Rafael Navarro-González of the National Autonomous University of Mexico in Mexico City, researchers added perchlorate to soil samples from Chile's Atacama Desert – long used as a terrestrial stand-in for Mars – and then heated the samples in a manner similar to Viking tests. Even though the desert soil contained plentiful organics, the test registered only the same two molecules that the Viking team had written off as contaminants. If perchlorate was present in the Viking samples, then the act of heating them initiated reactions that rapidly destroyed all organic molecules and, in the process, produced the presumed contaminants. "We can now say there is organic material on Mars, and that the Viking organics experiment that didn't find any had most likely destroyed what was there during the testing," Navarro-González said. In fact, because the perchlorate is inactive until heated, it could be present in the martial soil beside the organics for billions of years without affecting them. Ironically, it was our attempt to detect these molecules that resulted in their destruction. By the time this study was published, astronomers using ground-based telescopes had shown that one organic molecule, methane, sometimes puffed into the martian atmosphere in large quantities during the spring and summer and was destroyed faster than known processes could account for (see box, "A matter of methane").

The next step for Mars exploration will be another U.S. rover. Officially known as the Mars Science Laboratory and dubbed Curiosity, the six-wheeled vehicle is much larger than its predecessors, about the size of a compact car. It's scheduled to reach Mars in late 2012. Curiosity boasts an array of sophisticated instruments that will sniff the air, analyze rocks and soil, and image the terrain. Among Curiosity's experiments are liquid-based tests for organic molecules, which avoid the high temperatures that would set off perchlorate reactions. Additional U.S. and European missions are in the planning stages, too.

This is a time of unprecedented scientific study of the Red Planet; the brief survey above hardly does it justice. The complex story of Mars is gradually unfolding before our eyes.

Box 5.3 A matter of methane

One especially interesting organic molecule detected in the martian atmosphere is methane, the main component in natural gas. On Earth, more than 90% of methane is produced by living organisms, with the remainder emitted by geological activity like mud volcanoes. Since 2003, several research groups have detected substantial amounts of methane in Mars' atmosphere. Because ultraviolet sunlight breaks up methane molecules within about 400 years, the fact that the gas is detected at all indicates a recent release. On the other hand, methane survives long enough that winds should quickly mix it throughout the atmosphere.

But that's not what happens. In 2009, a team led by Michael Mumma at NASA's Goddard Space Flight Center reported the results from a 7-year study of the Red Planet using ground-based telescopes armed with instruments that could identify methane and water vapor. The observations covered 90% of the planet's surface. Their analysis shows that methane isn't smoothly distributed but occurs over specific areas, with concentrations increasing from the martian spring through the summer, and then rapidly declining. *Something* on Mars is actively releasing methane today.

The locations associated with the methane releases – Syrtis Major, Arabia Terra and Nili Fossae – are known to be rich in minerals altered by the past presence of water. Some minerals release methane easily when exposed to hot water and the planet's principal atmospheric gas, carbon dioxide. Underground methane produced this way could build up over time and then find its way to the surface during the spring and summer, when any icy barriers warm up and disappear. Scientists agree that the most likely explanation is geological, but they can't completely eliminate the possibility that pockets of microorganisms beneath the surface could produce the gas.

(continued)

The Martian Sky Show

Mars is farther from the Sun than we are, so it runs on an outer path of the solar system racetrack. Its captivating celestial movements are a result of Earth's catching up from behind, passing it and then pulling away on the inside lane. Figure 5.6

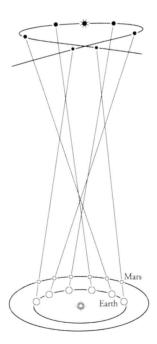

Fig. 5.6 The remarkable loop of Mars is less mystifying when seen from space. The loop at the top shows the path Mars appears to take through the stars as well as the planet's changes in brightness

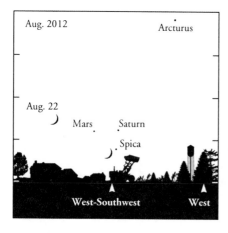

In mid-August 2012, look for Saturn (0.8), Mars (1.8) and Spica (1.0) lining up low in the southwest in the hour after sunset. By the time the Moon joins them on the 21st, the grouping has become a triangle

In February 2015, Venus (–3.9) is your guide to Mars (1.2). Look southwest about 45 min after sunset. The Moon joins the scene on the 19th, when the planets are closest

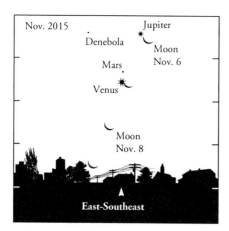

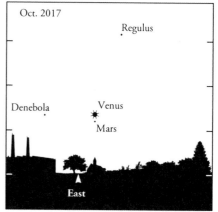

Mars (1.7) and Venus (–4.4) meet in early November 2015 while Jupiter (–1.8) looks on. Look high in the east-southeast 90 min before dawn. The waning crescent Moon slips through the lineup starting on the 6th

Mars (1.8) meets Venus (–3.9) in early October 2017. They separate quickly as Venus descends into twilight; the Moon sweeps past them after midmonth. Look east 45 min before dawn

Fig. 5.7 Mars is relatively inconspicuous when it appears near the horizon in the hour before dawn or after sunset. However, brighter planets and the Moon will help you find it. Note that the close morning conjunction between Mars and Jupiter in January 2018 is shown in Chap. 3 (Fig. 3.19)

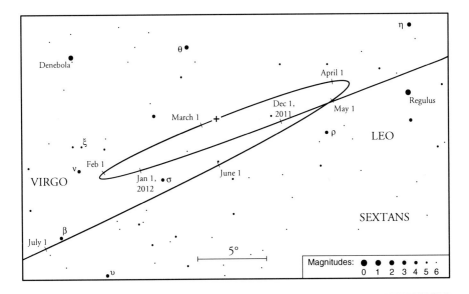

Fig. 5.8 Mars loop, 2011–2012. The Red Planet outshines Regulus, the brightest star in the constellation Leo, when it cruises past on Nov. 10, 2011; look for the pair high in the south in the hour before dawn. In late January, Mars resides near the 4th-magnitude star Nu (ν) in Virgo and then begins its westward reversal; the planet is now easily visible high in the south at 3 A.M. local standard time. Mars sweeps back toward Leo and reaches opposition (+) on March 3, when the planet is at its brightest and is visible all night long. When Mars halts again in mid-April, it remains easily visible high in the south in the hour after sunset. The Red Planet then courses eastward to finish out the apparition (Robert Miller)

provides an overview of how these motions generate the path Mars appears to take through the stars, but its detailed shape changes from apparition to apparition. Mars is truly at its best every 15–17 years, when it lies near the part of its eccentric orbit that brings it closest to the Sun. We last experienced one of these so-called perihelic oppositions in 2003; the next ones occur in 2018 and 2035. Because Mars is so physically small, it can fade to nearly second magnitude when seen near the opposite side of its orbit. Unfortunately, this is precisely when it's most likely to be involved in twilight conjunctions with other bright planets; the best of these arrangements for the decade are shown in Fig. 5.7.

The Red Planet's sky motions, which bedeviled all astronomers prior to Kepler, continue to give us a delightful exhibition of celestial geometry. Figures 5.8–5.12 on the following pages illustrate its track among the stars for each opposition this decade. As we follow Mars around the sky, we can reflect on the accomplishments of the Maya, the insights of Copernicus, the mathematical labors of Kepler, the fantasies of Lowell, the many scientific reversals – and the armada of spacecraft that may once again change how we think about the Red Planet (Table 5.2).

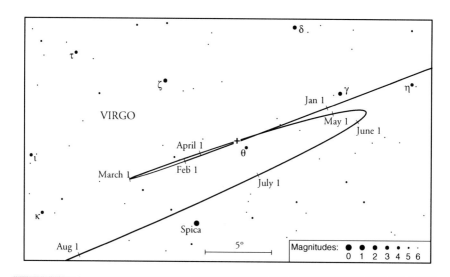

Fig. 5.9 Mars loop, 2013–2014. Mars executes more of a zigzag as it runs through the constellation Virgo. Mars is high in the south in the hour before sunrise when it passes the 4th-magnitude stars Eta (η) and Gamma (γ) Virginis in the latter half of December 2013, and in mid-January 2014 it passes Theta (θ). More notably, Mars stands within 5° of Spica, the constellation's brightest star, in early February; look for the pair high in the south at 4 A.M. local standard time. Mars begins its brightest phase of the apparition in March as Earth approaches; opposition (+) occurs near Theta Virginis on Apr. 8. By June, when Mars remains a bright object high in the south in the hour after sunset, it has begun the last leg of the apparition (Robert Miller)

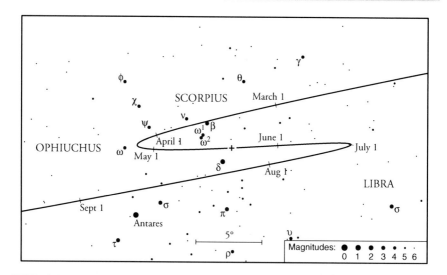

Fig. 5.10 Mars loop, 2016. Mars courses through the constellations Libra, Scorpius and Ophiuchus during this apparition, but the best stellar reference point is Antares, the brightest star in Scorpius. The star's name, which means "like Mars" comes from its orange coloring, but from spring through autumn, the Red Planet easily bests the star. Mars enters Scorpius for its first run through the constellation on Mar. 13, and skirts Beta Scorpii (β) 2 days later. At this point, Mars is high in the south two hours before dawn. By Apr. 2, Mars has entered Ophiuchus but halts its progress at midmonth, about 7° from Saturn (not shown) and less than 1° from 4th magnitude Omega (ω). As May opens, Mars crosses back into Scorpius for the run to opposition (+), which occurs May 22. A week later, the still-brilliant planet crosses into Libra. By July, when Mars is high in the south in the hour after sunset, the planet has resumed moving eastward for a final autumn sweep through these stars (Robert Miller)

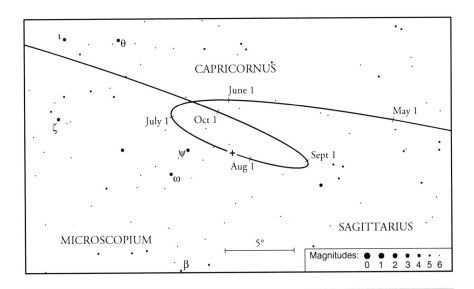

Fig. 5.11 Mars loop, 2018. This apparition is the Red Planet's brightest and closest since 2003 and it won't be as good again until 2035. In early April, look southeast in the hours before dawn for Mars and fainter Saturn separated by less than 2°. Mars crosses eastward into Capricornus in mid-May, easily outshining all of the stars on this chart, and executes a tight loop, heading westward in late June, when it shines high in the south at 3 A.M. local daylight time. In mid-July, as we rush toward the Red Planet, it passes 1.1° from the 4th-magnitude star Psi (ψ). Opposition (+) occurs on Jul. 27, when Mars is visible all night long. In late August, Mars changes course and runs eastward on a final swing through the constellation. By Halloween, the Red Planet is still shining brightly and stands high in the south 2 h after sunset (Robert Miller)

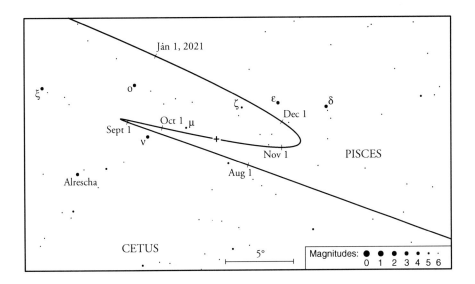

Fig. 5.12 Mars loop, 2020. As August opens, Mars is located in the constellation Pisces and is easily visible high in the south in the hour before dawn; it already far outshines every star on this chart. On Aug. 20, the planet passes 0.8° from 4th-magnitude Nu (ν). In early September, Mars begins its westward motion as Earth closes the distance. Opposition (+) occurs Oct. 13, when the Red Planet is visible all night long. In mid-November it changes course and resumes its eastward slide through Pisces. As 2021 opens, Mars is a bright tawny star high in the south 2 h after sunset (Robert Miller)

Table 5.2 Mars on the Web

Active missions

  2001 Mars Odyssey
  mars.jpl.nasa.gov/odyssey

  Mars Express
  mars.esa.int

  Mars Exploration Rovers
  marsrovers.jpl.nasa.gov

  Mars Reconnaissance Orbiter
  marsprogram.jpl.nasa.gov/mro
  hirise.lpl.arizona.edu

Future missions

  Mars Science Laboratory
  marsprogram.jpl.nasa.gov/msl

  Mars Atmosphere and Volatile Evolution (MAVEN)
  lasp.colorado.edu/maven
  www.nasa.gov/maven

Past missions

  Phoenix Mars Lander
  phoenix.lpl.arizona.edu

  Mars Global Surveyor
  www.nasa.gov/mission_pages/mgs
  www.msss.com/moc_gallery

  Mars Pathfinder
  mars.jpl.nasa.gov/MPF

  Vikings 1 and 2
  www.nasa.gov/viking

Interactive map

  Google Mars
  www.google.com/mars

References

Anon (1909) Science seeks to get into communication with Mars. N.Y. Times, May 2, 1909

Bricker HM, Aveni A, Bricker VR (2001) Ancient Maya documents concerning the movements of Mars. Proc. Nat. Acad. Sci. 98: 2107–2110

Carr MH (1981) The surface of Mars. Yale University Press, New Haven, Connecticut

Cohen IB (1985) Revolution in science. Belknap Press, Cambridge, Massachusetts

Dick SJ (1998) Life on other worlds: The 20th-century extraterrestrial life debate. Cambridge University Press, Cambridge, England

Dreyer JLE (1953) A history of astronomy from Thales to Kepler. Dover Publications, Inc., New York

Gibson EK, McKay DS, Thomas-Keprta KL, Clemett SJ (2010) Early Mars: A warm wet niche for life. http://www.lpi.usra.edu/meetings/abscicon2010/pdf/5062.pdf. Accessed 15 Oct. 2010

Gingerich O (2004) The book nobody read: Chasing the revolutions of Nicolaus Copernicus. Walker & Company, New York

Hecht MH, Kounaves SP, Quinn RC et al (2009) Detection of perchlorate and the soluble chemistry of martian soil at the Phoenix lander site. Sci. 325: 64–67. doi: 10.1126/science.1172466

Hoyt WG (1976) Lowell and Mars. University of Arizona Press, Tucson, Arizona

Irving T (2010) An up-to-date list of martian meteorites. http://www.imca.cc/mars/martian-meteorites-list.htm. Accessed 15 Oct. 2010

Kaufman M (2010) Not 'life,' but maybe 'organics' on Mars. Wash. Post. Sept. 4, 2010

Kerr RA (2010) Phoenix lander revealing a younger, livelier Mars. Sci. 329: 1267–1269. doi: 10.1126/science.329.5997.1267

Lapen TJ, Righter M, Brandon AD et al (2010) A younger age for ALH84001 and its geochemical link to shergottite sources in Mars. Sci. 33: 347–351. doi: 10.1126/science.1185395

Lefèvre F, Forget F (2009) Observed variations of methane on Mars unexplained by known atmospheric chemistry and physics. Nat. 460: 720–723. doi:10.1038/nature08228

Levin GV (2006) Modern myths of Mars. In: Hoover RB, Levin GV, Rozanov AY (eds.) Instruments, methods and missions for astrobiology. SPIE Proc 6309. http://mars.spherix.com/6309-12new.pdf. Accessed 15 Oct. 2010

Linton R (1922) The sacrifice to the Morning Star by the Skidi Pawnee. Leaflet No. 6, Field Museum of Natural History, Dept. of Anthropology, Chicago. http://www.archive.org/details/sacrificetomorni06lint. Accessed 15 Oct. 2010

Lowell P (1895) Mars. Houghton Mifflin Co., Boston

McKay DS, Gibson EK Jr, Thomas-Keprta KL et al (1996) Search for past life on Mars: Possible relic biogenic activity in martian meteorite ALH 84001. Sci. 273: 924–930

Mumma MJ, Villanueva GL, Novak RE et al (2009) Strong release of methane on Mars in northern summer 2003. Sci. 323: 1041–1045

Murray BC (1989) Journey into space: The first thirty years of space exploration. W. W. Norton & Co., New York

Navarro-Gonzalez R, Vargas E, de la Rosa J, Raga AC, McKay CP (2010) Reanalysis of the Viking results suggests perchlorate and organics at mid-latitudes on Mars. J. Geophys. Res. doi:10.1029/2010JE003599, in press

Rabin S, Nicolaus Copernicus. In: Zalta, E.N. (ed.) (2010) The Stanford encyclopedia of philosophy. http://plato.stanford.edu/archives/fall2010/entries/Copernicus. Accessed 15 Oct. 2010

Raulin-Cerceau F (2010) The pioneers of interplanetary communication: From Gauss to Tesla. Acta. Astronaut 67: 1391–1398. doi: 10.1016/j.actaastro.2010.05.017

Schafer EH (1977) Pacing the void: T'ang approaches to the stars. University of California Press Berkeley

Smith MD (2010) MGS TES surface temp extrema. Personal communication

Thurman M (1983) The timing of the Skidi-Pawnee Morning Star sacrifice. Ethnohist. 30:155–163

Thomas-Keprta KL, Clemett, SJ, McKay DS, Gibson EK, Wentworth SJ (2009) Origins of magnetite nanocrystals in Martian meteorite ALH8400. Geochim. Cosmochim. Acta. 73: 6631–6677 http://www.nasa.gov/centers/johnson/pdf/403099main_GCA_2009_final_corrected.pdf. Accessed 15 Oct. 2010

Treiman AH (1996) Fossil Life in ALH 84001? http://www.lpi.usra.edu/lpi/meteorites/life.html. Accessed Oct. 15, 2010

Touma J, Wisdom J (1993) The chaotic obliquity of Mars. Sci. 259: 1294–1297

Williamson RA (1984) Living the sky: The cosmos of the American Indian. Houghton Mifflin Co., Boston

Chapter 6

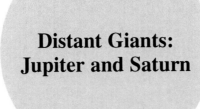

Distant Giants: Jupiter and Saturn

After following the frenetic wanderings of Mars, we turn now to planets that proceed through the sky at a much more leisurely pace. Jupiter and Saturn, the largest planets of the solar system, lie much farther from the Sun – and us – than Mars. Jupiter's lane of the solar system racetrack is about five times the size of Earth's, and Saturn's track is nearly twice as large as Jupiter's. Their wanderings through the constellations are much less dramatic than the splendid whirl of Mars, but the motions are similar, with retrograde loops centered on the time when they're opposite the Sun in our sky. As with Mars, Jupiter and Saturn come to opposition when our faster-orbiting Earth overtakes and passes them, and this is the time when they're closest, shine brightest and appear largest in a telescope. However, their slower orbital speeds translate to more frequent oppositions: Jupiter's recur every 13 months, while Saturn's happen a couple of weeks later each year (see Appendix D for details). Their plodding regularity results in seasonal appearances that approximate those of the background stars. Jupiter travels through roughly one constellation of the zodiac each year and Saturn tracks about half that, so skywatchers can count on seeing these planets only slightly later each successive year. A special treat occurs about every two decades, when Jupiter overtakes Saturn and the solar system's biggest worlds shine in tight formation.

Jupiter and Saturn represent a distinctly different class of planet than those we've discussed up to this point. They are the largest and nearest of the four gas giants of the outer solar system, the others being Uranus and Neptune. These mega-worlds are largely composed of hydrogen and helium, the two lightest elements in the cosmos. The inner solar system's small rocky planets, including Earth, have

F. Reddy, *Celestial Delights: The Best Astronomical Events through 2020*,
Patrick Moore's Practical Astronomy Series, DOI 10.1007/978-1-4614-0610-5_6,
© Springer Science+Business Media, LLC 2012

Table 6.1 Facts about Jupiter and Saturn

Jupiter	
Diameter (at 1 bar atmospheric pressure)	88,846 miles
	142,984 km
	11.21 times Earth's
Temperature (at 1 bar atmospheric pressure)	−163 °F (−108 °C)
Atmospheric composition	86.2% hydrogen (H_2)
	13.6% helium (He)
	0.1% methane (CH_4)
	< 0.06% water (H_2O)
	0.03% ammonia (NH_3)
Moons	63
Four largest moons, discovered in 1610 by Galileo:	
1.2% smaller than Mercury	Callisto
7.8% larger than Mercury, biggest moon in the solar system	Ganymede
10.2% smaller than Earth's Moon, possible subsurface ocean	Europa
4.8% larger than Earth's Moon, most volcanic body in the solar system	Io
Sidereal rotation period (length of day)	9.925 h
Obliquity (tilt of spin axis with respect to the orbital plane)	3.13°
Synodic period (time between successive conjunctions with the Sun)	1.093 years
Sidereal year (time for one full orbit around the Sun)	11.862 years
Tropical year (time between successive vernal equinoxes; i.e., the seasonal year)	11.857 years
Average distance from Sun	483.682 million miles
Light takes 43.3 min to travel this far.	778.412 million km
	5.2 Astronomical Units (AU)
Orbit inclined to Earth's	1.305°
Saturn	
Diameter (at 1 bar atmospheric pressure)	74,898 miles
	120,536 km
	9.45 times Earth's
Temperature (at 1 bar atmospheric pressure)	−218.5°F (−139.2°C)
Atmospheric composition	96.3% hydrogen (H_2)
	3.25% helium (He)
	0.45% methane (CH_4)
Moons	62
6.6% larger than Mercury, second-largest moon in solar system, atmosphere thicker than Earth's, liquid methane/ethane on surface	Titan
1/7th the size of Earth's Moon, ice volcano at south pole, source of E ring particles	Enceladus
Sidereal rotation period	10.66 h

(continued)

Table 6.1 (continued)	
Obliquity	26.73°
Synodic period	1.036 years
Sidereal year	29.457 years
Tropical year	29.424 years
Average distance from Sun	886.526 million miles
Light takes 1.3 h to travel this far.	1.427 billion km
	9.54 Astronomical Units (AU)
Orbit inclined to Earth's	2.485°

Fig. 6.1 Earth, Saturn and Jupiter compared (NASA images; montage by the author)

more in common with the largest moons of Jupiter and Saturn – some of which are as big as Mercury – than with these giants themselves. Everything about them is gargantuan. As a group, they hold 99.7% of the planetary system's mass. Jupiter alone possesses 318 times Earth's mass – and more than twice the combined mass of all of the other planets, their moons, and all of the asteroids and comets in the solar system. To put this into perspective, one could make the case that our planetary system consists of Jupiter, plus debris. Saturn comes in a distant second, weighing in at 95 Earth masses, but it possesses a superlative all its own. Saturn is surrounded by the brightest, most complex and most beautiful system of planetary rings ever seen. Additional statistics on both planets can be found in Table 6.1, and Fig. 6.1 offers a, visual comparison with Earth.

The regal pace at which Jupiter and Saturn drift through the constellations led to their association with the most powerful members of ancient pantheons. In Babylon, Marduk was regarded as the creator of the world and the god who established the cosmos' cyclic order by vanquishing Tiamat, one of two primordial creative elements whose chaotic energy dominated the early universe. For this, Marduk became the ruler of the gods. He took Jupiter (Neberu), the planet that most closely follows the Sun's path through the heavens, as his personal star. The number 12 had cultural importance in Mesopotamia, and in a time just shy of 12 years Jupiter makes the same trip through the zodiac that the Sun makes every 12 months. Jupiter could be seen as the sky's night watchman, dutifully performing celestial rounds, or as a kind of cosmic governor that maintained the stars and other planets at a safe and steady pace.

In Greece, Jupiter was associated with Zeus, chief god of the pantheon but a third-generation god. In one creation myth, Gaea (Mother Earth), arose from chaos and gave birth to Uranus (Sky), and together they created two races: the Titans and the one-eyed Cyclopes. Fearing the strength of the Cyclopes, Uranus exiled them, and in retaliation, Gaea persuaded the Titans to attack Uranus in a battle led by the youngest, named Kronos. As the defeated Uranus lay dying, he prophesied that Kronos, whose Roman equivalent is Saturn, would one day be similarly overthrown by his own children. Kronos hoped to avert the prophecy by eating his children, but his wife Rhea had had enough of this by the time Zeus was born. She put a stone in the child's place and secretly took the boy to the island of Crete. When Zeus matured, he confronted his father and slipped him a potion that made him vomit up the swallowed children. The regurgitated siblings pleaded with Zeus to lead an overthrow of the Titans, which he did and thereby assumed rule over the universe. Zeus was thought to make his presence known to mortals by hurling lightning bolts, and it's his Norse equivalent, Thor, who gives his name to the fifth day of the week.

The Egyptians of the New Kingdom (c. 1540 to 1070 B.C.) saw Jupiter and Saturn, together with Mars, as manifestations of the falcon-headed sky god Horus. Jupiter was "Horus who bounds the Two Lands," "Horus who illuminates the Two Lands," or "Horus who opens mystery," whereas Saturn was always "Horus the bull of the sky." The two lands were Upper and Lower Egypt, which were united around 3000 B.C.

To the ancient Chinese, Jupiter was Suixing, the Year Star. From about the 7th century B.C., Jupiter served a calendrical function based on its 12-year cycle around the sky. The sky regions of interest followed the celestial equator, an imaginary line midway between the north and south celestial poles, rather than the ecliptic. Jupiter comes to opposition 12 times in just over 12 years, and the twelfth opposition returns the planet to nearly the same stars as the first. The Chinese associated each of the 12 celestial regions with an animal, and as Jupiter moved from one zone to another, the year took on the characteristics of each of these animal realms. So the traditional Oriental cycle of year names – Year of the Rat, Year of the Ox, and so on – derives in part from Jupiter's dozen-year journey through the stars.

Saturn was Zhenxing (Quelling Star) or, following a later system that related the planets to fundamental elements, Tuxing (Soil Star). Among the Babylonians Saturn was Kayamanu and associated with Ninurta, originally a war god and hunter but also a fertility deity and the personification of the south wind. Saturn seems to have retained his association with agriculture when transported to the Italian peninsula. The Saturnalia, a week-long festival held in late December prior to the winter solstice, was among Rome's most popular celebrations. Today many of us may have similar feelings of release on the last day of the week, which gets its name from the god honored by the Saturnalia.

The Maya were aware of the retrograde loops of Jupiter and Saturn and saw particular importance in their "stationary" points, when motion against the starry background halts just before the planets reverse course. Inscriptions that herald events in the life of K'inich Kan B'ahlam II (635–702), the twelfth king of Palenque in what is now southern Mexico, suggest that the dates were chosen with Jupiter's retrograde motion in mind, and he may have regarded it as his own patron planet. On the evening of Jul. 19, 690, a waxing gibbous Moon joined an arc formed by Jupiter, Saturn and Mars in the head of the constellation Scorpius, and B'ahlam took advantage of the event to legitimize his rule. Scholars believe that the people of Palenque interpreted this celestial gathering as the First Mother joined by her three children, the Palenque gods known to archaeologists only as GI, GII, and GIII. The event initiated days of rituals that culminated with a small offering of B'ahlam's own blood, a symbolic reenactment of the First Mother's actions at what the Maya believed was the fourth and most recent cycle of cosmic creation. Both the rituals and the offering of royal blood established a bond between B'ahlam and the entire royal line at Palenque – and ultimately to the gods themselves. Two decades later, at the nearby Maya city of Yaxchilán, the ruler Itzamnaaj B'alam II (681–742) used a similar event to ensure his child's future claim to the throne. Jupiter and Saturn had a close conjunction just weeks after the child's birth. About 6 weeks later, on Oct. 28, 709, the planets stood 2.5° apart but had stopped moving against the stars – and that's when the king offered his blood sacrifice to the gods.

King of the Planets

Efforts to explain the motions of the planets ultimately gave humanity a new perspective on its place in the universe. Tycho's state-of-the-art observations of Mars played a pivotal role in the Copernican revolution, giving Kepler the data he needed to puzzle out the true structure of planetary orbits. Yet what really took the world by storm was the announcement of objects never before seen by any astronomer – "starlets" whirling near Jupiter that could only be detected with a brand new invention.

In 1608, a lens-maker in Holland named Hans Lipperhey (d. 1619) stumbled onto the fact that lenses, when combined in certain ways, could be used to make faraway objects appear much closer. In September he applied for a patent, but the application was denied because the device was considered too difficult to keep secret, and shortly thereafter Lipperhey's claim of priority was challenged by two of his countrymen. So while historians cannot claim Lipperhey as the telescope's inventor, his patent application remains the earliest documented evidence of a completed instrument in 1608. Within weeks, low-power telescopes were being made and sold in many northern European cities.

The following May, word of this wondrous new device reached the chair of mathematics at Italy's University of Padua, Galileo Galilei. In June he succeeded in building a telescope from glass lenses and a length of lead tube, producing an instrument capable of magnifying the size of distant objects by three times. Although it wouldn't be widely known until centuries later, Galileo had competition for astronomical discovery in a well-heeled English polymath named Thomas Harriot (1560–1621), who was able to purchase a Dutch-made telescope with about twice the magnifying power of Galileo's first effort. On the evening of Aug. 5 (Jul. 26 on the Julian calendar then used in England), Harriot turned his tube to the 5-day-old crescent Moon, sketched what he saw, and dated the drawing. It remains the oldest-known rendition of an astronomical body as seen through a telescope, beating Galileo by nearly 4 months. However, like most of Harriot's work, the drawing was never published in his time.

As summer turned to autumn, Galileo quickly advanced his design beyond version 1.0, and by November, "sparing neither labor nor expense," fashioned himself an excellent – by the standards of the day – 20-power telescope. He immediately undertook a series of lunar observations that lasted into December. But the big moment didn't come until twilight deepened on Jan. 7, 1610, with Jupiter, just a month past opposition, shining brilliantly above the nearly full Moon. Galileo turned his telescope to the planet and saw a trio of small, bright stars near it. "Though I believed them to be among the host of fixed stars, they aroused my curiosity somewhat by appearing to lie in an exact straight line parallel to the ecliptic," he wrote. But he found a different arrangement when he turned to the planet the following night: The stars were now all west of Jupiter, closer together, and spaced equally apart. On Jan. 13, a fourth starlet joined the other three, but by then Galileo had already figured out what he was seeing and fully realized its significance.

These four objects were orbiting around Jupiter, and he was watching that motion from the side. This was an unequivocal example that there was not one center of motion in the universe, but at least two – Earth and Jupiter. And if there were two, then the solution that made the most sense to Galileo is that all of the planets, including Earth, are actually moving around the Sun, as Copernicus had worked out. He rushed to tell the world about his findings and to encourage others to follow up. In March 1610, the booklet describing Galileo's telescopic work, *The Starry Messenger*, became an instant best-seller; the first press run sold out in a week, translations from the original Latin to vernacular Italian followed quickly, and within 5 years the main discoveries were published in Beijing. While this success was due in part to the specific and remarkable nature of the findings – mountains on the Moon, faint stars in the Milky Way too numerous to count, stars circling Jupiter – Galileo's direct, simple and succinct presentation helped, too. This was not a dry and deeply technical treatise like Copernicus' *On the Revolutions* or Kepler's *The New Astronomy*, but an exciting account of work in progress, of novel science being interpreted on the fly.

Galileo publicly referred to these moons as the "Medicean stars," a name that honored his patrons in Tuscany, but today astronomers refer to them collectively as the

Galilean satellites. In his notebooks, Galileo referred to them individually by number based on their maximum distance from Jupiter. In 1614, on a suggestion from Kepler, Simon Marius (1573–1624) proposed an improved naming system based on the many lovers of Zeus in mythology. This scheme wasn't finally adopted until the mid-1800s, when new discoveries around Saturn disrupted a similar scheme by finding new moons inside the orbits of those that were already known and numbered. To prevent further confusion, astronomers switched to a naming system for both planets. The Galilean moons, in order of distance from the planet, are Io, Europa, Ganymede and Callisto. As befits a giant planet, Jupiter's moons are supersized. Ganymede is slightly larger than the planet Mercury and is the largest satellite in the solar system. Callisto is slightly smaller, and both Io and Europa are about the size of Earth's Moon.

Improved telescopes soon revealed fascinating details on the planet. First astronomers noted cloud bands, and as telescopes improved these bands were resolved into swirls and festoons. The planet's most remarkable piece of meteorology, the Great Red Spot, was discovered in 1665 by Giovanni Domenico Cassini (1625–1712), who was looking for some kind of permanent marking to measure the planet's rotation period (Figs. 6.2 and 6.3). This amazing feature, which spans almost three times Earth's diameter, is an enormous high-pressure system that has remained on the planet for over three centuries. Using what he called the "Permanent Spot," Cassini measured the planet's rotation rate to nearly the modern value – an amazing 9.9 h. Cassini also noticed that features at different latitudes moved at different speeds around the planet, a phenomenon called differential rotation, and that Jupiter's shape is noticeably out-of-round and bulges at its equator. Spacecraft observations later showed that the Great Red Spot rotates once every 6 days and interacts with other, smaller oval storms at adjacent latitudes. The Great Red Spot's size and position vary slightly, as does its coloring, generally ranging from a pale orange to brick red; the source of those colors is unknown, with likely candidates being organic molecules, or compounds of phosphorus or sulfur.

Jupiter's atmosphere shows alternating light zones and dark bands, the largest of which are visible even in a small telescope. These are the visible manifestation of east–west jet streams that alternate direction from the equator to the poles, and they carry oval weather systems of all sizes. Jupiter's atmosphere has been especially busy lately. In 1998, two of three 60-year-old white oval storms located in a cloud band south of the Great Red Spot merged, and in early 2000 the third oval joined them. The resulting weather system, which is about a third the size of the Great Red Spot, was named Oval BA. In August 2005, amateur astronomers noticed that Oval BA was acquiring a reddish color, which gradually deepened until, by 2006, the blushing storm was nicknamed the Little Red Spot or Red Spot Jr. The following year, it was imaged by NASA's New Horizons spacecraft, which flew past Jupiter for a gravitational boost that will help it reach its target, Pluto, a little faster (the flyby is scheduled to occur in July 2015). Scientists suspect that the development of deeper colors in the red spots reflects increasing storm strength, perhaps by enabling them to dredge up coloring agents from lower in the atmosphere, but it's one of many mysteries about Jupiter's weather that remain unanswered.

Fig. 6.2 Pastel giant. Jupiter, the solar system's largest planet, is 11 times the size of our world. The large tawny oval is the famous and seemingly permanent Great Red Spot. It's a storm befitting a giant, large enough to engulf a pair of Earths. Jupiter's rapid spin powers fierce winds that shear its clouds, creating the banded appearance. They contain ammonia, hydrogen sulfide and water, which form cloud layers at different heights. Trace chemicals, possibly produced by reactions with sunlight, create the planet's color palette. NASA's Cassini spacecraft captured this view when it passed Jupiter in 2000 on its way to Saturn (NASA/JPL/ Space Science Institute)

Fig. 6.3 Jupiter's Great Red Spot (large oval, *top*) dominates this mosaic of images from Voyager 1's Jupiter flyby in March 1979. The storm spans more than 21,000 miles (33,800 km). First observed in telescopes about 350 years ago, the Great Red Spot is a persistent high-pressure system in Jupiter's southern cloud belts (NASA/JPL)

Impacting Jupiter

The most significant recent change to Jupiter's appearance occurred in 1994, when more than 20 large fragments from a disrupted comet named Shoemaker-Levy 9 (D/1993 F2) crashed into the planet's atmosphere. All of the short-period comets, that is, those with orbital periods of less than 200 years, show a decidedly jovian influence. Twice as many of them reach their closest point to the Sun within a semicircle centered on Jupiter's own perihelion than in one centered on the planet's aphelion. Sometime around 1930, to the best anyone can determine, this particular comet passed Jupiter in such a way that the giant planet was able to capture it into orbit and make it a temporary moon. On Jul. 7, 1992, the eccentric orbit carried the comet within 13,000 miles (21,000 km) of the planet's cloud tops – so close that jovian tides extensively cracked and weakened the nucleus; later, aided by the stress of its own rotation, the comet spun off fragments until it was completely disrupted. Astronomers remained unaware of the wrecked comet for another eight and a half months.

On the evening of Mar. 23, 1993, American astronomers Carolyn and Eugene Shoemaker and David Levy were imaging the area about 4° south of Jupiter as part of a program to discover new asteroids and comets. When examining the night's films a couple of days later, Carolyn saw something she described as a "squashed comet" – a bar of light capped with several faint tails that curved away to the north.

Follow-up observations revealed that the bar was embedded with numerous bright knots, icy fragments that stretched along the comet's orbit like pearls on a string. Calculations showed that these cosmic icebergs, with sizes up to 1.2 miles (2 km) across, were executing their final loop. They would meet a dramatic and fiery end the following summer when, one by one, they plunged into Jupiter's atmosphere at a speed of 134,000 mph (216,000 km/h), so fast that they could cover the distance from Earth to the Moon in less than 2 h.

This was the first collision of solar system bodies ever witnessed. Jupiter has no solid surface, so these impacts were "airbursts." As the comet chunks fell into the thickening atmosphere, they eventually experienced catastrophic stresses, broke apart, and converted their kinetic energy into heat so quickly that the event was indistinguishable from the detonation of high explosives. Although the fragments struck a location on the planet that was not directly visible from Earth, Jupiter's rapid rotation carried the sites into view minutes later. The best-placed observer was NASA's Galileo spacecraft, which was still 17 months away from its arrival at Jupiter. With a better angle on the impact zone than Earth-based telescopes, astronomers hoped that Galileo's instruments would be able to view the earliest light from some of the impacts. All eyes were on Jupiter when the first splinter, known as Fragment A, struck on Jul. 16, 1994.

"Prior to the event, no one knew for sure whether the collisions would produce phenomena detectable from Earth," recalled Heidi Hammel, who led a team of astronomers using the Hubble Space Telescope. Within half an hour of the Fragment A strike, initial reports of the impact's aftermath were coming in from ground-based observatories, but Hubble's images were delayed for two orbits before being radioed to Earth. The science team gathered in the basement control room at the Space Telescope Science Institute in Baltimore to watch as their first image appeared on the monitor. On the planet's limb, a small bright spot could be seen, but was this one of the moons emerging from behind Jupiter or an effect related to the impact? As the astronomers scrambled to account for the locations of the moons, a new image appeared and dispelled all doubt: Hubble had captured an enormous plume of ejected material rising thousands of miles above the planet's cloud tops. But one didn't need Hubble to see the circular brownish "scar" that formed around the site, produced as particles condensing in the impact plume rained back onto the planet.

Fragments kept arriving through Jul. 22. Infrared telescopes detected the heat of the entering fireballs, and Galileo detected visible flashes from some of them. Hubble watched additional plumes of hot gas rise above the planet's limb. Infrared observations revealed glowing spots as the plume material fell back into the atmosphere and rapidly heated it. Without a doubt, the physical and chemical effects of the impacts were far more complex than anticipated. The most spectacular scar came from the Jul. 18 impact of Fragment G. This "black eye" was the size of the Great Red Spot, easily visible in modest telescopes, and could be tracked into September. All told, the family of comet fragments known as Shoemaker-Levy 9 blasted Jupiter with the equivalent to 300 billion tons of TNT – millions of times the explosive energy of all weaponry humans have ever produced. The event raised awareness among both the public and scientists that large impact events still occur in the solar system, and it helped drive efforts to identify asteroids and comets that could be potentially hazardous to Earth.

At the time of the comet crash, astronomers figured Jupiter experienced an impact detectable from Earth every few hundred years. But exactly 15 years later, it happened again – and this time the likely culprit was a single asteroid. On Jul. 19, 2009, Australian amateur astronomer Anthony Wesley was imaging Jupiter from his home observatory when he noticed an unusually dark spot rotating into view in the planet's southern polar region. "It took another 15 minutes to really believe that I was seeing something new," he wrote. "I'd imaged that exact region only 2 days earlier and checking back to that image showed no sign of any anomalous black spot." He quickly emailed fellow amateur and professional astronomers.

Franck Marchis at the University of California, Berkeley, heard about the new discovery and wrote about it in his blog, requesting follow-up observations. Berkeley astronomer Paul Kalas, who was in Greece at the time, read the blog post, consulted with Marchis and other astronomers on the best way to observe the feature, and used his previously scheduled time on the giant Keck II Telescope in Hawaii to observe the area. The team's infrared image revealed a blemish about as big as the Pacific Ocean. Later Hubble observations showed that the new bruise exhibited slight color differences indicating that the invader was of asteroidal rather than cometary origin.

Incredibly, Wesley detected a 2-s-long flash from another, much smaller Jupiter impact on Jun. 3, 2010, while watching a live video feed from his telescope. At the same time, in the Philippines, amateur astronomer Chris Go was recording his own video feed and was able to confirm Wesley's observation. This time, even Hubble couldn't find debris, so astronomers believe Wesley and Go witnessed the flash of a brilliant meteor lighting up high above Jupiter's clouds. Astronomers recognize that they don't have a good handle on impact rates in the solar system, and it's possible that that the smallest detectable events at Jupiter may happen every few weeks. The watchful video eyes of amateurs provide a new capability for monitoring these small strikes.

It's tempting to regard Jupiter as a cosmic shield that protects the inner planets from asteroid and comet impacts, but in reality the planet tosses plenty of stuff our way. Asteroids that orbit the Sun in even multiples of Jupiter's 12-year circuit are said to be in an orbital "resonance" with it. For example, an asteroid that circles the Sun three times for every Jupiter revolution is in a 3:1 resonance: With every three orbits, the asteroid gets a tug from Jupiter at the same place in its orbit. These periodic nudges force a steady change in the asteroid's path, moving its perihelion closer to the Sun and its aphelion farther away. Over time the orbit may plunge deep into the inner solar system, crossing lanes occupied by the inner planets. There aren't many asteroids at the 3:1 resonance because Jupiter has cleared most of them away, and those that remain are temporary residents only. Close encounters with the inner planets or with Jupiter can further alter an asteroid orbit, sometimes quite drastically, and greatly enhance the possibility of a collision. What we know about the composition of asteroids that cross the orbits of Mars and Earth indicates that they come from every part of the asteroid belt. So collisions among asteroids knock fragments into resonance zones, and Jupiter eventually boots them out.

Jovian Close-up

The first space probes to Jupiter were Pioneers 10 and 11, which blazed a trail for NASA's spectacularly successful Voyager missions that followed. Launched in 1977, both Voyager spacecraft are still returning data as they cruise outward into interplanetary space. As of April 2011, Voyager 1 was more than 10.8 billion miles (17.4 billion km) away, so distant that its radio signals take more than 16 h to reach us. With a little luck, these spacecraft will serve as humanity's first interstellar probes in the next decade. The two Voyagers flew past Jupiter and Saturn, discovered dozens of moons, acquired awe-inspiring pictures and collected a wealth of data from which scientists still tease new discoveries. Voyager 1's planetary visits ended with Saturn, but thanks to a fortuitous celestial arrangement, some talented engineering and a bit of luck, Voyager 2 was able to return data as it flew past Uranus and Neptune before heading into deep space. "When I look back, I realize how little we actually knew about the solar system before Voyager," said Ed Stone, the mission's project scientist. "We discovered things we didn't know were there to be discovered, time after time."

NASA's Galileo mission followed the Voyagers to Jupiter; it arrived in late 1995, dropped a probe into the atmosphere, and spent more than 7 years orbiting the planet and mapping its moons. Additional flybys came from spacecraft using Jupiter for a gravity assist: the European solar probe Ulysses in 1992, the Saturn-bound Cassini-Huygens mission in 2000, and New Horizons in 2007. Next up will be NASA's Juno, an orbiter set to arrive in 2016.

All of the gas giant planets are essentially bottomless atmospheres, with Jupiter's being the deepest. Descending through its atmosphere, we first encounter hazes of hydrocarbons, like methane, propane and acetylene, produced by the action of ultraviolet sunlight. Below this layer, the pressure approaches that at Earth's surface, but the temperature is only $-230°F$ $(-146°C)$. At this level, we're immersed in a deck of bright white clouds formed by ammonia ice crystals. Plunging deeper, to twice the surface pressure of Earth's atmosphere (2 bars), the temperature rises to $-100°F$ $(-73°C)$, and it now increases steadily with depth. Here we encounter a yellow-orange cloud deck built from ice crystals of ammonium hydrosulfide. Usually we cannot see past these clouds, but spacecraft measurements and computational models indicate that the next cloud deck – located at pressures of about 2.3 bars and temperatures from freezing to $80°F$ $(27°C)$ – contains water ice crystals at the colder levels and water droplets lower down, just like clouds on Earth. Any familiarity pretty much ends there, however.

Scientists use minute frequency changes in spacecraft radio signals to map the structure of Jupiter's gravitational field, and this in turn enables them to describe the planet's internal structure below the clouds. Pressures and temperatures increase steadily, but the hydrogen atmosphere simply grows denser and hotter with depth and eventually, hundreds of miles beneath the clouds, the molecular hydrogen gradually starts to resemble a hot liquid. At depths ten times greater, when we've traveled only 20% of the way toward Jupiter's center, pressures approach a million bars and temperatures soar to $10,000°F$ $(5,700°C)$ – hotter than the Sun's surface. Here, the hydrogen becomes a more exotic substance called *metallic* hydrogen, an electrically conductive soup of protons and electrons that makes up most of Jupiter's mass and likely accounts for its potent magnetic field. Some 28,000 miles (45,000 km) farther down, about 80% of the way to the planet's center, the composition changes to a mix of water, methane, and ammonia under enormous temperatures and pressures. Another 4,400 miles (7,000 km) down and we're just 10% from the center, where the pressure rises to around 40 million bars and the temperature is some 40,000° F (22,000° C). Here, Jupiter's composition gradually morphs into a dense core of solid rock perhaps 12 times more massive than planet Earth. If this core exists, and not all scientists agree that it does, it may have been the gravitational seed that got Jupiter started. A major goal of NASA's Juno mission is to answer the many remaining questions about how Jupiter is put together.

Saturn's interior structure is similar to Jupiter's, although its lower mass and internal pressures lead to a smaller region filled with metallic hydrogen, which begins about halfway to the planet's center. Uranus and Neptune, which are even less massive, lack this substance altogether. Astronomers expect that many of the giant planets discovered in other solar systems, some of which are many times more massive than Jupiter, share its basic internal structure. Often called a "star that failed," the planet emits nearly 1.7 times as much energy as it absorbs from the Sun, but this internal heat is left from its formation 4.5 billion years ago; massive objects like Jupiter just can't rid themselves of this energy very quickly. It would take the mass of 12 more Jupiters to make a minimal star called a brown dwarf, and to fire up the nuclear reactions found in our Sun, you'd need to throw in about 68 more.

An extraordinary planet deserves extraordinary moons, and Jupiter's big Galilean satellites do not disappoint. A major surprise of the Voyager missions was their complexity and individuality, which scientists hardly expected from telescopic observations. "Those four orbiting moons are as unusual and different from one another as four bits of rocky debris orbiting close to the Sun – Mercury, Venus, Earth, and Mars. Voyager 1 had zoomed in on the four dark smudges near the gaudy telescopic image of Jupiter, revealing worlds and circumstances beyond our imaginations," wrote Bruce Murray, the director of NASA's Jet Propulsion Laboratory during the Jupiter and Saturn encounters.

There had long been hints that the big moons might be doing something interesting. Plots made by Galileo Galilei show that Europa and Io always have their conjunction on the side of Jupiter exactly opposite from where Europa and Ganymede have theirs. In the time it takes Ganymede to go around Jupiter once, Europa goes around exactly twice and Io goes around four times. This 1:2:4 resonance was likely set in place as the satellites first took shape around Jupiter. This gravitational interaction forces the orbits of the moons to maintain a slight eccentricity, and this in turn creates tides that flex the satellite shapes. Just as repeatedly bending a paper clip warms up the metal, these orbitally forced tides heat the interiors of Jupiter's big moons. The two most surprising effects of this heating occur at Io and Europa.

That tidal dissipation could melt a good portion of Io's insides was first realized by Stanton Peale at the University of California, Santa Barbara, and his colleagues. They worked out the details and submitted a paper to the journal *Science* in January 1979, presciently concluding: "Voyager images of Io may reveal evidence for a

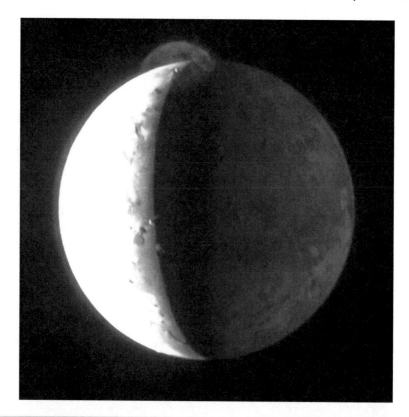

Fig. 6.4 A volcano near the north pole of Jupiter's moon Io lofts an eruptive plume (*top*) to a height of 200 miles (300 km). Illuminated by the setting Sun, the debris cloud rains down onto Io's nightside (*right*), made visible by sunlight reflected off of Jupiter. The source of the eruption is a volcano named Tvashtar, located at the brightest point within the plume and straddling the day-night line. The image was taken in February 2007 during the flyby of NASA's Pluto-bound New Horizons spacecraft (NASA/JHUAPL/SwRI)

planetary structure and history dramatically different from any previously observed." The paper was published on Mar. 2, just days before Voyager 1's closest approach to Jupiter. By then, images of the weirdly colored little moon, reminiscent of a pizza, was already making headlines. Scientists had expected Io to resemble our Moon, which is similar in size, but it lacked anything that looked like an impact crater and was coated with richly colored deposits that wouldn't look out of place around active fumaroles on Earth. The features looked so sharp that geologists agreed Io's volcanoes might still be active. On Mar. 8, Voyager 1's imaging system captured a crescent-shaped cloud – the plume of an active volcano – arcing high above the moon, making perhaps the speediest confirmation of a theoretical proposal in the history of science. Twenty-eight years later, the New Horizons spacecraft returned a strikingly similar image (Fig. 6.4). The outpouring of material on

Fig. 6.5 The face of Jupiter's icy moon Europa is remarkably smooth, bearing few impact craters. Enigmatic and sometimes strangely curved lines crisscross the surface. This view is a mosaic of images from Voyager 2's flyby in 1979. Beneath the ice, Europa may host a global ocean, one estimated to contain as much as twice the volume of water as all of the seas on Earth. The moon is a high-priority target for future space missions (NASA/JPL/USGS)

Io is so great that it has erased impact craters on the surface. With hundreds of volcanoes and constant eruptions, Io reigns supreme as the most volcanic body in the solar system.

The next farthest moon out from Jupiter, Europa, is also the smallest of the Galilean satellites. Its icy surface is highly reflective, curiously smooth – even the most prominent ridges rise no higher than the length of several football fields – and bears relatively few impact craters. Voyager, and later, Galileo, imaged an extensive network of dark linear fractures and strange curlicue ridges across the moon's frozen plains (Fig. 6.5). The curlicues, formally known as cycloidal fractures, are unique to Europa and provided the first evidence that a global ocean might exist beneath the moon's icy crust. These cracks likely develop at a slow pace, perhaps walking speed, when jovian tides flex the surface. As the moon moves along its orbit, the angle of the tidal stress changes and the crack propagates in a new direction, gradually tracing out an arc instead of a line. But jovian tides are too weak to crack the icy shell unless it sits atop a liquid ocean. The Galileo spacecraft also measured the moon's gravity and showed that Europa's internal structure consists of a rocky core and mantle capped by a shell of water some 90 miles (145 km) thick. Galileo magnetic measurements near Europa seem to be best explained if most of this water is in liquid form; similar evidence suggests liquid water layers beneath the surfaces of Ganymede and Callisto, too. The evidence remains circumstantial yet convincing, and scientists are faced with the fact that a body smaller than our own moon probably possesses an ocean containing roughly twice the amount of liquid water in all of the seas on Earth.

A major point of contention among planetary scientists is the thickness of the icy shell, with estimates ranging from as little as 2.5 miles (4 km) to many times more. Nearly half of Europa's surface consists of so-called chaos terrain. The Galileo mission's most detailed views of these regions showed that large plates of ice have broken apart, rotated into new positions and then re-froze, creating landscapes strikingly similar to rafting pack ice in Earth's polar oceans. The chaos terrain likely represents places where the crust has melted through, perhaps driven by concentrations of tidal heat. If so, the icy shell may be especially thin there – possibly thin enough that future space probes could gain access to fresh ice or maybe even the ocean. Europa ranks high in the list of places many scientists think might be hospitable for the development of some form of extraterrestrial life. Future visits to this moon are guaranteed.

Lord of the Rings

Despite familiar images from the Voyagers and Cassini, Saturn is still the object that makes the deepest impression through a small telescope. The planet looks unexpectedly unnatural, even downright fake. "There is something about Saturn that makes it hard to believe it's real," said Brad Smith, imaging team leader for the Voyager Saturn encounters. "You have to see it through a telescope. You never forget it – ever after, it remains a sort of focal point in the sky." Its bright, broad ring system reminds us that matter behaves differently in space than it does here on Earth. Although we now know that all of the giant planets have rings, Saturn's stand alone in terms of brightness, breadth and complexity.

In July 1610, Galileo made the first telescopic observations of Saturn, but his telescope was unable to resolve the iconic rings. Instead, he saw the rings as two smaller bodies on either side of the planet. He naturally assumed that they were moons, but they seemed unusually large – about a third the size of the planet itself – and they never seemed to move away from the planet the way Jupiter's satellites do. Adding to the mystery, the strange objects disappeared completely in 1612. "I do not know what to say in a case so surprising, so unlooked for and so novel," he wrote. Unknown to Galileo, the rings were oriented edge-on from Earth's perspective. As the angle improved again, he recovered the companions in 1616, but noted that they now appeared as "half ellipses." He didn't know what to make of them.

Over the next few decades, Saturn's odd and slowly changing shape continued to puzzle astronomers, but it would take a technological leap in telescope optics to reveal the planet's true form. It was the Dutch mathematician, astronomer and inventor Christiaan Huygens (1629–1695) who, working with his brother Constantyn, developed a new method of grinding and polishing lenses that greatly reduced distortions affecting image sharpness. In February 1655, they completed a new telescope with lenses made from the improved process, and Christiaan turned his attention to Saturn. He quickly noticed the moon we now call Titan – the first new planetary satellite seen since Galileo's Jupiter discoveries – and he tracked its course around the planet. In the fashion of the time, he crafted an anagram announcing the discovery and sent it to his scientific correspondents as a way to safeguard his finding while he prepared his publication. When properly solved, the anagram read: "A moon revolves around Saturn in sixteen days and four hours." The following year, Huygens published a brief tract that detailed his discovery and provided the solution to the anagram. He also included another anagram that encoded a new discovery.

During his initial observations, Huygens hit on the solution to Saturn's oddball features, and Titan was the key. He reasoned that all matter closer to Saturn than Titan must revolve faster than the moon. Because changes to the companions happened over years rather than days, they must take some form that is symmetrical around the planet. The fact that the companions came and went over the course of 15 years, half of Saturn's orbital period, indicated that Saturn was tipped on its axis much as Earth is, so astronomers had been seeing the planet and its companions at changing angles as it circled the Sun. When solved, Huygens' second anagram read: "It is surrounded by a thin, flat ring, touching it nowhere, and inclined to the ecliptic." Huygens then became involved in developing the first pendulum clock and writing a related book, but if Saturn wasn't foremost in his thoughts during this period, it wasn't far away. He analyzed previous drawings of the planet's appearance and continued telescopic observations to work out the tilt of the planet's spin axis. He finally published the details in his influential 1659 book, *The System of Saturn*. While he had solved the geometrical problem, the true nature of the ring, which he believed to be a single solid structure, would be debated for centuries.

Further discoveries came slowly. In 1671, 6 years after describing Jupiter's Great Red Spot, Cassini, now heading the new Paris Observatory, spotted one satellite, Iapetus, and he found another, Rhea, the following year. He then discovered a gap – today known as the Cassini Division – that splits the ring in two. Astronomers now refer to the outer one as the A ring and to the brighter inner one as the B ring. Cassini correctly proposed that the rings were not solid after all but were actually composed of large numbers of tiny satellites; he then discovered two more moons, named Tethys and Dione, in 1684.

The puzzle of the rings was finally solved in 1859, when physicist James Clerk Maxwell (1831–1879) published his analysis of its structure. He proved that the rings could not be solid or liquid and had to be made up of "an indefinite number of unconnected particles" independently orbiting the planet. By then, an additional interior ring, now labeled the C ring, three more satellites – Enceladus, Mimas and Hyperion – and another break in the A ring, now called the Encke gap, had been found. The American astronomer and mathematician Daniel Kirkwood was the first to suggest that any particle orbiting in the Cassini Division could experience a resonance with one of Saturn's moons. In much the same way Jupiter clears out lanes of the asteroid belt, a process also first described by Kirkwood, particles on resonant orbits with the moons would clear out of certain parts of the rings. Scientists now know that the moon Mimas (260 miles or 415 km) orbits once for every two orbits of a particle near the outer edge of the B ring. Every two orbits, particles at that location experience a tug from Mimas that gradually distorts their paths and sweeps them out of the Division. The 2:1 Mimas resonance is the strongest one operating in Saturn's rings, but there are many others at work, too.

Fig. 6.6 Ringworld. All of the solar system's giant planets have rings, but in grandeur and brightness Saturn's stand alone. This 2008 image from NASA's Cassini spacecraft also shows several moons, including Titan, the largest, at lower left (NASA/JPL/Space Science Institute)

Understanding the true complexity of the Saturn system had to await close-up spacecraft studies – the flybys of Pioneer 11 in 1979, Voyager 1 in 1980, Voyager 2 in 1981, and currently the orbiting Cassini mission, which arrived in 2004 (Figs. 6.6 and 6.7). The trillions of particles in the rings range from powder-sized grains to monolithic blocks, and all are composed of frozen water. Cameras on the Voyager spacecraft resolved each of the larger rings into thousands of tiny ringlets – some kinked or clumpy, others distorted by the presence of embedded moonlets. The images revealed density waves and corrugations rippling throughout the system, and electrically charged dust particles floating above the rings and creating the appearance of spokes. The dazzling close-ups from Voyager cameras forced scientists to push their dynamical theories to new limits. "Over and over, the spacecraft revealed so many unexpected things that it often took days, months and even years to figure them out," Ed Stone said. The journal *Science* summed up the flybys this way: "For sheer intellectual fun there has never been anything like the Voyager missions. Volcanoes on Io, ringlets around Saturn, braided rings – the observations are outrageous."

The "braided" ring is the faint, thin F ring, first seen in images returned by Pioneer 11. The F ring lies about as far from the outer edge of the A ring as New York does from Los Angeles and is at most a few hundred miles wide. Yet Larry Esposito, a planetary scientist at the University of Colorado, Boulder, and a member of the Pioneer, Voyager and Cassini teams, calls the F ring "one of the most dynamic objects in the solar system." The Voyager cameras revealed multiple strands of particles within the ring, and Cassini has shown that its features change in as little as a few hours. Two small potato-shaped moons discovered by Voyager 1,

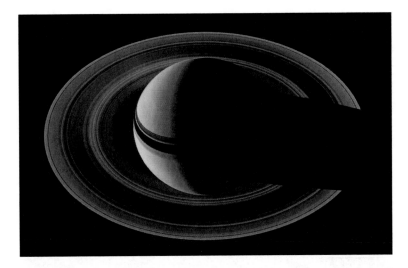

Fig. 6.7 This view of Saturn and its rings from the orbiting Cassini spacecraft provides a perspective unobtainable from Earth. We're hovering 39° above the unilluminated side of the rings and about 700,000 miles (1.1 million km) from the planet. The rings look relatively dark because only sunlight that passes through them is scattered toward Cassini's camera. The rings' shadows play across the planet as Saturn's shadow cuts across the rings (NASA/JPL/Space Science Institute)

named Prometheus and Pandora, orbit just inside and outside the F ring and were thought to "shepherd" the particles between them and keep them in place. But Cassini images show much more dramatic interactions – the moons peel off streamers and create dark wakes and other features (Fig. 6.8). Disturbances caused by the passage of Prometheus trigger the clumping of ring particles into loose aggregations – giant "snowballs" – up to 12 miles (20 km) across, but what the moon builds up, it later destroys in subsequent passes. Cassini images also indicate a moonlet embedded within the F ring and suggest that hundreds of icy chunks less than half a mile across lurk in the ring's core. Esposito and others suggest that there may be an ongoing cycle of creation and destruction in the F ring, with collisions between large icy bodies creating fragments that later reassemble into loose aggregations.

Such boom-and-bust cycles may help explain why Saturn's rings still exist at all. Mutual collisions grind the ring particles to dust, and the dust eventually falls into the atmosphere or is carried away by the planet's rotating magnetic field. Without replenishment, Saturn's glorious ring system is ultimately an evanescent phenomenon, with a life-span of tens of millions of years. The breakup of a wayward moon or comet probably created the rings, but it's very unlikely to have occurred this recently. On the other hand, interplanetary dust falling on the ring particles should gradually darken them over time, yet they remain youthfully

Fig. 6.8 Dark strands drape the inner edge of Saturn's dynamic F ring thanks to the gravitational influence of the moon Prometheus, which is just 63 miles (102 km) across and appears as the upper bright spot in this Cassini image. Another moon, Pandora (52 miles, 84 km) is visible at left. Prometheus orbits closer to Saturn than the F ring particles and shears out material when closest to the ring. Strips pulled out during earlier passes slowly blend back into the ring (NASA/JPL/Space Science Institute)

bright. Right now, planetary scientists can't explain the apparent longevity of the ring system.

Possible solutions include a variation on the recycling processes seen in the F ring, which could keep the particles looking fresh by constantly smashing them to pieces in collisions and reassembling them. It's also possible that scientists have grossly underestimated just how massive the rings are. Most planetary scientists agree that the mass of a Mimas-sized moon is sufficient to account for the main rings. But to avoid darkening by dust over the age of the solar system, the rings would need to possess as much as ten times more mass. Years from now, when the Cassini mission nears its end, planetary scientists hope to fly the spacecraft repeatedly between the rings and the planet in order to definitively weigh Saturn's rings.

Tale of Two Moons

Traveling from the F ring to the inner edge of the E ring involves about the same distance as circumnavigating Earth. The E ring is a vast, tenuous cloud of icy grains that envelops the orbits of the moons Mimas, Enceladus, Tethys, Dione and Rhea – a swath more than four times the width of Saturn itself. This makes it the biggest ring yet found in the solar system. After the Voyager encounters, which returned images of a snow-white and, in some areas, crater-free Enceladus, planetary scientists had little doubt that the 310-mile-wide (500 km) moon was the source of the ring's particles. Just how this might happen was anyone's guess. Everyone knew the culprit, but no one had a smoking gun.

Since 2005, Cassini has dramatically confirmed this idea by imaging and even flying through geysers blasting from the moon's south pole (Fig. 6.9). The total outflow is comparable to that of the Old Faithful geyser in Yellowstone National Park. Cassini has seen at least 30 individual geysers, all of which vent from fissures called "tiger stripes" that run across the south pole. Temperatures within the stripes are about −118°F (−83°C) – not exactly hot, but considerably warmer than the coldest parts of the moon's surface. Explaining how a body small enough to fit within the borders of New Mexico can generate enough heat to power these geysers

Fig. 6.9 Dramatic plumes spray out from locations near the south pole of Saturn's moon Enceladus, where vents blast icy particles, water vapor and organic compounds into space. *Inset*: The surface of Enceladus shows few impact craters but many fractures, folds and ridges – a tell-tale sign of recent geological activity. The moon is just 310 miles (500 km) wide. The prominent feature at top center is a 2,000-ft-deep (600 m) chasm (NASA/JPL/Space Science Institute)

remains a challenge. The same mechanism that warms the insides of Jupiter's Io and Europa – tidal heating – is the favorite. Water, energy and organic compounds are in abundance within the tiger stripes, so scientists can't completely dismiss the possibility that this little moon may even host some form of life. Deep inside the fissures, Enceladus might even possess environments at least as hospitable to life as the subsurface sea of Europa – but that's an investigation that will have to await future missions.

And then there's Titan, slightly larger than Mercury and the solar system's second-largest moon. Receiving only about 1% of the Sun's energy as Earth, Titan is incredibly cold, about −290°F (−180°C). Voyager images disappointingly showed only a big orange ball covered by an impenetrably thick haze, but other instruments revealed that Titan possessed a dense atmosphere – mainly nitrogen plus a small amount of methane – with a surface pressure half again as high as sea-level pressure on Earth. Ultraviolet sunlight converts the methane into complex, reddish organic molecules that form the haze, as well as hydrogen, which is so light it escapes into space. This means that the conversion of atmospheric methane is a one-way affair; if it were not somehow replenished by a source on or in the moon, the methane would be depleted in a few million years. Because methane and its primary photochemical byproduct, ethane, happen to be liquids under Titan's surface conditions, scientists imagined that the moon was awash in a vast sea of the two hydrocarbons. Yet radar and other studies from Earth found no evidence to support the idea.

Once Cassini arrived at Saturn in 2004, it began making repeated Titan flybys, viewing the moon in infrared wavelengths as well as with radar, both of which can penetrate the haze, but there were no surface liquids in sight. In late December, Cassini released its very important passenger, the European Space Agency's Huygens probe. Huygens rushed into Titan's atmosphere on Jan. 14, 2005, deployed its parachute, and drifted down through the atmosphere for two-and-a-half hours, sniffing the air, tasting haze particles, and acquiring hundreds of images. The landscape captured by Huygens looks eerily familiar – and more earthlike than many images of Mars. Photochemical goo drifting down from the haze layers should coat and darken the entire surface, but the icy highlands are bright and clean, as if recently washed by some form of precipitation. Meandering channels wind through these rugged hills and suggestively terminate in dark lowlands (Fig. 6.10). Huygens touched down in what looks like a cobble-strewn plain, and while it spotted no fluids, it measured a methane humidity of 45%. "We now have the key to understanding what shapes Titan's landscape," said the University of Arizona's Martin Tomasko, principal investigator of the Huygens imager. "Geological evidence for precipitation, erosion, mechanical abrasion, and other fluvial activity says that the physical processes shaping Titan are much the same as those shaping Earth."

Yet the fluids were nowhere in sight. The following month, Cassini's radar imager found numerous areas in Titan's northern polar region that looked suspiciously like lakes. Months later, a similar oval feature, named Ontario Lacus for its similarity to the size of Lake Ontario in North America, was identified near the

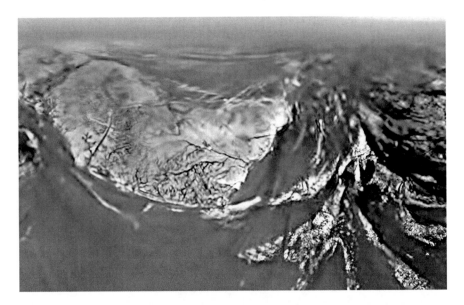

Fig. 6.10 As the Huygens probe descended by parachute through Titan's thick, hazy atmosphere, it captured images of an eerily earthlike landscape. This computer-processed mosaic shows the view from a height of 6 miles (10 km), similar to the cruising altitude of an airliner. The bright highlands, which lie north of the landing site, contain a network of branching channels. Steeply sloped ridges between the channels rise as high as 650 ft (200 m). A rain of liquid ethane and methane scours the channels as it flows downhill toward the dark floodplains (ESA/NASA/JPL/University of Arizona)

south pole. In 2007, Cassini instruments confirmed that Ontario Lacus contained liquid ethane, which meant that the hundreds of similar northern features were also likely lakes. Then, 2 years later, the spacecraft's imager caught a conclusive glint of sunlight reflecting off of the surface of Kraken Mare, a large lake in the north. The total surface area of the northern lakes nearly equals that of France and is 40% larger than the Caspian Sea. Using updated information about Titan's atmosphere provided by instruments aboard Huygens, planetary scientists now can estimate in greater detail what fills these lakes. It's a noxious brew of ethane, propane (cooking fuel), methane, butane (lighter fluid), acetylene (welding gas), with a pinch of highly toxic hydrogen cyanide.

Ethane and methane seem to play the same role on Titan as water does on Earth – raining from storm clouds, collecting in streams that wind through highlands, carving out channels, spilling onto floodplains and pooling into lakes. Each hemisphere seems to have a winter rainy season, which explains why there are now few lakes in the south and hundreds in the north. As Saturn lumbers in its orbit and the long seasons gradually change, Titan's monsoons will shift to the south, and with them will go the "land of lakes."

Dance of the Giants

Each year, Jupiter gains on Saturn by about 18°, so after about two decades it comes full circle and catches up to Saturn again. The two planets were opposite each in the sky during 2010, so throughout the decade they move progressively closer, as shown in Fig. 6.11. In December 2020, the giant planets shine at their closest in evening twilight, but long before then they also fly formation with Mercury, Venus and Mars. The best of these groupings are shown in Figs. 6.12 and 6.13.

A planet's distance from Earth significantly impacts its brightness at opposition. For much of the decade Jupiter dims steadily as it comes to opposition at ever greater distances from us until 2017. Saturn is also moving outward and reaches its maximum distance in 2018, and the planet's slow pace means that it won't match the brightness of its 2003 opposition, which occurred near perihelion, until 2031.

Saturn's rings add an icy twist to its brightness changes. Twice during Saturn's 29-year orbit, during its equinoxes, the rings appear edge-on as seen from Earth,

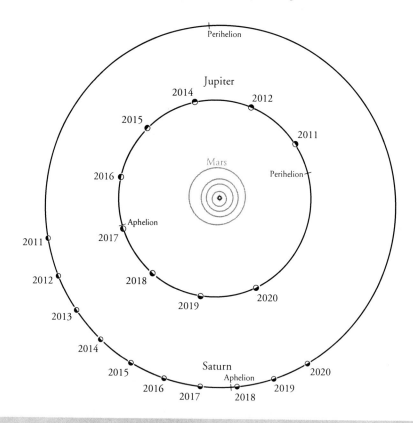

Fig. 6.11 This plot shows the relative positions of Jupiter and Saturn at each of their upcoming oppositions. Jupiter gains on Saturn each year throughout the decade, which culminates in their close conjunction in December 2020

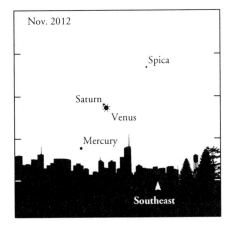

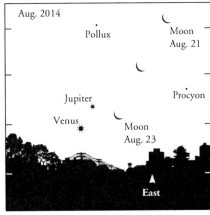

On Nov. 27, 2012, Saturn (0.6) passes bright Venus (–3.9) in the hour before dawn. Look for nearby Spica, the brightest star in Virgo, and then, closer to dawn, try finding Mercury (0.4)

In late August 2014, Jupiter (–1.8) shines with Venus (–3.8) low in the east in the hour before dawn. Jupiter, which is ascending into morning twilight, meets descending Venus on Aug. 18

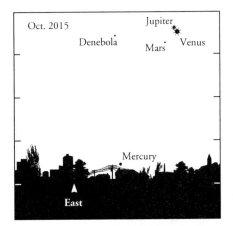

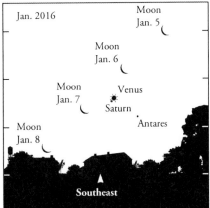

Look for a well-placed planetary mashup in late October 2015, when Venus (–4.5), Jupiter (–1.8) and faint Mars (1.7) rise 3 h before the Sun. Jupiter and Venus are closest on Oct. 26. Try spotting Mercury (–0.9) in the half-hour before dawn

As 2016 opens, Venus (–4.0) takes a morning meeting with Saturn (0.5) above the star Antares. The Moon sweeps past the converging planets before their Jan. 9 closest approach. Look southeast in the hour before sunrise. Scan for Mars (1.2) high in the south and Jupiter (–2.2) shining in the southwest

Fig. 6.12 The best twilight gatherings of Jupiter and Saturn, 2012 through 2016

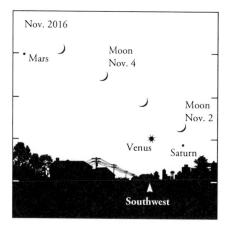

The waxing Moon guides you through the evening planets as November 2016 opens. Look for Saturn (0.5), Venus (−4.0) and Mars (0.4) in the southwest 45 min after sunset. Saturn and Venus lie 3° apart a few days earlier, on Oct. 29

As 2019 opens, descending Venus (−4.4) converges toward ascending Jupiter (−1.8). The planets are closest on Jan. 22. Look southeast in the hours before dawn to find the brilliant pair

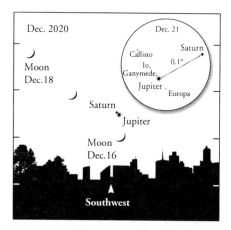

In December 2020, Jupiter (−2.0) drifts ever closer to Saturn (0.6). Look southwest in the hours after sunset. *Inset:* On Dec. 21, when Jupiter passes 6 arcminutes from Saturn, a small telescope will reveal both planets, as well as Jupiter's four largest moons

Fig. 6.13 The best twilight gatherings of Jupiter and Saturn, 2016 through 2020. See also Fig. 3.19 for Jupiter's Jan. 2018 conjunction with Mars

Table 6.2 Jupiter and Saturn on the Web

Active missions
   Cassini (Saturn orbiter)
   saturn.jpl.nasa.gov
   ciclops.org

   New Horizons (Jupiter, Pluto and beyond)
   pluto.jhuapl.edu

   Voyagers 1 and 2 (Jupiter, Saturn and beyond)
   voyager.jpl.nasa.gov

   Juno (Jupiter orbiter)
   www.nasa.gov/juno

Past missions
   Huygens (Titan probe)
   saturn.esa.int

   Galileo (Jupiter orbiter)
   solarsystem.nasa.gov/galileo

   Pioneers 10 and 11 (Jupiter, Saturn and beyond)
   www.nasa.gov/centers/ames/missions/archive/pioneer.html

that is, we pass through the ring plane. This happens because Saturn spins on an axis tilted about 27° to its orbit, which gives us different perspectives on the ring system as the planet makes its way around the Sun. Because the rings are made of icy material, they bounce a lot of sunlight back to us and make a significant contribution to Saturn's overall brightness. When the rings are edge-on, as in 2009, they make no contribution, but for most of the coming decade the rings will be opening up; they reach their maximum angle in 2017. So as Saturn moves to more remote portions of its orbit during the decade, the rings tip toward us and make the planet a little bit brighter than it otherwise would be.

Jupiter and Saturn are the most distant worlds easily seen with the naked eye, so they're the final planets we'll discuss. Their slow march through the heavens gave them a regal air, but it took the invention of the telescope to reveal how kingly these worlds really are. The resources listed in Table 6.2 provide a starting place for deeper exploration of Jupiter and Saturn. Surrounded by dozens of satellites, some of which are truly planet-sized, we can think of the two giants as the centers of their own miniature solar systems. Even as astronomers catalog new planets around distant stars, we have only begun to explore the surprising and fascinating analogs to them right in our back yard: the biggest moons of the biggest planets.

References

Drake S (1957) Discoveries and opinions of Galileo. Doubleday and Co., Inc., Garden City, New York

Boslough MBE, Crawford DA (2008) Low-altitude airbursts and the impact threat. Int. J. Impact Eng. 35: 1441–1448. doi: 10.1016/j.ijimpeng.2008.07.053

Burrows WE (1990) Exploring space: Voyages in the solar system and beyond. Random House, Inc., New York

Chapman A (2009) A new perceived reality: Thomas Harriot's Moon maps. Astron. Geophys. 50: 1.27–1.33. doi: 10.1111/j.1468-4004.2009.50127.x

Chodas PW, Yeomans DK (1996) The orbital motion and impact circumstances of Comet Shoemaker-Levy 9. In: Noll KS, Weaver HA, Feldman PD (eds.): The collision of Comet Shoemaker-Levy 9 and Jupiter. Cambridge University Press, New York

Cook JR (2010) Saturn then and now: 30 years since Voyager. visit. http://www.nasa.gov/mission_pages/voyager/voyager20101111.html. Accessed 11 Nov. 2010

Cooper HSF Jr (1983) Imaging Saturn: The Voyager flights to Saturn. Holt, Rinehart and Winston, New York

Cordier D, Mousis O, Lunine JI et al (2009) An estimate of the chemical composition of Titan's lakes. Astrophys. J. 707:L128–L131

Crawford DA (1997) Comet Shoemaker-Levy 9 fragment size estimates: How big was the parent body? Ann. NY Acad. Sci. 822: 155–173. doi: 10.1111/j.1749-6632.1997.tb48340.x

Cuzzi JW (2010) An evolving view of Saturn's dynamic rings. Sci. 327, 1470–1475. doi: 10.1126/science.1179118

Esposito LW (1993) Understanding planetary rings. Annu. Rev. Earth Planet Sci. 21: 487–523

Falorni M (1987) The discovery of the Great Red Spot. J. Brit. Astron. Assoc. 97: 215–219

Greenberg R (2010) The icy Jovian satellites after the Galileo mission. Rep. Prog. Phys. 73: 1–20. doi:10.1088/0034-4885/73/3/036801

Hammel H (1999) How astronomers use The Astronomical Almanac. In: Fiala AD, Dick SJ (eds.): Proceedings of the Nautical Almanac Office Sesquicentennial Symposium. U.S. Naval Observatory, Washington

Hammel H, Wong MH, Clarke JT et al (2010) Jupiter after the 2009 impact: Hubble Space Telescope imaging of the impact-generated debris and its temporal evolution. Astrophys. J. Lett. 715: L150. doi: 10.1088/2041-8205/715/2/L150

Kargel JS (2006) Enceladus: Cosmic gymnast, volatile miniworld. Sci. 311:1389–1391

Levy DH (1998) The collision of Comet Shoemaker-Levy 9 with Jupiter. Space Sci. Rev. 85: 523–545

Lovett L, Horvath J, Cuzzi J (2006) Saturn: A new view. Abrams, New York

Marchis F (2009) Urgent: Possible impact on the southern pole of Jupiter? http://www.cosmicdiary.org/blogs/nasa/franck_marchis/?p=391. Accessed 15 Oct. 2010

Morrison D, Samz J (1980) Voyage to Jupiter. NASA SP-439. NASA, Washington, D.C

Murray B (1989) Journey into space: The first three decades of space exploration. W. W. Norton and Company, Inc., New York

O'Connor JJ, Robertson EF (1997) Christiaan Huygens. http://www-history.mcs.st-andrews.ac.uk/history/Biographies/Huygens.html. Accessed 15 Oct. 2010

O'Connor JJ, Robertson EF (2003) Giovanni Domenico Cassini. http://www-history.mcs.st-andrews.ac.uk/history/Biographies/Huygens.html. Accessed 15 Oct. 2010

Peale SJ, Cassen P, Reynolds RT (1979) Melting of Io by tidal dissipation. Sci. 203: 892–894

Schiele L, Freidel D (1990) Forest of kings: The untold story of the ancient Maya. William Morrow and Company, Inc., New York

Van Helden A (1995) Hans Lipperhey. The Galileo Project. http://galileo.rice.edu/sci/lipperhey.html. Accessed 15 Oct. 2010

Van Helden A (1995) Satellites of Jupiter. The Galileo Project. http://galileo.rice.edu/sci/observations/jupiter_satellites.html. Accessed 15 Oct. 2010

Van Helden A (2006) Huygens' ring, Cassini's Division and Saturn's children. Smithsonian Institution Libraries, Washington, D.C

Wesley A (2009) Impact mark on Jupiter, 19th July 2009. http://jupiter.samba.org Accessed 15 Oct. 2010

Whitehouse D (2009) Renaissance genius: Galileo Galilei and his legacy to modern science. Sterling Publishing, New York

Chapter 7

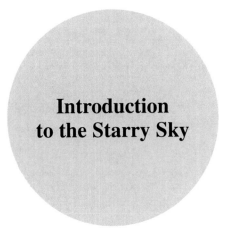

Introduction to the Starry Sky

A little infinity is good for the soul. Like the vastness of the sea or some ancient natural landscape, a dark sky filled with myriad stars has the capacity to touch us on the deepest level. Every person ever born has lived and died on the surface of a modest planet orbiting a modest middle-aged star, one of perhaps 400 million in a modest spiral galaxy that is itself one of innumerable galaxies in the vast cosmos. Few other experiences communicate the enormity of space and time than that of gazing under a starlit sky. "When we are chafed and fretted by small cares," wrote astronomer Maria Mitchell in 1866, "a look at the stars will show us the littleness of our own interests."

Until now, we've mostly focused on the sky's seven moving lights – the Sun, the Moon, and the five naked-eye planets – with only passing reference to the starry backdrop against which we view their motions. Now the stars and constellations take center stage. This overview provides a working knowledge of the brightest stars visible to the unaided eye, their most recognizable patterns, and their cyclic appearances throughout the year. The first glimpse of a star-studded sky can be a little disconcerting; with so many stars, finding even the brightest and most famous constellations can be frustrating. For beginners, there's actually a positive benefit to the light-fogged skies of the suburbs and small cities. By washing out the faintest stars, scattered light makes it easier to identify the main patterns. For best results, view from areas shielded from glaring sources like streetlights or billboards that shine light directly into the eye. Parks often offer enough relief from glaring lights to provide a nice view of at least some portion of the starry sky, especially if they're located near a large lake.

Because the human eye takes a few minutes to adapt to darkness, observing time is another important factor in successfully exploring the starry sky. The human eye contains two types of light-sensitive cells: cones and rods. Cones operate best under

F. Reddy, *Celestial Delights: The Best Astronomical Events through 2020*,
Patrick Moore's Practical Astronomy Series, DOI 10.1007/978-1-4614-0610-5_7,
© Springer Science+Business Media, LLC 2012

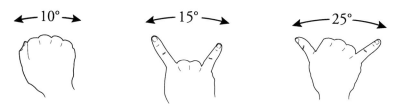

Fig. 7.1 A fist held at arm's length spans the bowl of the Big Dipper, about 10° across. The "rock on" sign extends across the Great Square of Pegasus (15°), and "hang loose" (25°) almost covers the Dipper's full length

bright light and provide color vision, while rods excel at detecting faint light but can't render color. The eye switches from one system to the other as needed, but there's a noticeable lag when moving into poorly illuminated environments. This occurs because bright light bleaches out the key purple-red photosensitive protein stored inside rod cells. When moving into darkness, the pupil opens to admit more light, the eye flips from cone mode to rod mode, and the photosensitive protein slowly regains its color, giving rods their full night-vision sensitivity in about 30 min. Because any exposure to light bleaches some of the protein, cities and suburbs aren't dark enough for this process to complete. Dim red light actually has the least impact on night vision, which is why red-filtered flashlights are a token of stargazing. For the purpose of identifying basic constellations during, say, an evening stroll or a run to take out the trash, 5 or 10 min away from glaring lights should be sufficient. But the longer you look the more you'll see.

Another thing to keep in mind: Think big, because constellations sprawl across the sky. The Big Dipper, which is the most noticeable part of the constellation Ursa Major (the Greater Bear), stretches nearly 26° from end to end, and the entire figure, measured from the final star of the Dipper's handle to the stellar tip of Ursa's nose, is 44° across. Diagrams in this chapter show angular distances between prominent stars, and you can use your hand to estimate them in the sky. As shown above, a fist held at arm's length corresponds to a sky angle of about 10°, which is about the width of the Dipper's bowl. Useful guides for estimating larger angles comes in the form of two well-known hand gestures: "rock on" and "hang loose" (Fig. 7.1).

The constellations represent humanity's first efforts to organize the sky. Most of these groupings, which form "star pictures" memorializing figures and objects from Greek myth, have been known since antiquity. In the modern sense, the constellations are 88 formally delineated regions of the sky that enclose these traditional star patterns. Yet even in the first decades of the twentieth century, there was no general agreement where one constellation ended and another began. The fuzziness of these boundaries became increasingly troublesome as the pace of new discoveries increased, particularly for astronomers concerned with variable stars – stars that changed brightness, sometimes dramatically – because the official variable names included the constellation. In fact, prior to 1922, there wasn't even a definitive list of constellations recognized by all astronomers. That year, the newly formed International Astronomical Union (IAU) officially adopted 88 constellations and standardized their names. The IAU resolved the boundary issues a few years later, in 1930, by establishing borders finessed so that all existing variable stars remained in the constellations for which they were named.

But if Ursa Major is a constellation and it contains the Big Dipper, then what's the Big Dipper? It's an "asterism," an easily recognized star pattern that is well known but has no official standing. Other examples: the Summer Triangle, which includes stars from three constellations; the Sickle in Leo; the Great Square, part of the constellation Pegasus; the Teapot of Sagittarius; the Pleiades star cluster in Taurus; and the Heavenly G, which is built from stars in six winter constellations.

Of the 88 constellations, more than half were described by Claudius Ptolemy in his *Almagest*, which was written around A.D. 150 and reflected established Greek celestial traditions dating back centuries. These are described in a popular poem, *The Phaenomena*, composed by Aratus around 275 B.C., that passes down constellations described in a lost work of the same name written by Eudoxus (366 B.C). The constellations not supplied by the Greeks were added by European astronomers in the 16th and 17th centuries, sometimes to fill in gaps between already established figures but more often to create new ones in the far southern sky, which had not been visible from Greece. Some of these new additions are, in the words of astronomer Ken Croswell, "so dim and pointless that astronomers and casual stargazers alike have complained about them ever since." And no wonder. Despite classical-sounding Latinate names, creations like Antlia (Air Pump), Caelum (Sculptor's Chisel) and Norma (Carpenter's Square) make parts of the southern sky seem about as romantic and inspirational as a tool shed.

The Greek constellations and their associated mythology were fully formed when Eudoxus compiled it in the fourth-century B.C. But there's no evidence for such an elaborate system when Homer composed *The Iliad* and Hesiod penned his farmer's almanac four centuries earlier. Both works mention a pair of constellations, (Orion and the Great Bear), a pair of star clusters (the Pleiades and the Hyades) and a pair of individual stars (Sirius, Arcturus), and in *The Odyssey* Homer additionally mentions Boötes the Herdsman. There's no mention of other constellations in any surviving Greek sources dated before the fifth century. It's possible that Homer and Hesiod recorded the only constellations known by the Greeks in their time. If so, then where did the others come from?

Bradley Schaefer, an astronomer at Louisiana State University, has examined this problem, with some notable results. By studying both the southern void, where no Greek "star pictures" exist, and the writings of Aratus, Schaefer found celestial clues that can be used to pin down both the geographic latitude and the origin time for the constellations. The void reflects the southernmost portion of the sky visible from the latitude of origin, and this limit is associated with a particular time because of the wobbly motion (precession) of Earth's axis. Astronomical clues in Aratus, which appear to have originated from Eudoxus, likewise put similar constraints on time and place. Schaefer's results give a position within about 62 miles (100 km) of 36° north latitude and within about a century of 1130 B.C. This puts the Greek mainland too far north but provides a good match for Qal'at Sherqat in modern Iraq. The site is better known as Ashur, the capital of the Assyrian empire from the fourteenth to the ninth century B.C.

The earliest account of constellations in this region comes from a frequently copied document known by its first words: *Mul.Apin*, "The Plow." This trio of cuneiform tablets contains lists of information such as the rising and setting of Mesopotamian star patterns, many of which have direct correspondence to Greek star groups, for example, "The True Shepherd of Anu" is Orion, "The Great Twins" is Gemini, and the "Bull of Heaven" is Taurus. Others, such as our Aries ("The Hired Man") and Triangulum ("The Plow") use the same stars but with different names. The oldest surviving copy of *Mul.Apin* dates from 687 B.C., but the observations go back much farther. After analyzing the *Mul.Apin* statements much as he did the Greek ones, Schaefer found that they describe the sky as seen within a century of 1370 B.C. at a location within 81 miles (130 km) of 35° north latitude. Given the overlaps in content, place and time, he concludes that it seems likely a significant part of our star lore originated with Assyrian priests and that much of the transfer to Greece occurred after 500 B.C.

A completely different tradition of star groupings emerged in ancient China. About a millennium after Babylonian scribes made the oldest extant copy of *Mul.Apin*, an unidentified Chinese astronomer was creating what would become the world's oldest known set of star charts. Carefully plotted on a long scroll of mulberry paper, the Dunhuang Star Atlas offers a unique snapshot of the sky as it was organized during the T'ang dynasty. Dunhuang is located in western China's Gansu province. By the fourth century it was a bustling oasis town in the Gobi Desert, a strategic frontier watering hole along the Silk Road, one of the trade routes that linked China with India and points west. It was at about this time that Buddhist monks began transforming hundreds of nearby natural grottoes into the beautifully decorated temple complex known as the "Caves of the Thousand Buddhas." Sometime around 1036, as invaders threatened the region, someone sealed up a cave stocked with textiles, paintings and about 45,000 manuscripts – including both the star atlas and *The Diamond Sutra*, the world's oldest complete printed book. The entrance was then camouflaged with a painting, and the makeshift vault was forgotten.

Five hundred years later, when the Silk Road was abandoned by the Ming dynasty, oasis towns like Dunhuang fell into decline, and with few pilgrims traveling the route, so did the caves. In the 1890s, a Daoist monk named Wang Yuanlu arrived at the caves, appointed himself their caretaker and began restoration efforts. In 1900, workmen stumbled onto the secret entrance of the "library cave." Yuanlu was quick to appreciate the importance of the discovery and hoped that it would enable him to generate government interest in preserving the cave's contents, but it didn't work out that way. When the British archaeologist Aurel Stein visited the area in 1907, he negotiated a small reimbursement with Yuanlu in return for removing many of the items. Others followed and made similar deals, enriching institutional and private collections around the globe while they depleted the grottoes and even damaged some of them. In any case, that's how the Dunhuang Star Atlas traveled through time and space to arrive at its present location, the British Library in London.

According to a 2009 study, the star atlas dates to the seventh century. Its maps plot the locations of more than 1,300 stars grouped in 257 small constellations, most associated with objects or persons of the empire. All stars are shown as the same

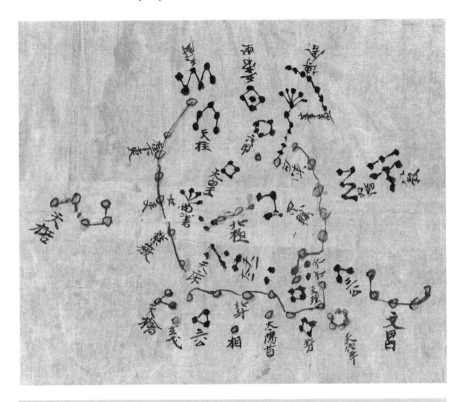

Fig. 7.2 This ancient map of the northern sky is part of the Dunhuang Star Atlas, the oldest complete star atlas known, dated to the late seventh century. Chinese constellations are much smaller, more numerous, and incorporate fainter stars than their Western counterparts. This map shows the sky around the north celestial pole. Star positions in the hand-drawn atlas were found to be accurate to within a few degrees. This map of the north polar region shows a recognizable Big Dipper (bottom), a part of the modern constellation Ursa Major (British Library)

size, regardless of brightness, a curiosity of Chinese star maps that has never been explained; evidently, the patterns were paramount. While the maps are clearly plotted from stellar observations, the document is primarily astrological in nature, perhaps a field guide for interpreting what various celestial events meant to China. The atlas is designed around Jupiter's long march through the stars, which is rounded to a 12-year cycle by Chinese tradition. The sky is divided into a dozen Jupiter stations, each of which is mapped in the atlas and associated with a state within the empire whose fortune is supposedly affected by Jupiter's presence. A thirteenth chart shows the stars associated with the area around the north celestial pole, centered on the Celestial Emperor and his Purple Palace and surrounded by constellations representing his family, servants, military officers and their associated housing. Although most modern star patterns are difficult to quickly discern, the Big Dipper easily stands out (Fig. 7.2).

Personal Stars

Even people who have never before observed under the starry sky probably know at least a single constellation by name – the one associated with their astrological sign of the zodiac. These signs actually correspond to real constellations located along the ecliptic. If glimpsing your "personal stars" provides the motivation to get out under a dark sky, that's all the better as far as we're concerned.

The word zodiac stems from the Greek *zodiakos kuklos*, "the circle of animals," a kind of celestial bestiary containing a variety of persons and creatures. Some are mythological, like Gemini the Twins, others are fanciful, like Capricornus, a half-fish, half-goat. The patterns were well established in Greece by the fourth century B.C. and were later adopted by the Roman world (Fig. 7.3).

Unfortunately, most of the zodiacal constellations are rather subtle, and a few are downright invisible to suburban skywatchers. But because the zodiac stars follow the path of the ecliptic, the planets Mars, Jupiter and Saturn can serve as bright guides to your personal constellation. Table 7.1 lists the locations of the three planets in the zodiac through 2020. Even from very dark sites, don't expect to see a recognizable shape in the stars. This is connect-the-dots writ large, and most constellations, especially those of the zodiac, require an almost hallucinogenic imagination to see anything resembling their namesakes.

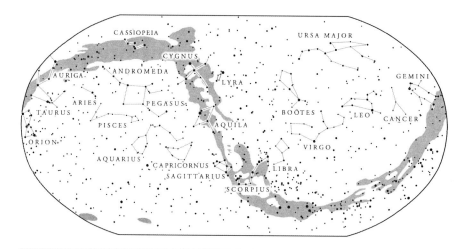

Fig. 7.3 The signs of the zodiac originated from twelve constellations that lie along the ecliptic, the Sun's apparent annual path through the stars. The arc of constellations from Taurus and Aries (*left*) through Sagittarius and Scorpius (*bottom center*) to Cancer and Gemini (*right*) comprises the zodiac stars; additional prominent constellations are also labeled. This map shows the entire celestial sphere, but stars much farther south than the tail of Scorpius cannot be seen from mid-U.S. latitudes, such as Denver, St. Louis, or Washington, D.C. The shaded band is the Milky Way, the faint glow of innumerable stars that marks the plane of our galaxy

Table 7.1 Planets in the zodiac through 2020. Near opposition, the planets Mars, Jupiter and Saturn guide you to every zodiacal constellation but Aquarius. However, there's a bonus — Ophiuchus, whose modern boundaries intercept the ecliptic

Constellation	Mars	Jupiter	Saturn
Aries	–	2011	–
Taurus	–	2012	–
Gemini	–	2014	–
Cancer	–	2015	–
Leo	2012	2016	–
Virgo	2014	2017	2011–2012
Libra	–	2018	2013–2015
Scorpius	2016	–	–
Ophiuchus	–	2019	2016–2017
Sagittarius	–	2020	2018–2020
Capricornus	2018	–	–
Aquarius	–	–	–
Pisces	2020	–	–

One oddity of the modern sky is that the ecliptic makes a short run through the constellation Ophiuchus (OFF-ee-YOO-kus) the Serpent Bearer, which was never in anyone's zodiac. This reflects the general fuzziness of constellation boundaries prior to their formal definition by the IAU, and it was further enhanced by the rule that all cataloged variable stars known at the time remain in the constellations for which they were named.

Now, turn to the section in this chapter that matches the current season and start getting acquainted with the rest of the starry sky. Each section is accompanied by a finder chart that will help guide you to the most obvious constellations. At the end of the chapter you'll find bimonthly charts covering the entire sky visible from middle-northern latitudes.

Winter's Gems

We typically spend a lot less time outdoors during the cold winter months, but the early nights often attract skyward gazes from many who otherwise might not give stars a passing thought. Even as we rush from warm car to warm home, we're often compelled to pause and admire the brilliant stars that glitter overhead. The stars seem so much brighter on a crisp winter evening – and in fact, they really are. Table 7.2 lists the brightest stars, nearly half of which are located in the winter sky. Two of these stars lie in a single constellation, Orion (oh-RYE-un) the Hunter, so it will be our guide; refer to Fig. 7.4 as you investigate the crystalline winter sky.

Table 7.2 The brightest stars. This listing ranks stars by their visual magnitude, which includes companion stars too close to see without a telescope. The constellations Carina, Centaurus, Crux and Eridanus aren't labeled on the star charts; they're either too dim for urban skies or not completely visible from middle-northern latitudes

Name	Distance (light years)	Magnitude	Constellation
Sun	0.000016	−26.74	–
Sirius	8.58	−1.46	Canis Major
Canopus	309	−0.72	Carina
Rigel Kentaurus (Alpha Centauri A & B)	4.36	−0.29	Centaurus
Arcturus	37	−0.04	Boötes
Vega	25	0.03	Lyra
Capella	43	0.08	Auriga
Rigel	860	0.12	Orion
Procyon	11.43	0.37	Canis Minor
Hadar	392	0.61	Centaurus
Achernar	140	0.46	Eridanus
Alpha Crucis	325	0.52	Crux
Betelgeuse	642	0.70	Orion
Altair	16.7	0.77	Aquila
Aldebaran	67	0.85	Taurus
Antares	550	0.96	Scorpius
Spica	250	1.04	Virgo
Pollux	34	1.14	Gemini
Fomalhaut	25	1.16	Piscis Austrinus
Mimosa	280	1.25	Crux
Deneb	1,425	1.25	Cygnus
Regulus	79	1.35	Leo

On any clear night at the beginning of the year, we can step outside at about 10:30 P.M. local time, face south, and see a swarm of bright stars. Scan the sky from horizon to zenith (directly overhead) and a distinctive row of three stars becomes readily apparent. This is the belt of Orion. To the upper left of the belt is a decidedly reddish star – Betelgeuse (BET-el-joos) – and to the lower right is the bright, bluish star Rigel (RI-jel). Because the Hunter is facing you, these stars represent his right shoulder and left knee, respectively. Scan to the right of Betelgeuse to find Bellatrix (bel-LA-triks), and to the left of Rigel to find Saiph (safe), about where you'd expect the other shoulder and knee. To the west of Bellatrix is a dim string of stars that may be faintly visible depending on your observing circumstances. This is the giant's shield, raised in some renditions to fend off the rush of nearby Taurus, the celestial bull.

A trio of stars dangles from Orion's belt, making up his sword. You may notice an odd misty glow from the middle star of the sword, and with a pair of binoculars,

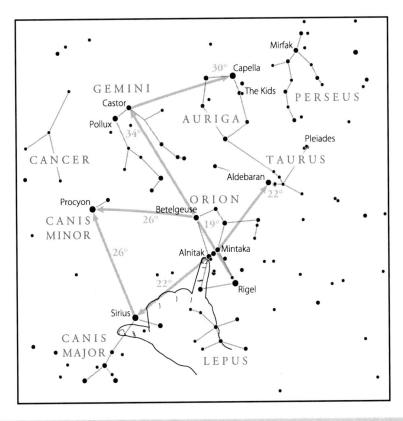

Fig. 7.4 Orion will guide you to several of the bright winter constellations. Highlights of the season include Sirius, the brightest star in the sky, and the compact Pleiades star cluster

you'll see that this is no optical illusion – there really is something strange going on there. In fact, that middle star is not a star at all, but the famous Orion Nebula, a vast cloud of dust and gas that has formed thousands of stars (Fig. 7.5). The hottest of these young stars produce intense but invisible ultraviolet light. The gas absorbs this energy and reradiates it at visible wavelengths in a process called fluorescence, which creates the misty glow we see. The Orion Nebula lies about 1,350 light-years distant – that is, the light we see has taken 1,350 years to reach us – and is about 25 light-years across. It's by far the closest star-forming region. As big as it is, the nebula is really only the brightest portion of a far larger cloud complex that encompasses more than half of the constellation.

One reason the winter stars are so bright is that many are young, and any collection of young stars has a disproportionate number of hot, bright members. For stars, mass is destiny. Massive stars burn hottest and brightest, but run through their nuclear fuel the fastest. A star like Rigel may shine for a mere 10 million years, whereas our Sun has shone for nearly 5 billion years, which is about half its expected

Fig. 7.5 The second "star" in Orion's sword is the nearest stellar nursery and deserves a look through binoculars or a telescope. Its hottest young stars produce ultraviolet radiation that sets the gas aglow (NASA/ESA/M. Robberto (STScI) and the Hubble Space Telescope Orion Treasury Project Team)

lifetime. In other words, Rigel didn't exist when dinosaurs walked the Earth. Astronomers observing the Orion Nebula with the Hubble Space Telescope have found young stars enveloped in protoplanetary disks of gas and dust, which astronomers have nicknamed "proplyds." Many of these disks seem to be evaporating under the intense ultraviolet radiation and strong outflows (called stellar winds) from the nebula's hottest stars.

Turn now to Taurus (TOR-us) the Bull, a constellation whose stars might be classed as adolescents. Taurus is a zodiacal constellation whose main stars are easy to find. Follow Orion's belt upward and to the right (northwest) to the bright orange star Aldebaran (al-DE-bar-an). Aldebaran marks one tip of a V-shaped cluster of stars known as the Hyades (HI-uh-deez) that mark the face of the Bull; his horns

Fig. 7.6 Sapphire sisters. The distinctive Pleiades is the sky's brightest and most famous star cluster, a winter gem that is easily visible even from light-fogged cities. We view the cluster through a scrim of intervening dust that shows up as a bluish haze in long-exposure images. (Antonio Fernandez Sánchez)

stretch 17° to the northeast. The most noticeable stars in Taurus, however, are the Pleiades (PLEE-uh-deez), a compact star cluster about 14° from Aldebaran. The Pleiades are known as the Seven Sisters in Greek mythology, although few people can spot more than six stars with the naked eye. A pair of binoculars will reveal more than a dozen or so, and the cluster totals more than 500 (see Fig. 7.6).

The Pleiades, the mythical offspring of Atlas and Pleione, and the Hyades, the daughters of Atlas and Aethra, are scientific as well as mythological relatives. Both are classified as open clusters, which are physically related groups of stars loosely bound by their mutual gravitational attraction. The Hyades, at a distance of about 150 light-years, is one of the closest, composed of stars born some 625 million years ago. Farther away, at about 440 light-years, is the Pleiades cluster, whose stars are about 115 million years old.

The Pleiades is such a distinctive pattern that its worldwide prominence in folk-lore should come as no surprise. German researcher Michael Rappenglück makes an interesting case that specific dot patterns on the walls of Paleolithic caves in France, dating back some 17,000 years, are intended as representations of the Pleiades. At the opposite extreme, the Pleiades is Subaru in Japanese folklore – a name now shared by an automobile company with a stellar logo.

The Navajo and Blackfeet of North America used the cluster as the basis of a stellar calendar. Crops needed to be planted between the time the Pleiades vanished in evening twilight (late April) and when they reappeared before the Sun (mid-June); the cluster's arrival in the evening sky (late September) warned that the first frost was near. The principal Navajo deity, Black God, even used the cluster to adorn his forehead.

Among tribes in the Amazon valley, the cluster's appearance marked the start of the rainy season and the migration of birds. The Guaranis of Paraguay begin their year when the Pleiades make their first appearance in the predawn sky. Colonial sources recorded many Inca names for the cluster – Collca, Larilla, Oncoy – and in one case is described as being the mother of all stars. Farmers in drought-prone areas of Andean South America still monitor the Pleiades for changes in apparent brightness around the time of their winter solstice. The goal is to try to predict summer rainfall amounts and accordingly adjust the planting dates for potatoes, their most important crop. In a 2000 study, scientists found that poor visibility of the Pleiades in June was caused by increases in light cirrus cloud cover that signaled an active year for the Pacific ocean-warming phenomenon known as El Niño. This, in turn, means reduced rainfall during the growing season several months later, so the farmers delay planting by up to 6 weeks. Similarly, many aboriginal peoples of Australia associated a rainy period with the cluster's appearance, and cursed it if rain failed to follow.

The Aztecs of Mesoamerica believed that the Pleiades would herald the end of the world and began their most important ceremony on a date when the cluster – known to them as Tianquiztli (Marketplace) – crossed overhead at midnight. At the end of their 52-year calendrical cycle, when the dates of their secular and sacred calendars coincided, priests watched anxiously as the cluster approached the zenith. If the Pleiades stopped moving as it passed overhead, then the world would come to an end; if not, it would continue on through another 52-year cycle.

Before turning from Taurus, look for the star that marks the tip of the Bull's lower horn, the one nearest Orion. About the width of a finger tip at arm's length (1°) to the upper right (northwest) of this star lies the remnant of a star that, nearly 960 years ago, shone brightly enough to be seen in the daytime. Although it's now invisible to the naked eye, this debris is the Crab Nebula, an expanding "star wreck" that marks the site of a spectacular explosion known as the supernova of 1054. Such explosions end the brief lives of massive stars when they run out of fuel and collapse under their own weight, a phenomenon discussed in greater detail in Chap. 9.

The next stop in the winter sky is a pentagon of stars above the horns of Taurus – Auriga (aw-RYE-guh) the Charioteer. The constellation's most obvious member is the bright star Capella. To the ancient Greeks, Capella represented Amalthea, a she-goat that suckled the exiled Zeus. Knowing this is about the only way to make sense of the name given to the triangular asterism just south of Capella – The Kids. For the beginning observer, Auriga isn't a terribly interesting constellation. A small telescope reveals a number of star clusters like the Pleiades and Hyades, though much more distant and hence smaller and fainter.

Returning now to Orion, we'll trace the line of his belt away from Taurus to the brightest nighttime star visible from Earth. This is Sirius, a star weighing in with twice the Sun's mass but whose brightness stems largely from its proximity to

Earth, a relatively nearby 8.6 light-years. Sirius is the Dog Star, the brightest in Canis Major, the Greater Dog. This star's appearance in the east just before sunrise in early August was thought to trigger the onset of the hottest days of the year, often referred to as the "dog days." To the ancient Egyptians, the first appearance of Sirius in the morning sky was of great importance, for it signaled the start of the vital Nile flood, a coincidence that led them to the discovery of a 365-day year by 2,800 B.C. The star was associated with the goddess Isis, and Orion's belt, which points toward her, was associated with Osiris. Canis Major is one of the few constellations whose rough outline actually suggests the shape of its namesake, and most of its stars are visible from suburban skies; its head, however, is composed of faint fourth-magnitude stars.

Box 7.1 What's in a star name?

A star by any other name would shine as bright, but how stars and other astronomical objects receive their names is a little-appreciated aspect of astronomy. A new object might be named for the person who discovered it or who caught it doing something unusual. Examples include Hubble's Variable Nebula, which Edwin Hubble found brightening and fading thanks to shadows cast by moving dust clouds, or Barnard's Star, a nearby red dwarf E. E. Barnard discovered that boasts the fastest motion across the sky – about half the Moon's apparent width in a century. Most often, though, sky objects receive names and numbers as a result of detections in various types of surveys.

Consider Sirius, the brightest star in the sky and known since antiquity. It became Alpha Canis Majoris in the Greek-letter-plus-Latin-possessive mashup introduced by Johann Bayer in his 1603 star atlas, and it became 9 Canis Majoris in John Flamsteed's 1725 work. More prosaic appellations followed: BD −16 159 in the 19th-century's *Bonner Dürchmusterung* list, HD 48915 in the 1920s-era *Henry Draper Catalog*, and SAO 151881 in the Smithsonian Astrophysical Observatory catalog of the 1960s. The star's detection by NASA's Einstein X-ray satellite provided the sobriquet 1E 064255–1639.4, while the Two Micron All Sky Survey contributed 2MASS J06450887–1642566; in these cases, the numbers correspond to the star's position in astronomical coordinates. There's also HIP 32349, GCRV 4392, ADS 5423 A, CCDM J06451–1643A, USNO 816 and many more. But it'll always be Sirius to us.

The imposing task of keeping all these designations straight is the job of the Astronomical Data Center at Strasbourg Observatory in France. Computers there host SIMBAD, which is *the* reference for the identification of astronomical objects. By 2011 SIMBAD's amassed information included nearly five million objects with more than 14 million identifiers. From there, the database will only grow, as a new generation of sensitive surveys begins scanning the skies.

Animals seem to come in pairs in the heavens, and the Greater Dog has a lesser companion to the northeast. Trace a line left through the shoulders of Orion until you come to the next bright star, Procyon (PRO-see-on). This star is Canis Minor's main claim to fame, though beginning stargazers may also appreciate the constellation's simplicity: Only one other star is required to complete its standard figure. Canis Minor and Canis Major are often regarded as the hounds that accompanied Orion on his hunts. Both Procyon and Sirius make it into lists of the brightest as well as the nearest stars (see Tables 7.2 and 7.3).

Our final winter stop is another constellation of the zodiac – Gemini the Twins. The bright stars that mark the heads of the Twins can be found along a line extended from Rigel through Betelgeuse. The brighter of the two stars is Pollux, and his twin Castor lies closer to the north celestial pole. Although the stars have been part of a Twins constellation for millennia, Castor is distinctly the fainter of the two. Gemini is about equal to Orion in length and extends toward the Hunter from the "twin" stars. Castor and Pollux would stand out in any other season, but they pale before the competing winter stars.

Table 7.3 The nearest stars. Few of the brightest stars are nearby, and few of the nearest stars are visible at all. Most of the galaxy's stars are faint red dwarfs like Proxima Centauri, which outnumber stars like our Sun by nearly ten to one

Name	Distance (light-years)	Magnitude	Constellation
Sun	0.000016	−26.74	–
Proxima Centauri (Alpha Centauri C)	4.24	11.05	Centaurus
Rigel Kentaurus (Alpha Centauri A & B)	4.36	−0.29	Centaurus
Barnard's star	5.98	9.57	Ophiuchus
Wolf 359	7.78	13.53	Leo
Lalande 21185	8.29	7.47	Ursa Major
Sirius	8.58	−1.46	Canis Major
Luyten 726–8 A & B (BL and UV Ceti)	8.72	12.06	Cetus
Ross 154	9.67	10.44	Sagittarius
Ross 248	10.30	12.29	Andromeda
Epsilon Eridani	10.50	3.73	Eridanus
Lacaille 9352	10.69	7.34	Piscis Austrinus
Ross 128	10.93	11.16	Virgo
EZ Aquarii A, B & C	11.26	12.30	Aquarius
61 Cygni A & B	11.4	4.66	Cygnus
Procyon A & B	11.43	0.37	Canis Minor
GJ 725 A & B	11.49	7.86	Draco
GJ 15 A & B	11.65	8.01	Andromeda
Epsilon Indi A, B & C	11.81	4.68	Indus
DX Cancri	11.82	14.90	Cancer
Tau Ceti	11.90	3.49	Cetus

The winter sky also boasts a giant asterism composed of the brightest stars of the constellations we've discussed. It's called the Heavenly G, and it runs from Capella to Castor and Pollux, then to Procyon and Sirius, Rigel, Aldebaran and, for the jog that completes the letter's horizontal, Betelgeuse.

Before departing the winter sky, there's an aspect of it often goes unnoticed but becomes fairly obvious when pointed out: the colors of the winter stars. These stars are just bright enough to stimulate the color receptors in the eyes. They stand out particularly well when comparing stars of contrasting colors, such as ruddy Betelgeuse and bluish Rigel in Orion. Aldebaran in Taurus is another colorful star, running perhaps a bit more toward the orange than Betelgeuse.

Colors are an important clue to the physical nature of stars because they indicate surface temperatures. Contrary to everyday experience, the bluest stars are the hottest. Rigel's surface temperature runs about 20,200°F (11,200°C) or more than twice as hot as the surface of the Sun. On the other hand, the surface of Betelgeuse runs about 6,100°F (3,400°C), or two-thirds as hot as the Sun's. The white and blue-white stars are in their prime, but redder stars like Betelgeuse and Aldebaran are nearing the ends of theirs. They've reached the so-called red giant phase, a period marked by a dramatic expansion that spreads the star's energy over a larger surface area while reducing its surface temperature. More massive than our own Sun, Betelgeuse actually qualifies as a red supergiant. If placed at the Sun's position in the solar system, Betelgeuse would engulf all of the inner planets and extend almost to Jupiter's orbit.

The Stars of Spring

After the brilliant stars of winter, the spring sky may seem a bit drab, but compensation comes in the form of more tolerable temperatures for those who want to linger outdoors. The major constellations are quite easy to identify thanks to a handy and notable guiding asterism – the Big Dipper. Refer to Fig. 7.7 for an overview on finding your way around the spring sky.

The familiar Big Dipper, also known as the Drinking Gourd in North America and in Europe as the Plough or Wagon, is the brightest part of Ursa Major, the Greater Bear of the north. The Big Dipper lies relatively close to the north celestial pole and is one of a handful of circumpolar star groups that never set from middle northern latitudes. If you live north of the line of latitude that runs through New York City and Salt Lake City (about 41°), then the Big Dipper is always above the horizon and fully available as a guide to the sky.

As twilight deepens in early March, look for the Dipper standing on its handle low in the northeast. The asterism's stars are visible from all but the most brightly lit urban sites, and it should be fairly easy to identify its distinctive bowl and curving handle. As the night wears on, the Dipper arcs higher but appears increasingly upside down relative to the south-facing view shown in Fig. 7.7. By 10:30 P.M. local daylight time in early April, the Dipper is nearly overhead as the sky darkens.

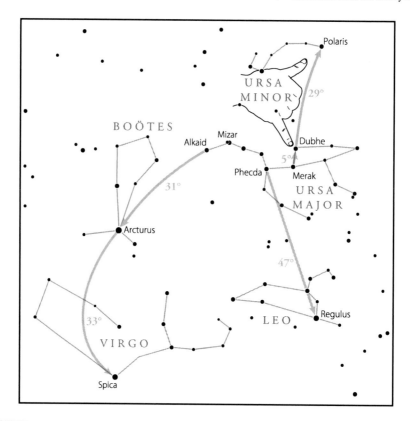

Fig. 7.7 The Big Dipper, the most famous of spring star groups, shows the way to several constellations. Mizar and Alcor, in the Dipper's handle, also serve as a test of your vision

Once you've established how the Dipper is oriented during your viewing session, rotate the finder chart to match it.

Take a good look at the second star from the end of the Dipper's handle. This is Mizar, located about 83 light-years from us, and it's mainly known for its faint fourth-magnitude companion, named Alcor, hovering just 12 arcminutes away. Alcor can easily be seen by those with good eyesight and reasonably dark skies, but try using binoculars if you're having trouble. Mizar and Alcor are the most famous example of a double star, that is, two stars that appear close together from our viewpoint on Earth. Double stars can result from chance alignments of stars that actually have no physical association, but in the case of Mizar and Alcor the two probably orbit each other; even now, astronomers can't say for certain. If they do, they're joined by many others. Alcor is a binary too, and four additional stars orbit Mizar, making this a sextuple system.

With the exception of Alkaid, the end star of the handle, and Dubhe (DU-bee), the top tip of the bowl, the Big Dipper's stars are traveling together through space as part of the same star cluster, the nearest one to Earth. The stars occupy a line about about 30 light-years long at an average distance from us of about 80 light-years, and are of similar age – about 250 million years old.

Once you've found the Big Dipper, the next challenge is to see how much of its constellation, Ursa Major, you can find. In myth, the Greater Bear is usually depicted as being perpetually tracked by a hunting party represented by the three stars in the Dipper's handle. What's striking is how consistently the mythical bear appears in cultures as diverse as ancient Greece, Siberian tribes, and native American groups such as the Cherokee, Zuni and Tlingit. Bradley Schaefer thinks that the most logical explanation for the uniformity of the bear story is that it arrived in the New World at the same time humans did – some 12,000 years ago, when Paleolithic hunters migrated from Asia into North America across the Bering Strait during the last ice age. Of course, the star tale must be far older than this. "However it came to be," wrote Schaefer, "the Great Bear is quite likely one of the oldest inventions of humanity."

Now use the ancient constellation to find the single star closest to the north celestial pole – the North Star, the Pole Star, Polaris. The two front stars of the Dipper's bowl, Dubhe and Merak, form a line segment about 5° long. Imagine extending this line northward by another six times and you'll pass near a single moderately bright second-magnitude star. This is Polaris, and facing this star means you're looking due north.

Polaris lies almost directly above Earth's North Pole, so it forms the pivot around which the celestial vault revolves. The Pawnee of North America called it "the star that does not walk around" and groups in Europe and Asia referred to it as the "nail of the world" or the "pillar of heaven." Polaris marks the end of the handle of the Little Dipper, the most prominent part of Ursa Minor, the Lesser Bear. Most of the constellation's stars are too faint to see from cities, but Polaris and the stars constituting the bottom of the Little Dipper's cup are easy targets from the suburbs. While Polaris seems unmoving in the night sky, it's actually about three-quarters of a degree from the north celestial pole and in long-exposure photographs shows a noticeable trail. While the Pole Star is often seen as a symbol of steadfastness and reliability, the long, slow wobble in Earth's spin known as precession causes our planet's rotational axis to execute a wide circle around the sky. Although the axis happens to point close to Polaris now, this isn't permanent. When the Pyramids were being built in Egypt, the star Thuban in Draco was closest to the north celestial pole, and 12,000 years ago it was near the bright summer star Vega. However, Polaris still has a significant amount of job security because Earth's spin axis will move ever closer to it over the next 90 years.

Return to the Big Dipper to locate some additional spring constellations. Following the arc of the handle stars, imagine extending a long curve to the southeast until you spot a bright, yellow-orange star. This is Arcturus (arc-TOOR-us) in Boötes (bo-OH-teez) the Herdsman. Arcturus is the brightest star in the spring sky and is one of the closest, about 37 light-years away. Its name, which means "the Bear Watcher," is one of the earliest astronomical references in ancient Greek literature. The relatively faint stars of Boötes form an elongated kite, about a hand-span high, extending north of the constellation's brightest star.

The maneuver from the Big Dipper's handle is best remembered as the "arc to Arcturus." Extend this same arc through Arcturus and you'll eventually reach another star of similar brightness. This is Spica (SPI-kuh), the brightest star in the Virgo, a zodiac constellation associated with the harvest. So in addition to the "arc to Arcturus," you "speed on to Spica." The name, which refers to "an ear of grain," comes from Rome, but its identical Greek meaning is clearly a continuation of a concept originating in Mesopotamia. Spica is actually a pair of hot blue stars, each many times more massive than the Sun, with orbital separations only about a third of the distance between Mercury and the Sun. Even at a distance of 250 light-years, Spica shines as one the sky's brightest stars.

Virgo is one of the largest constellations and is among the most conspicuous of the zodiac. Looking toward the constellation directs our view out of the populous, dusty plane of our home galaxy, the Milky Way, and into the wider universe. Distant galaxies are scattered throughout the spring sky, but Virgo contains an astounding concentration called the Virgo Cluster of Galaxies. It contains some 1,300 galaxies positively identified as members and lies 54 million light-years away. Small amateur telescopes can reveal perhaps a hundred of these galaxies in the region about a hand-span north of Spica. Although this patch of sky appears empty to the naked eye, it teems with literally trillions of distant Suns.

Our final spring target is Leo, the heavenly Lion, another of the constellations that line the ecliptic and thus a member of the zodiac. Return to the Big Dipper and this time use the two stars representing the back of the bow, Megrez and Phecda. Imagine a southward arc that passes through these stars and continues for on for about two hand spans. This brings you to Regulus (REG-yuh-lus), the "little king" that marks the celestial lion's heart. The constellation's most distinctive feature is "The Sickle," an asterism shaped like a backward question mark that outlines the lion's head and terminates at Regulus. The star lies very close to the ecliptic, and as seen in drawings elsewhere in this book, is frequently visited by the Moon and planets.

Regulus contains about 3.4 times the Sun's mass, with a correspondingly high surface temperature and bluer color. It also rotates exceptionally fast, its equatorial region whirling at about 700,000 mph (1.1 million kilometers/h), or about 160 times faster than the Sun's. As a result of this rapid spin, the star is flattened into an oblate shape is about a third larger across its equator than across its poles. This spin also affects the star's surface temperature, which is 50% hotter at the poles than it is on the equator. Regulus is actually a quadruple star system, and interactions with an younger version of the closest companion, an evolved star called a white dwarf, likely spun up the lion's heart.

Summer's Starclouds

Summer can be a bit frustrating for stargazers. The temperatures are congenial, but the nights start late, humidity runs high and insects can be a nuisance. The reward for overcoming these difficulties is a fascinating look into our home galaxy (Fig. 7.8).

The best place to start your explorations of the summer sky is with the Summer Triangle, an asterism formed from the brightest stars of three constellations.

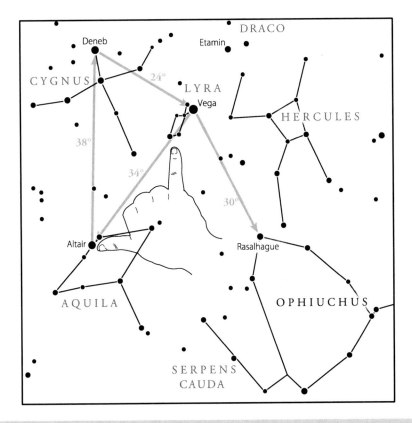

Fig. 7.8 Start your explorations of the summer sky with the Summer Triangle of Vega, Altair and Deneb. Farther south lie two distinctive zodiac constellations: Sagittarius and Scorpius

In early June, face northeast as twilight fades and look for the three brightest stars. The highest and brightest is Vega (VAY-guh or VEE-guh), the lowest and farthest south is Altair and the dimmest and northernmost is Deneb (DEH-nehb).

Vega, the luminary of the small constellation Lyra (LYE-ruh) the Lyre, is one of the brightest stars and, at a distance of just 25 light-years, among the closest. The lyre is the harp of the luckless Orpheus, the Greek who used his musical skills to charm the guardians of Hades so that he could rescue his beautiful wife Eurydice from the realm of the dead. He lost her as they exited the underworld when, contrary to instructions, Orpheus looked back too soon to see if she was still behind him.

A Chinese tale casts Vega as Zhinu, a divine princess, the youngest daughter of the Queen of Heaven, who fell in love with the mortal herdsman Niulang, identified with Altair. The queen didn't think this a proper match and ultimately separated them forever by a Silver River (the Milky Way). Moved by their love, the queen allowed them to meet once a year. On the seventh day of the seventh lunar month on the traditional calendar, all of the world's magpies fly up to the heavens and form a bridge across the river so that the lovers may be reunited. This is the story behind the Double Seven Festival, sometimes called the Chinese Valentine's Day, which is said to date back to the Han dynasty (206 B.C.–A.D. 220). Festivals in both Japan (Tanabata) and Korea (Chilseok) celebrate the same folk tale.

Vega shines with a white color, is about 350 million years old, and has a little more than twice the Sun's mass. Like Regulus, Vega is flattened by rapid rotation and its poles are cooler than the equatorial bulge. Interestingly, the star's spin axis points almost directly toward us, which made understanding the rotational effects all the more difficult. Infrared studies in the 1980s identified a vast cloud of infra-red-emitting dust around the star, and in 2005 NASA's infrared-sensing Spitzer Space Telescope showed that this debris disk consists of fine dust particles being blown outward by the pressure of Vega's starlight. The total mass of the disk is only about a third that of the Moon, so astronomers believe it to be a temporary structure that probably formed within the last million years, a result of asteroid or comet collisions close to the star. Now, Vega's light is gradually pushing this material out into the galaxy.

Altair, the herdsman of the Chinese folk tale, is the brightest star of Aquila the Eagle. It's another nearby white star and another rapid rotator distorted by its spin. The name Altair meant "the Eagle star" to the Mesopotamians, and the modern name derives from the Arabic for "the Flying Eagle." Aquila bears at least a vague resemblance to its namesake. With Altair at the head, the Eagle flies northeast with broad wings more than a hand-span across.

Deneb, the faintest of the Summer Triangle stars, is by far the most remarkable, holding its own against Vega and Altair despite that fact that it's situated 57 times farther away, at a distance of about 1,425 light-years. Beginning its energy-producing career with at least 15 times the mass of the Sun, Deneb is a live-fast/die-hard type of star, fated to explode as a supernova at some future time. It is some 50,000 times as luminous as the Sun, and astronomer James Kaler has noted that if Deneb were as close as Vega, it would easily be visible in broad daylight, shining some 15 times brighter than Venus at its best. Deneb, which means "tail," is also part of an avian constellation: Cygnus (SIG-nus) the Swan. The body of the Swan stretches about 22° from Deneb with an even larger wingspan. Urban stargazers will have trouble locating the full extent of the constellation, but its brightest stars form the more easily noticed asterism of the Northern Cross, which is topped by Deneb.

Summer brings the very best views of our galaxy, which we see as the ragged silvery band of the Milky Way. Spend some time under the stars any clear, dark night away from urban sky glow and you'll notice a misty lane arching across the sky. Recognized since antiquity, this glow was often associated with a heavenly path or celestial stream by ancient peoples. The Romans, fond of road construction, called it *Via Lactea*, or the Milky Way. To the Greeks it was *galaxias kuklos*, the milky circle, and it is from this phrase that our word galaxy derives. But a name used by the San people in southern Africa's Kalahari Desert better hints at the Milky Way's true nature: "The Backbone of the Night."

For the Milky Way is nothing less than the structure of our galaxy seen edgewise. As scientists began to realize this, the Roman name for a cosmic highway was transferred to our galactic home. The first step in understanding our galaxy began in 1610, when Galileo turned his spyglass to the Milky Way and revealed its luminous glow to be individual stars "so numerous as almost to surpass belief." He wrote: "Upon whatever part of it the telescope is directed, a vast crowd of stars is immediately presented to view. Many of them are rather large and quite bright, while the number of the smaller ones is quite beyond calculation."

The summer Milky Way is worth a trip out to the countryside to ensure a good look. Once your eyes have adapted to the darkness, the silvery path exhibits surprising variations of light and dark. The region near the head of Cygnus is particularly interesting, as suggested in Fig. 7.3. This is the start of the Great Rift, an elongated dark patch that splits the Milky Way down to the horizon. Villagers in modern Peru count among their constellations animal shapes formed by dark patches in the southern Milky Way; these "dark constellations" are probably an astronomical relic of the great Inca empire. Binoculars make a great addition to any observing session where you can make out the Milky Way.

If we followed the imaginary flight of Cygnus southwest along the Milky Way, we would come upon a distinctly reddish star named Antares (an-TAH-reez), the brightest member of the zodiac constellation Scorpius the Scorpion. Literally translated from the Greek, Antares simply means "like Mars," a reference to their similar colors, but we've always preferred the hint of competition implied in the less accurate translation, "rival of Mars." As 2016 opens and nearby Mars and Antares face off, the star easily bests the planet. But a few weeks later, the situation reverses as Mars begins approaching Earth; the planet remains on top for the remainder of the year.

Antares, the heart of the celestial Scorpion, is a cool red supergiant, similar in evolutionary state, mass, and size to Betelgeuse in Orion. Scorpius is a relatively bright constellation, but it's less obvious than it deserves to be because it hangs low in the south as seen from middle-northern latitudes. Its sinuous chain of stars and especially hooked tail actually takes little imagination to see as a connect-the-dots scorpion.

East of the Scorpion's tail is another group of stars that resembles a teapot. About the angular size of an open hand, the Teapot asterism is the brightest part of the constellation Sagittarius the Archer. Seen from dark skies this distinctive pattern seems lost within the glow of the Milky Way's thickest star clouds. Looking toward the Teapot's spout, we're gazing toward the very center of our galaxy, which lies about 28,000 light-years away. We're located about halfway between the center and our galaxy's outer edge, which means the galaxy spans about 100,000 light-years. Unfortunately, intervening dust, gas and stars prevent us from seeing the innermost galaxy with our eyes, but satellite observatories and ground-based telescopes have mapped the region in radio, infrared and X-ray wavelengths.

Astronomers know that the center of the galaxy hosts a giant black hole with a mass about four million times greater than the Sun's. In fact, most big galaxies harbor big central black holes, and the Milky Way's is by no means the most massive. But unlike many of these galaxies, our black hole seems to be relatively quiet for the moment. When large amounts of matter fall into a black hole, the result is intense radio, X-ray and gamma radiation, along with jets of particles traveling near the speed of light – in other words, it's the kind of thing that makes an impression. In 2010, astronomers using data from NASA's Fermi Gamma-ray Space Telescope found striking evidence that the Milky Way's central region experienced some kind of eruption as recently as 10 million years ago. Gamma rays are highest-energy form of light. What the researchers found were two never-before-seen gamma-ray-emitting bubbles that extend 25,000 light-years north and south of the galactic center. Invisible to human eyes, the bubbles occupy more than half of the sky, stretching from the constellation Virgo to the constellation Grus. Studies show that the bubbles formed quickly from some sort of impulsive event in the galactic center, and they may turn out to be the remnant of an eruption from the supersized black hole at the center of our galaxy. As you reflect on the Milky Way's silent beauty, try to imagine the incredible energies at work in its unseen heart.

Autumn's "Watery" Sky

As Earth continues its journey around the Sun, the planet's night side is directed away from the galactic center and toward a region of the sky that seems almost starless by comparison. As in spring, we're largely looking out of the galaxy's disk. The shortening days make the Summer Triangle an obvious feature of the evening sky, its prominence enhanced by the dearth of bright fall constellations. In particular, the "watery" constellations of the zodiac – Capricornus the Sea-goat, Aquarius the Water Bearer, and Pisces the Fishes – are as faint as they are famous. Nevertheless, the fall sky does offer some treats, including one of the year's very best. Figure 7.9 illustrates the autumn highlights and the best routes for locating them.

Our first stop is a regal constellation, Cassiopeia (CASS-ee-uh-PEE-uh). As evening twilight darkens in mid-September, look low in the northwest for the Big Dipper. Use the Big Dipper's pointer stars to locate Polaris, and then continue extending the imaginary line until you encounter a group of stars in the northeast that resemble a rather lopsided numeral 3. Depending on the time of year and the hour, the Queen's stars may be better described as an M, W or E as well. Cassiopeia appears deceptively small on our maps but in the sky spans about 13° – a bit less than the angle between index and little fingers held at arm's length.

Many of the best-known constellations associated with the autumn sky appear to have been home-grown creations of the ancient Greeks rather than Mesopotamian imports, at least according the Bradley Schaefer, who believes they were probably in place by 500 B.C. These constellations are Cassiopeia, the matriarch of a legendary family that includes her husband Cepheus (SEF-ee-us) and daughter Andromeda, the heroic Perseus, Pegasus the Winged Horse and Cetus (SEE-tus) the Sea Monster.

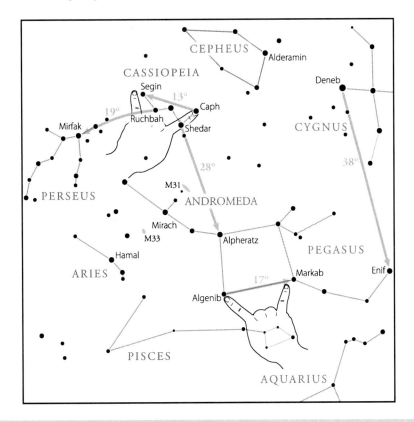

Fig. 7.9 The Great Square of Pegasus is the keystone of the autumn sky, though you may need Cassiopeia's assistance to find it. Hop along Andromeda's chains to locate the jewel of the autumn sky: M31, the Andromeda Galaxy. From a dark site, both M31 and the fainter and more distant galaxy M33 are visible to the naked eye

The trouble all began when vain Cassiopeia, the Queen of Ethiopia, boasted that she was more beautiful than the daughters of the sea god Poseidon. To punish her, Poseidon flooded the land and sent the monster Cetus to ravage the country. King Cepheus consulted an oracle about how to end the god's rampage but learned that nothing could stop Cetus until he chained his daughter Andromeda to a rock as a sacrifice to it.

Meanwhile, Perseus was having his own troubles, having accepted the task to kill Medusa – best known for her serpent hair-do and ability to turn any who looked upon her into stone – and return with her head. By viewing her reflection in his shield, Perseus was able to do the deed and, as he decapitated Medusa, the winged horse Pegasus sprang from her blood. On his way home to complete his task, Perseus stopped in Ethiopia and saw Andromeda chained to the rocks. Using the head of Medusa, Perseus slew the monster by turning it into stone, thereby winning Andromeda's hand in marriage. All of the tale's *dramatis personae* are on display in the autumn sky, but we'll focus on four: Cassiopeia, Pegasus, Andromeda and Perseus.

Use Cassiopeia to locate the asterism known as the Great Square, which is part of the constellation Pegasus. Draw an arc from Cassiopiea's central star, which lacks a proper name and goes by its Greek-letter designation of Gamma, south to Shedar. Then continue following the arc for another 28° until you come to a star about as bright as Shedar. This is second-magnitude Alpheratz, a star with dual citizenship in the sense that it links the figures of Pegasus and Andromeda, to whom it officially belongs. From Alpheratz, drop south to Algenib (al-JEE-nib), then west to Markab – the longest dimension of the "square," about 17° – then north to Scheat (SHE-at). An eastward run back to Alpheratz, where you started, closes out the asterism. A square hardly suggests a horse, of course, but the western edge boasts some equestrian features. Extending from Markab is an arc of stars representing the horse's neck and face that ends with Enif (EE-nif), the brightest star in Pegasus. Stars trailing from Scheat give the impression of forelegs. Those looking for wings, however, are on their own.

To follow Andromeda the Chained Maiden, we'll return to Alpheratz and explore points to the northeast. Two trails of stars extend to the northeast in a slender V shape. This is essentially all there is to Andromeda, and in light-fogged urban skies only the brighter southerly star trail will be noticeable. Hovering near the northern trail is a soft oval glow that turns out to be the closest large galaxy to our own. This is the Andromeda Galaxy, which is also known as M31 after its designation in Charles Messier's eighteenth-century catalog of showcase telescopic sights. Similar in size, mass and type to our own galaxy, M31 lets us see from the outside what the summer Milky Way shows us from within. From dark country skies, M31 is easily noticeable, and tenth century Persian astronomers made the first textual reference to the object, which they referred to simply as the "little cloud." The light from that oval glow began its journey 2.5 million years ago, long before modern humans walked on Earth, which makes the Andromeda Galaxy one of the most distant objects visible to the naked eye (Fig. 7.10).

To locate the galaxy, start at Alpheratz and hop two stars northeastward along the trail to reach second-magnitude Mirach (MY-rak), and then jump to the fourth-magnitude star Mu Andromedae about 4° above it. Extend the imaginary line from Mirach another 3° to reach fourth-magnitude Nu Andromedae. If you can see Nu at all, then the sky may be dark enough for you to notice a milky patch of light nearby; if not, try binoculars.

Before the 1920s, astronomers referred to galaxies like M31 as spiral nebulae, generally assuming them to be in the same class of object as gas and dust complexes like the Orion Nebula. In 1912, Henrietta Leavitt (1868–1921) at Harvard College Observatory made an important discovery about a class of stars called Cepheid variables, named for the first member of the class, which is located in the constellation Cepheus. These stars undergo cyclic brightness changes that were related to their true luminosity: The longer it takes a Cepheid to complete its cycle, the brighter it is. By comparing a star's actual luminosity to its apparent brightness,

Fig. 7.10 The misty oval of M31 transforms into a grand galaxy when imaged through a telescope. The Andromeda Galaxy is a giant spiral galaxy much like our own. It's located 2.5 million light-years away and is among the most distant objects visible to the naked eye (NOAO/AURA/NSF/B. Schoening and V. Harvey)

astronomers can determine its true distance. Today, these stars remain important cosmic yardsticks.

On Oct. 5, 1923, American astronomer Edwin Hubble (1889–1953) targeted M31wth the 100-in. telescope at California's Mount Wilson Observatory and took a long-exposure image of the galaxy, part of a series of images of the galaxy he took that autumn. When Hubble later examined the glass photographic plate, he identified three brightened stars that he believed were novae – a type of stellar eruption – associated with the nebula. But one of these stars was actually a Cepheid, and the period-luminosity relationship for these stars showed that the

Andromeda Nebula was truly an "island universe" in its own right, a galaxy far beyond our own. Astronomers have now established that M31 lies 2.5 million light-years away.

Another, fainter galactic neighbor lies about as far from Mirach as M31, but in the opposite direction. This is the Triangulum Galaxy, a spiral located about 2.9 million light-years away. Seeing its subtle glow with the naked eye requires excellent observing conditions; if yours aren't up to the task, try using binoculars to scan for it once you'd found M31.

The last constellation we'll consider in the autumn tableau is Perseus, the hero who slew Medusa and rescued Andromeda. To locate these stars return to Gamma Cassiopeia, the W's central star. Draw an imaginary line from Gamma to Ruchbah and extend the line about 19° to second-magnitude Mirfak, the brightest star of Perseus, which is located at the fork in the constellation's vaguely Y-shaped figure. Mirfak, also known by its Greek letter designation of Alpha Persei, is a yellow supergiant with about 5,000 times the Sun's luminosity and about seven times its mass. Mirfak and many of the constellation's stars are physically related as part of the Alpha Persei star cluster, which is about 580 light-years away and only about 50 million years old.

Take to the Stars

With a little effort, anyone can spot the stars and constellations briefly discussed in this chapter. In time, you'll find that they've become familiar companions anticipated with the change of seasons. The maps on the following pages provide more comprehensive views of the seasonal changes and allow you to locate additional constellations. The charts show the entire sky as seen from middle northern latitudes at different times of the year (Figs. 7.11–7.16). We've included just the stars one can expect to see from a reasonably dark urban site, such as a large park, however, we've also included all of the lines used in standard constellation figures, so there may be no star symbol where the constellation lines imply one exists.

To use the charts, rotate them so that the direction shown at the bottom of the map matches the direction you're facing. Look for key constellations to help you get your bearings and to show you how the chart images scale up to the sky. The circular outer boundary of these charts represents idealized horizon, so trees and buildings may block your view in certain directions. The center of the map is the zenith, the point directly overhead. The directions east and west may at first appear reversed on the maps, but they're correct because with sky charts we're looking up at the sky. To preserve your night vision while referring to the charts, cover the lens of your flashlight with a red filter or several layers of grocery-bag to dim and redden the light.

Do your initial stargazing from the back yard or a suburban park, where the general sky glow washes out the faintest stars and simplifies the sky. Once you've mastered a few constellations, look for darker skies to test your skills so you can appreciate the subtler wonders mentioned in this chapter. And when the clouds roll in, as they inevitably will, use the resources in Table 7.4 to continue your exploration of the starry sky.

January and February Stars

Early January	10:30 P.M.
Mid January	9:30 P.M.
Early February	8:30 P.M.
Mid February	7:30 P.M.

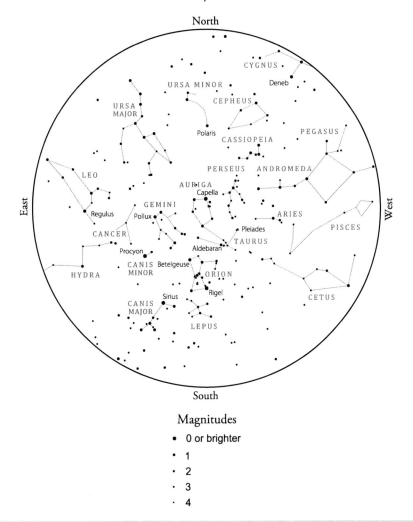

Magnitudes

• 0 or brighter
• 1
· 2
· 3
· 4

Fig. 7.11 Late winter evenings gleam with brilliant stars. Use Orion as a guide to help you hunt other prominent constellations

March and April Stars

Early March	11:30 P.M.
Mid March	11:30 P.M.
Early April	10:30 P.M.
Mid April	9:30 P.M.

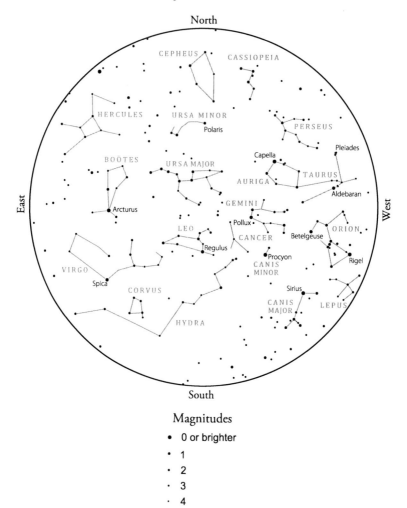

Fig. 7.12 The bright stars of winter can still be seen in the western sky as spring arrives, but they drift ever closer to the glow of evening twilight. Ursa Major's Big Dipper is now prominent and serves as a guide to the spring sky

May and June Stars

Early May	12:30 A.M.
Mid May	11:30 P.M.
Early June	10:30 P.M.
Mid June	9:30 P.M.

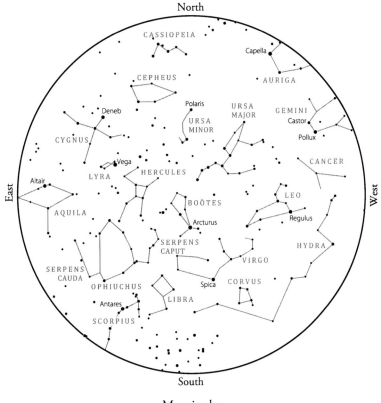

Magnitudes

- • 0 or brighter
- • 1
- · 2
- · 3
- · 4

Fig. 7.13 As spring turns to summer, the last of the winter constellations drift into evening twilight and a new group of bright stars – Deneb, Altair, Vega, and Antares – gleams in the east. The first three form the asterism known as the Summer Triangle

July and August Stars

Early July	12:30 A.M.
Mid July	11:30 P.M.
Early August	10:30 P.M.
Mid August	9:30 P.M.

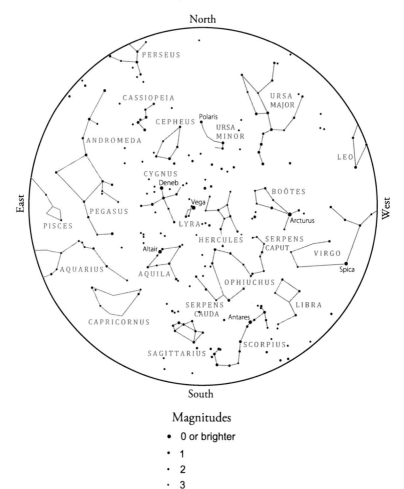

Magnitudes

- • 0 or brighter
- · 1
- · 2
- · 3
- · 4

Fig. 7.14 The trio of stars marking the Summer Triangle are most visible now. From a reasonably dark location, the faint band of the Milky Way can be seen arcing from Cygnus to Scorpius

September and October Stars

Early September 11:30 P.M.
Mid September   10:30 P.M.
Early October   9:30 P.M.
Mid October   8:30 P.M.

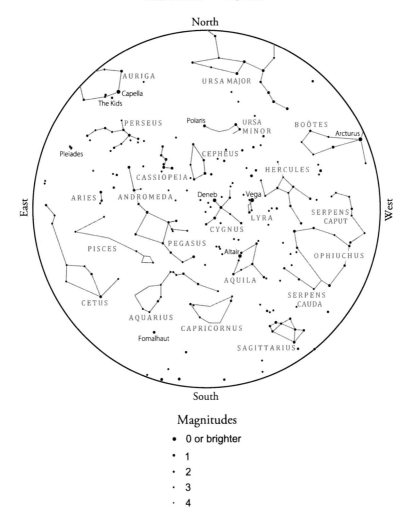

Magnitudes

● 0 or brighter
· 1
· 2
· 3
· 4

Fig. 7.15 The Summer Triangle stars gleam high in the western sky even as the faint constellations of autumn emerge in the east. The distinctive shape of Cassiopeia is easy to spot and serves as a guide to the Great Square asterism in Pegasus

November and December Stars

Early November 11:30 P.M.
Mid November   9:30 P.M.
Early December   8:30 P.M.
Mid December   7:30 P.M.

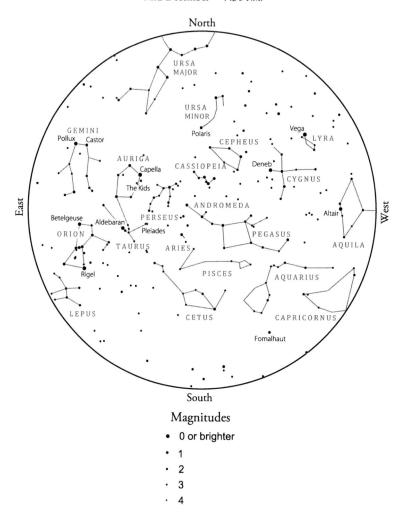

Fig. 7.16 The stars of summer, now low in the western sky, gradually yield to the stars of autumn and winter. Cassiopeia, Pegasus and Andromeda ride highest in the sky now, and the bright winter constellations of Orion, Taurus and Auriga emerge in the east

Table 7.4 Stars and constellations on the Web

General Information
Astronomy magazine
www.astronomy.com
Jim Kaler's Skylights
stars.astro.illinois.edu/skylights.html
Sky & Telescope magazine
www.skyandtelescope.com

Images
Astronomy Picture of the Day
apod.nasa.gov
The Constellations
www.iau.org/public/constellations
The World at Night
www.twanight.org

Interactive maps and tools
Google Sky
www.google.com/sky
WorldWide Telescope
www.worldwidetelescope.org
Sky View Café
www.skyviewcafe.com

Myths and legends
Ian Ridpath's Star Tales
www.ianridpath.com/startales

References

Bauer BS, Dearborn DSP (1995) Astronomy and empire in the ancient Andes: The cultural origins of Inca sky watching. University of Texas Press, Austin

Bonnet-Bidaud J, Praderie F, Whitfield S (2009) The Dunhuang Chinese sky: A comprehensive study of the oldest known star atlas. J. Astron. Hist. Herit. 12: 39–59

Croswell K (2002) The brightest red dwarf. http://kencroswell.com/thebrightestreddwarf.html. Accessed 15 Oct. 2010

Harper GM, Brown A, Guinan EF (2008) A new VLA-Hipparcos distance to Betelgeuse and its implications. Astron. J. 135:1430–1440. doi:10.1088/0004-6256/135/4/1430

Holberg JB (2007) Sirius: The brightest diamond in the sky. Springer-Praxis, Chichester, UK

Kaler JB (2006) Cambridge encyclopedia of stars. Cambridge University Press, New York

Kaler JB (2009) Deneb. http://stars.astro.illinois.edu/sow/deneb.html. Accessed 15 Oct. 2010

Kunitsch P, Smart T (2006) A dictionary of modern star names: A short guide to 254 star names and their derivations, second edition. Sky Publishing, Cambridge, Massachusetts

Research Consortium on Nearby Stars (2010) The one hundred nearest star systems. http://www.recons.org/TOP100.posted.htm. Accessed 15 Oct. 2010

Norris R (2009) Searching for the astronomy of aboriginal Australians. In: Vaiskunas J (ed.) Astronomy & cosmology in folk traditions and cultural heritage. Klaip da Univ. Press, Klaip da, Lithuania

Orlove BS, Chiang JCH, Cane MA (2000) Forecasting Andean rainfall and crop yield form the influence of El Niño on Pleiades visibility. Nat. 403: 68–71

Rappenglück M (1999) Paleolithic timekeepers looking at the golden gate of the ecliptic: The lunar cycle and the Pleiades in the cave of La-Tête-Du-Lion (Ardèche, France), 21,000 BP. Earth, Moon and Planets, 85–86:391–404. doi: 10.1023/A:1017069411495

Rochberg F (2004) The heavenly writing. Cambridge University Press, New York

Schaefer BE (2002) The latitude and epoch for the formation of the southern Greek constellations. J. Hist. Astron. 33: 313–350

Schaefer BE (2006) The origin of the Greek constellations. Sci. Amer., Nov., 96–101

Schaefer BE (2007) The latitude and epoch for the origin of the astronomical lore in Mul. Apin Bull Amer. Astron. Soc. 38: 157

Schafer EH (1977) Pacing the void: T'ang approaches to the stars. University of California Press, Berkeley

Su KYL, Rieke GH, Misselt KA et al. (2005) The Vega debris disk: A surprise from Spitzer. Astrophys. J. 628:487–500

Su M, Slatyer TR, Finkbeiner DP (2010) Giant gamma-ray bubbles from Fermi-LAT: active galactic nucleus activity or bipolar galactic wind? Astrophys. J. 724:1044–1082. doi:10.1088/0004-637X/724/2/1044

Toulman S, Goodfield J (1961) The fabric of the heavens: The development of astronomy and dynamics. Harper and Row, New York

Chapter 8

Meteors and Meteor Showers

Now and then, we're dramatically reminded that our planet doesn't orbit the Sun all by itself. As evening twilight deepened at the tail end of rush hour on Monday, Jan. 18, 2010, scores of people in Pennsylvania, New Jersey, Delaware, Maryland and Virginia glimpsed a brilliant fiery object streaking across the sky. It lasted only seconds, but a smoke trail marked its passage for many minutes, gradually turning from linear to serpentine as high-level winds sheared it. At the time, physician Frank Ciampi was finishing up paperwork at his family practice in Lorton, VA, before heading home. Moments after commuters saw the streaking object, he heard a crash in another room that was so loud he thought a bookcase had toppled over. When he investigated, he found bits of wood, plaster and insulation scattered outside one of the examining rooms. There was more debris inside and an obvious hole in the ceiling – plus fragments that together formed a mango-sized chunk of rock. After realizing that the rock might be a meteorite, Ciampi alerted the Smithsonian Institution in Washington, D.C., which collected the mysterious visitor and confirmed that it was indeed a rock from space.

Earth is under constant bombardment. Each day, more than 100 billion meteoroids larger than 1 μm – about a third the mass of a small sand grain – enter the atmosphere. Particles about 0.003 in. (0.06 mm), or about half the thickness of a U.S. dollar bill, do so about every second. Each day, our planet accumulates about 73,000 lb (33 metric tons) of interplanetary material, most of it in the form of fine dust particles. Fortunately, the number of meteoroids – the general term for small interplanetary debris – decreases rapidly as their sizes increase, so while there's lots and lots of dust, there are relatively few objects big enough to punch through ceilings, and far fewer still that are capable of causing the devastating damage associated with a large impact crater. The smallest particles simply drift down from the top of the atmosphere. Larger ones blaze a brief visible trail across the sky,

F. Reddy, *Celestial Delights: The Best Astronomical Events through 2020*,
Patrick Moore's Practical Astronomy Series, DOI 10.1007/978-1-4614-0610-5_8,
© Springer Science+Business Media, LLC 2012

sometimes called a "shooting star" but officially known as a meteor. A couple of times a year, a rock as big as a baseball tumbles toward us and flames out as a brilliant meteor called a fireball. As the fireball travels, observers may see it flare several times and report it trailing "sparks." The flares are detonations – audible as thunderous sounds if the breakup happens at a low-enough altitude – and the sparks are fragments of the main body. Many of these pieces may reach the ground along the fireball's ground track, possibly to be retrieved as meteorites.

The most dramatic recent event of this kind occurred over the Chicago suburb of Park Forest, IL, on Mar. 26, 2003. About 10 min before local midnight, a bright fireball streaked across the north-central U.S. It was seen by observers in Illinois, Indiana, Michigan and Missouri and captured by security and police-car video cameras. Residents of Chicago's southern suburbs were awoken by thunderclaps from detonations, and in Park Forest meteorites struck parked cars, punched through ceilings and smashed windows. Some 40 fragments were recovered there, with hundreds more found elsewhere along a 6-mile-long (9.7 km) area. This is the most densely populated region to experience an event like this in modern times. It is also perhaps the most unusual, because Park Forest had its own resident expert – Steven Simon, who studies meteorites at the University of Chicago. Throughout the night, residents called police, who took reports and carted away specimens as evidence. The Park Forest police station then became Meteorite Central, with police encouraging residents to bring in any specimens they found and forwarding calls about possible meteorites to the university. All told, scientists collected about 66 lb (30 kg) of rock fragments – but they estimate the fireball's pre-entry mass could have been as high as 7 metric tons, or roughly the mass of a large elephant.

Between myriad microscopic dust particles and objects like the Park Forest fireball, there are small bits of debris that create meteors we can observe any night of the year, provided we're patient enough. Under a dark sky, any observer can expect to see between two and seven meteors each hour every night. These are sporadic meteors, small unlucky pieces of rock swept up by our planet as it orbits the Sun. Each hour, Earth sweeps out a tunnel through space about 67,000 miles (109,000 km) long, so the sporadic meteoroids big enough to make visible trails are truly few and far between. However, there are certain times of year when the numbers of visible meteors run much higher, which allows observers to optimize their meteor-watching time (Fig. 8.1). That's what this chapter is about.

Visible meteors typically flame out at altitudes between 44 and 75 miles (70 and 120 km), where the atmosphere rapidly thickens. As air molecules smash into the rushing meteoroid, its surface heats up and a thin layer of minerals begins to evaporate. Scores of atoms are ejected with each impact, creating a cloud that surrounds and travels with the particle. This increases its effective surface area, which in turn leads to more collisions, more heating, more ablation. The meteoroid is now a meteor, glowing ever brighter as it descends. As the air density rises, there comes a point where the vapor cloud in front of the meteor becomes thick enough that its atoms begin colliding with each other, creating a shock wave. Together, collisions between air, meteoroid vapor and the rock itself create a trailing column of ionized atoms. The excited gases give off characteristic wavelengths of light in much the

Fig. 8.1 The 1998 Leonid "attack of the fireballs" as seen by an all-sky camera at the Modra Astronomical Observatory at Comenius University in Bratislava, Slovakia. About 150 fireballs brighter than magnitude –2 are recorded. Follow the converging trails to locate the constellation Leo (*left*). At least three bright meteors pass through Orion's torso (Modra Astronomical Observatory)

same way that light is generated by a neon light or fluorescent lamp, and this glowing gas produces most of a meteor's light. So for the fastest showers, a particle with a mass of only 0.005 oz (0.13 g) – equivalent to a large sand grain – can produce a meteor as bright as zero magnitude. The glow stops either when the body of the meteoroid has been scoured away completely or when it has slowed to speeds no longer capable of creating an ionized trail.

There are two phenomena to watch for even after a meteor fades out. The first is a dimly glowing greenish wake that lasts a few seconds. It's caused by excited oxygen atoms and is commonly seen with fast meteors. The other is called a persistent train, a faint billowing trail associated with bright meteors that can last many minutes. Meteor trains somewhat resemble the contrails seen behind high-altitude aircraft. They glow because iron oxide and sodium stripped from the meteoroid take part in reactions that form oxygen and ozone.

Several times during the year, Earth encounters swarms of small particles that greatly enhance the number of meteors. The result is a meteor shower, during which observers may see dozens of meteors every hour. Clumps and filaments within these swarms may produce better-than-average displays in some years, with rates of hundreds per hour. And every now and then, we're treated to a truly spectacular display that produces thousands of visible meteors for a brief period – a meteor storm, defined as more than 1,000 meteors per hour. In fact, the impetus for the development of meteor astronomy came from storms that erupted from the normally tepid November shower called the Leonids.

The Lion Roars

The first hint that meteor showers were capable of something truly unusual came during the wee hours of Nov. 12, 1799. At Cumaná, Venezuela, the German naturalist and explorer Alexander von Humboldt and the French botanist Aimé Bonpland were still in the first few months of a trip that would last 5 years. Bonpland awoke, decided to enjoy the fresh air under a starry sky, and beginning at about 2:30 A.M. was treated to an extraordinary sight. "Thousands of bolides and falling stars succeeded each other for four hours," wrote von Humboldt. "Mr. Bonpland relates that, from the beginning of the phenomenon, there was not a space in the firmament equal in extent to three diameters of the full Moon that was not filled at every instant with bolides and falling stars."

That same morning, Andrew Ellicott, an American surveyor charged with defining the border between the U.S. and the Spanish territories in Florida, was sailing off Key Largo and also witnessed the shower. "The phenomenon was grand and awful; the whole heavens appeared as if illuminated with skyrockets, flying in an infinity of directions, and I was in constant expectation of some of them falling on the vessel," he wrote. "They continued until put out by the light of the Sun after daybreak."

As vivid as these accounts are, they attracted little attention until decades later. There isn't enough information to determine a reliable number for the peak rate of meteors, but it must have been a truly spectacular show because the faintest trails would have been washed out by light from a nearly full Moon.

As Table 8.1 shows, the Leonids are normally a lower-tier shower, much less impressive than the always reliable Quadrantids, Perseids and Geminids. The visible trails of shower meteors all seem to radiate from a common point in the sky, and it's the location of this so-called radiant that gives the shower its name; for the Leonids, it lies within the Sickle asterism that marks the Lion's head. The radiant is a perspective effect. The particles travel in roughly parallel paths as Earth plows into them, so they converge in the distance just like highway lines or railroad tracks (Fig. 8.2). A meteor shower is usually named either for the constellation where its radiant appears or for the nearest bright star.

Determining the peak hourly rate of a meteor shower from visual observations is a tricky business, even more so when dealing with sometimes very sparse historical records. The goal is to find a number called the Zenithal Hourly Rate, or ZHR, which represents the greatest number of meteors seen every hour under absolutely

Table 8.1 The year's best meteor showers. Dates of maximum activity vary from year to year; check Appendix A for the correct date in any given year. Hourly rates given reflect the average number of meteors expected for an observer under ideal conditions (clear, dark, moonless sky) with the radiant overhead; actual rates will be lower. Shower rates may also fluctuate from year to year, and outbursts do not necessarily coincide with the peak date of the annual shower

Shower	Max. date (UT)	Typical rate (per hour)	Notes	Source	Possible significant outbursts
Quadrantids	Jan. 3–4	120	Moderate speed	2003 EH$_1$	
Lyrids	Apr. 21–22	15–20	Fast. Outburst of 50 per hour in 1982	C/1861 G1 (Thatcher)	2042
Eta Aquariids	May 5–6	50+	Very fast, often leave trains, but best seen from the southern hemisphere	1P/Halley	
Delta Aquariids	Jul. 29–30	15–20	Moderate speed; best seen from the tropics and southern hemisphere	96P/Machholz	
Perseids	Aug. 12–13	100	Fast. Outbursts in 1993, 2004 and 2009 and modest outburst likely in 2016	109P/Swift-Tuttle	2016
Draconids	Oct. 8–9	Variable	Slow. Strong storms (exceeding 5,000 per hour) in 1933 and 1946, strong outbursts (500 per hour) in 1995 and 1998 and (350 per hour) in 2011	21P/Giacobini-Zinner	2018 2025
Orionids	Oct. 21–22	25	Very fast, often leave trains	1P/Halley	
Leonids	Nov. 17–18	20	Very fast, often leave trains, some lasting many minutes. Outburst in 1998, storm in 1999, outburst in 2000, storm in 2001, storm in 2002, additional modest outbursts since	55P/Tempel-Tuttle	2031– 2035
Geminids	Dec. 13–14	120	Moderate speed	3200 Phaethon	
Ursids	Dec. 22–23	10	Moderate speed. Strong outbursts in 1945 and 1986	8P/Tuttle	2020

ideal conditions – a clear dark sky on a moonless night with the radiant directly overhead. Observed rates are always lower than this value because every location differs from the ideal in one way or another, but with this number in hand astronomers can compare the activity of different meteor showers or track the changing activity of a single shower from year to year.

After 1799, the Leonids lapsed back to their usual trickle. Then, in 1831, a few observers in Spain, France and the eastern United States counted a meteor every minute on the morning of Nov. 13. The best was yet to come. The following year, meteors rained through the predawn sky from the Ural Mountains to the eastern shore of Brazil. The display was intense in spite of interference from a waxing

Fig. 8.2 A straight road converges into the distance, and so do the observed paths of meteors. The convergence point is called the radiant. Leonid meteors radiate from a point within the constellation's Sickle asterism

gibbous Moon right near the shower radiant. Observers reported many bright fireballs and enjoyed a wonderful display with a peak estimated at 2,000 Leonids per hour.

And then, in 1833, the Leonid shower outdid itself. Along the east coast of North America, from Canada to Mexico, anyone under a clear sky in the hours before dawn saw hundreds of meteors every minute; the peak rate was around 60,000 per hour. Native Americans referred to the event as "the night the stars fell." An Annapolis, MD, observer described the meteors as falling "like snowflakes." An observer near Augusta, GA, reported that "the stars descended like a snowfall to the earth" and that the brightest meteors left trains that "would remain visible … for nearly fifteen minutes." "The scene was truly awful," wrote a cotton planter in South Carolina, "for never did rain fall much thicker than the meteors fell towards the earth; east, west, north, and south, it was the same." The display was again rich in bright fireballs and meteors were seen even after sunrise.

In the weeks that followed, wild theories involving electrified air or flammable gases filled the popular press. According to some accounts, a "black cloud" hung overhead and it was from this that the meteors seemed to descend, but the cloud was an optical illusion caused by the greatly foreshortened trails of meteors closest to the radiant. Others noticed that the meteors seemed to stream from the constellation Leo and that its source rose with the stars as the night wore on.

Accounts of the storm witnessed by Humboldt in 1799 and of the 1832 storm drew renewed attention, and many newspapers commented on the coincidence that all three meteor showers had occurred near the same date.

Yale College mathematics professor Denison Olmsted (1791–1859) caught the last hour of the 1833 storm. "The flashes of light, though less intense than lightning, were so bright as to waken people in their beds," he wrote. Intrigued by what he saw, Olmsted analyzed reports of the phenomenon from around the country. He established the shower's geographic extent and estimated that over 207,000 meteors were seen that night. Olmsted determined that the meteors had originated beyond Earth and that they had entered the atmosphere traveling in parallel paths. He concluded that the meteors were part of a nebulous body of unknown nature orbiting the Sun and that the shower was caused by Earth's passage through this object.

Astronomers watched the 1834, 1835 and 1836 displays with heightened interest, and while the Leonids showed outbursts that would be considered unusually good in most years – with ZHRs between 60 and 150 – they could only be an anticlimax after the great storm. The following year, the German physician and astronomer Heinrich Olbers (1758–1840) examined the available information and concluded that the main Leonid swarm returned about every 33 years. "Perhaps we shall have to wait until 1867 before seeing this magnificent spectacle return," he wrote. Meanwhile, the shower fell back to its usual modest numbers.

In the 1860s, as the possibility of more extraordinary meteor activity approached, Yale professor Hubert Newton (1830–1896) whet appetites by combing Chinese, Arab and European records for evidence of previous Leonid returns.

He turned up evidence for 11 great meteor showers dating from the year 902. An account in 934 associated the meteors with a traumatic terrestrial event: "There was an earthquake in Egypt … and flaming stars struck against one another violently." A Japanese report from 970 noted that "all over the sky myriad stars streamed from east to west, scattering without interruption, their shapes like knives and swords."

Modern orbital calculations show that comet 55P/Tempel-Tuttle, the source of the Leonid swarm, passed within 2.13 million miles (3.43 million km) of Earth – less than nine times the distance to the Moon – on Oct. 26, 1366. This was the second-closest known approach to our planet by any comet; the record goes to comet Lexell in 1770, which swept past us at only three times the Earth-Moon distance. The dates of historical showers fall earlier than the modern one because Tempel-Tuttle's whole orbit slowly rotates around the Sun (a phenomenon already discussed with respect to Mercury's advancing perihelion). The same year of the comet's closest approach, the Leonids made quite an impression to a Portuguese chronicler:

> And afterward [the stars] fell from the sky in such numbers, and so thickly together, that as they descended low in the air, they seemed large and fiery, and the sky and the air seemed to be in flames, and even the earth appeared as if ready to take fire…. Those who saw it were filled with such great fear and dismay, that they were astounded, imagining they were all dead men, and that the end of the world had come.

Newton determined that the interval between Leonid storms was 33.25 years and predicted its next return on the night of Nov. 13–14, 1866. After a "warm-up" outburst of one meteor a minute in 1865, a small Leonid blizzard returned on schedule, delighting observers from England to India with a peak hourly rate of about 8,000. Careful meteor counts from the United Kingdom enabled astronomers to determine the moment of peak activity and estimate the width of the cloud of material responsible for the storm. A modern reduction of the data gives the peak at 1:12 UT on Nov. 14, with the shower above half the maximum value for 50 min. This means that the swarm of particles responsible for the shower spread across some 55,000 miles (89,000 km) of space.

Another fine display occurred the following year, with rates around 1,200 per hour despite interference from a nearly full Moon. The 1868 shower occurred under a new Moon and was widely observed in Europe and North America. The heavens obliged with an unusual outburst – about 400 per hour – displaying a broad period of heightened activity that lasted many hours but showed no clear peak.

The decade also brought a solution to the mystery surrounding the origin of the Leonids and of other meteor showers then being recognized. Orbital calculations of the Perseid meteor stream by Giovanni Schiaparelli, better known for his later work on Mars, revealed a very strong resemblance to the orbit of a bright comet discovered in 1862 and concluded that all meteor showers were caused by the disintegration of comets. The idea was quickly accepted, thanks to additional associations between comet orbits and meteor streams. Using information from the 1866 Leonid shower, Urbain Le Verrier published its orbital characteristics, and astronomers immediately pointed out the similarity between the orbit of the Leonid swarm and that of the newly discovered comet Tempel-Tuttle.

Comets are now thought to be "dirty snowballs" containing a mixture of dust and frozen gases. They only become visible near their closest approach to the Sun, when icy regions on the comet's surface become warm enough to evaporate. This creates jets of gas that erode the comet's surface and carry away any solid matter mixed in with the original ice (Fig. 8.3). At each pass near the Sun, the comet ejects a stream of material. These particles receive small accelerations from the pressure of sunlight, from the outflow of gas called the solar wind, and from the gravitational pull of the planets. Over time, their orbits become slightly modified, and the swarm becomes a little more diffuse with every orbit, eventually forming a broad band of material that Earth encounters every year. A meteor shower occurs on the date in the year when the Earth passes nearest to the band of material associated with a comet's orbit. The clumps evident in the Leonid swarm are an indication of its youth: The various streams of particles simply haven't had time to become uniformly distributed along Tempel-Tuttle's orbit. The Leonid shower puts on its most impressive displays when Earth passes near the populous cores of streams ejected during past returns of the comet. Each stream takes its own slightly different path around Sun, and we aren't guaranteed an encounter with them during any given return of the comet.

The new understanding of the origin of meteor showers, coupled with the terrific track record of the Leonids, bolstered astronomers' confidence that another extraordinary meteor storm would fill the skies in November 1899. Newspapers in Europe and America spread the word. In what many hoped would be a prelude to the main

Fig. 8.3 The dusty jets of *Halley's comet* create two of the year's best meteor showers – May's Eta Aquarids and October's Orionids. The European Space Agency's Giotto probe flew past the comet's icy nucleus in 1986. Bright jets of gas and dust stream away from the comet's sunward side, while the night side is silhouetted against the background of bright dust. The bright large spot at the center of the nucleus is believed to be a sunlit hill about 1,600 ft (500 m) high (Max Planck Institut für Aeronomie)

event, the 1898 shower provided an outburst of 100 meteors an hour. But astronomers didn't understand the Leonids as well as they thought, and the well-advertised spectacle simply failed to materialize. This was "the worst blow ever suffered by astronomy in the eyes of the public, and has indirectly done immense harm to the spread of science among our citizens," wrote Charles Olivier, an astronomer at the University of Virginia. In fact, some astronomers had calculated that the swarm's close encounters with Jupiter and Saturn after 1869 had slightly altered its course, bringing it well inside Earth's orbit; this eliminated any possibility of a meteor storm.

The new century brought strong outbursts in 1901 and 1903, and then the Leonid showers slumbered again. More modest outbursts occurred in the early 1930s, but after 1934 rates dropped to their usual trickle. It was looking as though the once-great Leonid shower had finally begun to fizzle out. Judging by the 1899 return, a close approach to a dense portion of the Leonid stream no longer seemed very likely; astronomers held out little hope for a big event in 1966. In his book on meteor science published in 1961, Canadian astronomer D. W. R. McKinley gloomily concluded "it is highly improbable that we shall ever again witness the full fury of the Leonid storm."

Beginning that year, the Leonids exhibited annual enhancements that never exceeded 130 per hour. And that's probably the rate most of the world witnessed on the morning of Nov. 17, 1966. But die-hard observers in the central and western United States watched, often through partly cloudy skies, with the hope that the shower's best years were far from over. The 4-day-old Moon had set well before midnight; Jupiter and Mars bookended Leo. Observers began noticing a better-than-usual shower after midnight; it only improved as the radiant rose ever higher, and it rates increased dramatically around 3 A.M.

Astronomer James Young in Wrightwood, CA, awoke to check the cloud cover and saw numerous meteors through the thinning cloud deck. He headed to nearby Table Mountain Observatory to try to capture it on film – something no one had bothered to prepare for. "The shower peaked around 4 A.M. with some 50 meteors falling per second," he recalled. "We all felt like we needed to put on hard hats! To further understand the sheer intensity of this event, we blinked our eyes open for the same time we normally blink them closed, and saw the entire sky full of streaks – everywhere!" Estimates of the shower's peak exceeded a ZHR of about 100,000. There was no doubt about: The Leonid shower was back!

The Millennium Storms

The most recent suite of Leonid storms never reached the level seen in 1966, but they certainly did not disappoint. In 1998, relatively chunky material ejected by the comet in the 1300s produced a brilliant display of fireballs. A full-blown meteor storm, with a rate reaching 3,700 per hour, materialized briefly over the eastern Mediterranean on Nov. 18, 1999. A strong outburst occurred in 2000, hampered by a bright Moon, but the real show-stopper came in 2001.

Earth passed through several Leonid streams composed of material ejected in the 17th and 19th centuries. The result was a period of storm-level activity lasting several hours, with the best display – about 3,000 per hour – occurring over China and Mongolia. However, the sky show over North America did not disappoint, as observers from Florida to Hawaii were treated to spectacular celestial fireworks, with numerous fireballs and many meteors sporting red, blue and green colors.

At the time, Bill Cooke, the lead for NASA's Meteoroid Environments Office at Marshall Space Flight Center in Huntsville, AL, was one of many astronomers refining models to predict the timing and peak rates of the shower. Expecting that the 2001 Leonids would exhibit a single peak visible from Hawaii, he was part of team deployed to record the event at the U.S. Air Force's Maui Observing Station

(AMOS) atop the island's Mount Haleakala. NASA has a significant stake in meteor outbursts, with an occupied space station, piloted rocket flights, and numerous satellites dedicated to observing Earth and space. Cooke's office functions to provide forecasts that alert satellite operators to meteor threats so that they can take steps to mitigate potential damage, such as feathering solar panels into the shower's direction and turning sensitive components toward Earth. It's the orbital equivalent of battening down the hatches.

However, rather than a single peak, the 2001 Leonids displayed two, and Cooke was roughly in between them. "The 2001 Leonid display was undoubtedly the best meteor shower I have seen," he recalled. AMOS is an Air Force facility used in space surveillance, and as soon as it was dark, multiple lasers – used to help telescopes counteract the turbulent effects of the atmosphere so that they acquire the best images – shot skyward from domes housing various equipment. "On a normal night, this would have been plenty spectacular, but on the night of Nov. 17, the laser show was nothing compared to the celestial fireworks of the Leonids," Cooke wrote. The shower was rich in brilliant fireballs. "Each fireball across the sky was accompanied by the cheers and approving shouts from the large crowd of people gathered in the state park below. Never have I heard a crowd so excited by a meteor shower, and their excitement was matched by my own." The show continued through dawn.

Astronaut Frank Culbertson saw the display from a unique vantage point – the orbiting International Space Station (ISS). "It looked like we were seeing UFOs approaching the Earth flying in formation, three or four at a time," he said. "There were hundreds per minute going beneath us, really spectacular. It's like being in the middle of a hailstorm." Shields on the ISS protect it from meteoroid impacts.

Just as the Leonid displays of the 1830s gave birth to meteor astronomy, the recent showers prompted a resurgence of the field. The increased power of desktop computers in the mid-1990s had finally enabled astronomers to implement an idea first proposed by Irish astronomers a century before – computing the orbits of "fictitious" Leonid meteoroids ejected from comet Tempel-Tuttle on each of its passages through the inner solar system while accounting for the gravitational effects of the planets and other forces that affect the particle positions. The computational task is impressive: Track up to one million imaginary particles as they orbit the Sun over hundreds to thousands of years. Particles intercepting Earth indicate that an outburst, and the relative numbers of those Earth-striking meteoroids provide information used to forecast meteor activity. Much like weather predictions, these forecasts can be a mixed bag. The times of the 1999, 2000, and 2001 Leonid activity peaks were predicted to within a few minutes, but the intensities of the displays were either underestimated (1999) or overestimated (2001). "Obviously, the meteor forecast models still need a bit of work," Cooke said, "but the present state of the art is sufficient for NASA and spacecraft operators to determine times and strategies for safegaurding their orbiting vehicles."

Concern for the safety of satellites represents another hallmark of this new era of meteor science. During past Leonid storms, few if any spacecraft were in Earth orbit; now there are hundreds, some of which perform functions vital to our way of life. The material that creates meteor showers is a fluffy conglomerate of dust, too insubstantial to survive passage through the atmosphere and reach the surface. Meteor showers do not increase the chances of a meteorite impact. Nevertheless, whenever a meteorite does fall, media reports always helpfully include the name of whatever meteor shower is active at the time. Don't buy it. "We occasionally see kilogram-sized Leonid fragments burning up in Earth's atmosphere," Cooke said, "and they appear as very bright fireballs that disintegrate completely before hitting the ground."

Satellites, on the other hand, are quite vulnerable to meteoroids. The enormous speeds of the particles more than make up for their fragility. Investigators agree it's likely that a single strike by a particle from the Perseid meteor shower led to the demise of a communications satellite named Olympus.

A more rugged satellite, our Moon, is repeatedly struck by meteoroids. Between 1970 and 1977, seismic stations deployed there by Apollo astronauts picked up impacts during several meteor showers, including the Leonids. In addition, something never before recorded took place during the 1999 storm. Independent observers monitoring the Moon's night side videotaped six brief flashes no brighter than a fourth-magnitude star. These flashes were caused by the impact of relatively large Leonid fragments.

Although a Leonid particle is composed of very fragile material, it hits the lunar surface at speeds a hundred times faster than a rifle bullet. A fragment weighing a few pounds can create a blast as powerful as one from a few hundred kilograms of high explosives. The impact vaporizes the meteoroid as well as some of the lunar surface within a few feet of the impact point. This creates a brief flash detectable through telescopes on Earth and can hurl tons of lunar soil and rock on ballistic trajectories. The real danger to astronauts and equipment on the Moon is not from a direct or nearby meteoroid hit, but from the spray of "shrapnel" associated with more distant strikes. By 2008, more than 100 lunar impacts had been detected, and current models indicate that the Moon is struck by a meteoroid with a mass greater than 2 lb (1 kg) more than 260 times a year. Scientists admit that the flux of meteoroids is so uncertain that this number could be several times higher.

Observing Meteor Showers

Once, when out under the stars observing the Perseid meteor shower, there was a burst of conversation followed by long lull in the chatter. Suddenly, my stepson Chris, a 9-year-old first-time meteor watcher, likened the experience to something he knew better: "It's kind of like fishing," he said. That's a pretty fair description, and it's one I've used ever since: You're out in the presence of nature, enjoying the company of friends and a free fireworks show. To maximize the experience, dress warmly, set down a blanket or lawn chair at a dark site, and get comfortable. But instead of turning to your companions during the conversation, keep your eyes on the stars. Meteors will appear in every part of the sky, but interference from streetlights or the Moon will screen out the fainter and more frequent ones, so the darker the site, the better. Put bright light sources behind you or locate your observing site so that nearby buildings shield you from as many as possible. Remember, you want to keep your eyes from unnecessary exposure to light as long as possible so they can reach the maximum sensitivity conditions will allow.

On any night of the year, meteors appear faster, brighter, and more numerous after midnight. That's when your location has turned into the Earth's direction of motion around the Sun and plows into meteor particles nearly head-on, rather than having them catch up from behind. The peak activity of a meteor shower occurs in the hours when the Earth passes closest to the orbit of the shower particles. The ideal circumstance for any observer is for the shower to peak at a time when its radiant is highest in the sky during the morning hours; most of the year's best showers have the potential to meet these criteria. Table 8.1 lists information on the major meteor showers, including the approximate date of maximum activity, typical hourly rates at peak, and the general appearance of the meteors. The calendar date for a given shower's peak varies a day or so due to leap years. Refer to Appendix A for the correct date and hour in a given year. The hour is only really important for the clumpy Draconids, Quadrantids, and of course the Leonids. Most meteor streams are diffuse enough that looking at the actual hour of peak activity will have little impact in what you'll see. A more important consideration is the the altitude of the meteor shower's radiant, because the higher it is, the more meteors you'll see.

Most of the meteors seen during the annual showers arise from fluffy particles not much more massive than sand grains. As the particle enters the atmosphere, it collides with gas atoms and molecules and becomes wrapped in a glowing sheath of heated air and vaporized material boiled off its own surface. Many of the faster, brighter meteors may leave behind persistent trains. Occasionally, you'll see a fireball – a spectacular meteor so bright that it outshines even Venus. More rarely, a fireball will fragment, undergoing flares along its path and possibly trailing "sparks." Such a fireball is often called a bolide. The following section provides background on the major meteor showers so that you can become more acquainted with their individual characters.

Quadrantids (kwo-DRAN-tidz): A strong shower well placed for northern observers, cold temperatures prevent the Quadrantids from being more widely appreciated. With meteors visible between Dec. 28 and Jan. 12, the Quadrantids exhibit a narrow peak around Jan. 3 and a top ZHR of 120, with reported rates varying from 60 to 200. Roughly 5% of the meteors leave trains.

Quadrantid particles hit the atmosphere at speeds of nearly 92,000 mph (148,000 km/h), which, perhaps incredibly, is middling as meteors go. The moderate speed is a result of the stream's orbit, which intersects Earth's orbit quite steeply (Fig. 8.4); the particles practically fall straight down on us. When the shower was first recognized as annual in 1839, the radiant occurred in a constellation that is now no longer recognized – Quadrans Muralis (the Wall Quadrant) – that has since been divided between Hercules, Boötes and Draco. Now, the radiant occurs in Boötes and is located north of the familiar kite-shaped pattern, roughly midway between Hercules and the last star in the Big Dipper's handle.

Until late 2003, this was the only major meteor shower without a known "parent" body. In March of that year, astronomers found a near-Earth asteroid

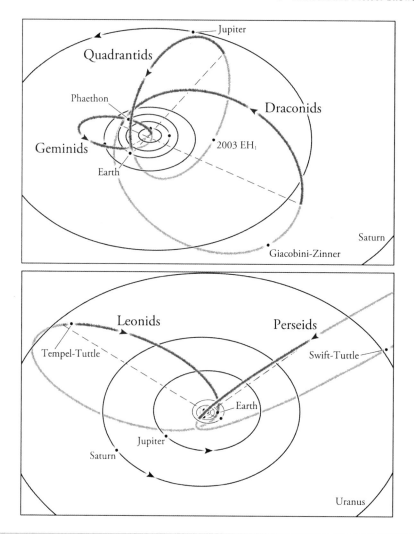

Fig. 8.4 This illustration shows the orbits of several well-known meteor showers and their parent bodies. Lighter shading indicates where the orbits pass beneath the ecliptic plane. Object positions are correct for Dec. 31, 2020. *Top:* Geminid particles hit Earth near the midnight side, making it one of the few showers best seen in the late evening. Similarly, Quadrantid meteors plunge steeply downward onto Earth's orbit. The Draconid swarm actually catches us from behind, making its meteors among the slowest. *Bottom:* Earth runs nearly head-on into Perseid and Leonid particles, which makes their meteors among the fastest

designated 2003 EH$_1$. As further observations clarified the object's orbit, Peter Jenniskens, an astronomer at NASA's Ames Research Center in California, noted its similarity to the meteoroid stream's orbit. He had previously shown that the Quadrantids must be a relatively young shower, created within the last 500 years. The reason is that the shower's farthest point to the Sun occurs just inside Jupiter's orbit. Each time Jupiter approaches the stream, its gravity perturbs some of the meteoroids and rapidly disperses them, so to be as compact and active as it is the stream must be no older than this. Jenniskens predicted that the parent might have been hiding as a asteroid-like object in a high-inclination orbit. He suspects that this object is really a fragment from the breakup of a comet, that this event gave birth to the Quadrantid shower, and that more big fragments may yet be found within the stream.

Lyrids (LYE-ridz): Although the Lyrids were not recognized as annual occurrence until 1839, Chinese records of its activity date back to Mar. 23, 687 B.C.: "7th year of Duke Zhang of Lu … During the night, stars fell like rain, together with the rain." This remains the earliest observation that can be connected to a known meteor shower.

The Lyrids appear from Apr. 16 to 25 and peak around Apr. 22; the radiant lies near Lyra. Despite the low annual rate, the Lyrids have the capacity for impressive outbursts of over 50 meteors per hour. These displays happened in 1803, 1922 and 1982, which suggests a 60-year period. In 1867, the shower was linked to its parent comet (C\1861 G1). The Lyrid meteors are bright and rather fast: 110,000 mph (176,000 km/h). About 15% of the meteors leave persistent trains.

Studies by the Audrius Dubietis and Rainer Arlt of the International Meteor Organization indicate that the Lyrids are not especially well behaved. The time of the maximum can vary by nearly half a day, the length of the maximum changes too, and there are additional year-to-year variations in the ZHR.

Eta Aquariids (EY-tuh Ah-KWAIR-ee-idz): This is the first of two showers derived from *Halley's comet*. The Eta Aquariids occur from Apr. 19 to May 28, with a broad peak centered on May 5–6. The shower was discovered in 1870 and linked to Halley 6 years later. The radiant is located near the Y-shaped asterism in Aquarius known as the Water Jar and is named for one of its stars.

This shower is best for observers in the southern hemisphere, where the hourly rate can exceed 50; from northern latitudes, expect fewer than half of this. The shower's ZHR may rise as high as 85 in years when a 12-year variation caused by Jupiter's influence brings the stream closest to Earth; however, this cycle is on the wane throughout the first half of the decade. The meteors are among the fastest – 148,000 mph (238,000 km/h) – and typically faint; the brighter ones have a yellowish color, and perhaps 30% leave trains.

Southern Delta Aquariids (DEL-tuh Ah-KWAIR-ee-idz): This is the busiest of a group of streams active at the same time and, as the name suggests, it's best seen in the southern hemisphere. They may be seen between Jul. 12 and Aug. 23 and peak near Jul. 30. The radiant is located near the star Skat, otherwise known as Delta Aquarii, the third brightest star in the constellation. These meteors hit the atmosphere at moderate speed – 92,000 mph (148,000 km/h) – and they tend to be faint, leaving few trains behind. The stream is debris from comet 96P/Machholz.

Perseids (PUR-see-idz): The best-known of all meteor showers, the Perseids never fail to put on a good show and, thanks to its late-summer peak, are usually widely observed. The earliest record comes from China in July of A.D. 36: "At dawn, more than 100 small meteors flew on all four sides." The shower was first recognized as annual in 1835, and in 1862 it was the first meteor shower to be linked to a comet. Generally visible from Jul. 17 to Aug. 24, the speed – 132,000 mph (212,000 km/h) – brightness and high fraction of persistent trains (nearly half) easily distinguish Perseid meteors from other showers active at this time. The shower peaks around Aug. 12, with a ZHR of about 100. Outbursts are known to occur when the parent comet is in the vicinity and as part of a 12-year cycle caused by Jupiter's influence, which brings the stream particles a little closer to Earth. The last Jupiter-related outburst occurred in 2004, so 2016 could bring another.

Perseid rates gradually increased from a mid-1960s ZHR of 65 to about 90 in the early 1980s. The rising rate buoyed hopes that comet 109P/Swift-Tuttle would soon return; it had last been seen in 1862 was thought to orbit every 130 years. Outbursts in 1991 and 1992 heightened this expectation, and astronomers finally recovered the inbound comet in late 1992. The possibility of a large outburst in August 1993 led to a first-ever NASA decision to postpone a scheduled launch based on meteor shower activity. The agency decided to postpone the launch of the space shuttle Discovery in order to minimize the risk to the vehicle and crew. The outburst materialized with a ZHR of about 300 Perseids an hour.

Russian cosmonauts aboard the orbiting Mir space station, who could watch the fireworks in the atmosphere below them, also took precautions. During the shower's peak, they took refuge in the Soyuz spacecraft docked to the station and closed the hatches, ready for a quick getaway in the event Mir became heavily damaged. As many as 60 particles reportedly hit Mir, many of which were audible as "pings" to the cosmonauts, but the most serious damage turned out to be a few holes in the station's numerous and easily damaged solar panels.

The 1993 Perseids hold the dubious distinction of causing the first probable "kill" of a functioning satellite by a meteoroid. The satellite, named Olympus, was an $850 million telecommunications satellite operated by the European Space Agency. In fact, it was the largest civilian telecom satellite at the time of its 1989 launch. While leaving work on Aug. 11, the program's spacecraft manager, Doug Caswell, heard a colleague joke about the Persied threat: "Olympus dies at midnight!" As luck would have it, something *did* happen just before the midnight hour, when the Perseid rate crested 100 per hour. The satellite unexpectedly began spinning out of control and subsequently broke contact with Earth. The satellite carried enough fuel for normal operations, but early problems had left it with no margin for anything unexpected. To stop the twice-per-minute spin, controllers rapidly depleted the remaining fuel. Once control over the satellite was reestablished, so little of the attitude-control fuel remained that ground control was left with no choice but to declare the satellite dead, place it in a graveyard orbit, and shut it down.

Investigators concluded that one of the satellite's solar panels had probably sustained a hit from a Perseid meteoroid. "It was not absolutely confirmed," Caswell said, "but there is no better explanation." Just seconds before the strike, controllers had executed a planned firing of thrusters. The impact vaporized the meteoroid into a cloud of ionized gas. The meteor vapor, augmented by the thruster exhaust gases, together provided a conduit that permitted electrical charges on the surface of the spacecraft to penetrate inside it, and a strong jolt damaged sensitive attitude-control electronics. Olympus demonstrated the importance of meteor-induced electrostatic discharges in causing spacecraft anomalies.

Another incident involving the Perseids occurred over Alaska during the 2009 outburst. At 5:23 UT on Aug. 13, NASA's Earth-observing Landsat 5 satellite began tumbling out of control. Within hours, the flight operations team was able to stabilize the satellite, and normal operations resumed after 2 days. Later analysis showed that the Perseid radiant was visible to the satellite at the time it lost control. This, plus the general similarity to the Olympus incident, suggests that Landsat 5 took a hit from a piece of the Perseids.

Draconids (DRAK-uh-nidz): Draconid activity occurs between Oct. 6 and 10, with a maximum on Oct. 8. Draconid meteors move slowly, with encounter speeds of less than 45,000 mph (72,000 km/h), and are typically faint. The radiant is located in the head of the constellation Draco (DREY-koh) the Dragon and never sets from northern hemisphere locations north of Las Vegas, NV, and Nashville, Tenn. Better yet, it's highest in the early evening in early October. In 1933 and 1946, the shower produced brief but intense meteor storms exceeding 5,000 per hour. In 1998, the shower spiked to a rate of about 500 over eastern Europe, and in 2005 a modest outburst produced about 40 meteors an hour. Preliminary data from the moonlight-hampered 2011 outburst shows a peak of about 350 per hour.

The Draconid shower is periodic rather than annual. Most activity on record occurred in years when the stream's parent comet, 21P/Giacobini-Zinner, made its closest approach to the Sun, which happens every 6.6 years. Astronomers modeling the comet's dust streams successfully predicted a strong display in 2011 of at least 350 per hour, and they anticipate similarly strong outbursts in 2018 and 2025. More modest activity, at a level akin to that seen in 2005, is expected in 2012, 2014, 2019, 2023 and 2030.

Orionids (ohr-EYE-uh-nidz): This is the sister stream of the Eta Aquariids, as both arise from comet 1P/Halley. Although discovered as annual in 1864, the connection to Halley wasn't made until 1911. The radiant lies in northern Orion near its border with Gemini. Orionid meteors can be found between Oct. 2 and Nov. 7, with a typical peak rate of about 25 around Oct. 21. Orionid meteors are among the fastest, hitting the atmosphere at about 148,000 mph (238,000 km/h), and are generally faint. About 20% leave trains that persist a second or so. The observed rates show fluctuations of more than 10% due to a 12-year Jupiter-induced cycle. They were higher than normal from 2008 to 2010, so Orionid returns will decline out to 2015 before rebounding later in the decade.

Leonids (LEE-uh-nidz): Leonids are active from Nov. 10 to 23, with a maximum around November 17 of about 20 meteors per hour. Because Earth runs into the orbiting particles almost directly head-on, Leonid meteors travel faster than those of any other shower – 159,000 mph (256,000 km/h) – and as a result roughly half of them leave persistent trains, some of which can be seen for several minutes. The shower's most notable feature, of course, is its habit of producing dramatic outbursts and meteor storms when we intercept streams of dense material ejected at previous returns of the parent comet, 55P/Tempel-Tuttle.

Our planet passed through such streams annually from 1998 to 2002, but the Leonids will not rise to this level of activity for decades. Several modest outbursts have been observed since 2002, but these are faint echoes of the Lion's roar, with ZHRs similar to what one can see annually in the Geminid and Perseid showers. The outbursts actually are even less impressive visually because they're rich in small particles that produce faint meteors; only the smallest meteoroids have drifted far enough from the comet's orbit to intersect Earth. Worse yet, most computer models suggest that Jupiter's tug on the dense Leonid streams will cause them to miss Earth on their next pass, making full-fledged meteor storms impossible until at least 2098. Outbursts are likely after 2030, however, and at least one modeler suggests that a meteor storm is possible in 2034.

Fig. 8.5 Bullet sky. The annual Geminid meteor shower peaks in mid-December and never disappoints. This all-sky view from Mount Uludag, Turkey, compresses two hours of the 2009 Geminid shower into a single image by merging 150 short exposures. The winter stars share the sky with twenty-six bright Geminid meteors. Lights from the nearby city of Bursa wash out the northwestern sky (Tunç Tezel)

Geminids (JEM-uh-nidz): The Geminid shower is one of the year's best and most reliable displays, with numbers exceeding the August Perseids and a radiant, located near the star Castor, that is well placed for evening viewing (Fig. 8.5). Active from Dec. 7 to 17, the Geminids reach maximum near Dec. 13 with a ZHR of 120. The shower's meteors strike the atmosphere at medium speed – 78,000 mph (126,000 km/h) – are often bright, and appear white and yellow-white.

The parent body of the Geminids, discovered only in 1983 and named 3200 Phaethon, is classed as an asteroid, not a comet. This is really another clue that the distinctions made by astronomers based on limited observational information aren't necessarily reflected in nature. Comets and asteroids are really parts of a continuum of bodies with different histories of formation and evolution. For example, several objects in the main asteroid belt between Mars and Jupiter exhibit comet-like behavior, sprouting tails of gas and dust when nearest the Sun. Eventually all of the ice that gives rise to cometary behavior evaporates, leaving only the rocky material behind. These bodies could remain intact to form an "extinct comet" class of asteroids, and planetary scientists suggest that perhaps as many as a third of the asteroids whose orbits cross Earth's actually may be defunct comets.

Ursids (UR-sidz): The poorly observed Ursid shower has the disadvantage of occurring between two impressive ones, the Geminds and the Quadrantids. The Ursid radiant, which lies near the second-magnitude star Kochab in Ursa Minor, never sets for most northern observers and is best placed for observation in the early evening. The shower produced significant outbursts, with ZHRs of about 50, in 1945 and 1986. Ursid meteors are active Dec. 17 to 26 and peak around Dec. 22, usually with about ten meteors an hour.

It's clear from the brief descriptions above that we know relatively little about the streams of asteroid dust and comet litter that our planet occasionally passes through. The only way to learn more about what's really out there is to monitor what happens when Earth encounters this debris. Much of the observational work is done by dedicated amateur groups trained and coordinated by the non-profit International Meteor Organization. The results of these observational campaigns are made available to professional astronomers around the world. Just as weather observers supplied the ground truth that helped improve weather forecasts, meteor observers help astronomers improve their computational models by witnessing and reporting on what really happens in the sky. For more about meteor showers on the Web, see the resources in (Table 8.2).

Look and Listen

Rarely, something truly surprising will accompany the streak of a bright meteor: sound. One method for producing meteor sound is well understood, the other … not so much.

When a large fireball penetrates very low into the atmosphere, it's accompanied by shock wave that rapidly decays into sound. Just as we can hear aircraft shock

Table 8.2 Meteors on the Web

American Meteor Society
www.amsmeteors.org

Gary Kronk's Meteor Showers Online
meteorshowersonline.com

International Meteor Organization
www.imo.net

Lunar impact monitoring news
www.nasa.gov/centers/marshall/news/lunar

NASA's Leonid MAC site
leonid.arc.nasa.gov

Spaceweather.com
www.spaceweather.com

waves as sonic booms, a distant rumble may follow a deeply penetrating fireball by a few minutes. Traveling from the altitude of the meteor, the sound waves take several minutes to reach the ground and observers hear them long after the meteor has disappeared.

What has been more puzzling are persistent reports of seemingly impossible sounds heard simultaneously with a fireball's passage. A Chinese record from 817 described a meteor "which made a noise like a flock of cranes in flight." Edmond Halley noted the phenomenon among the eyewitness accounts he studied when reconstructing the path and altitude of a fireball seen over England in March 1719. Halley found that the meteor's height "exceeded sixty English miles" with a speed of "above 300 such miles in a minute." But he concluded:

Of several Accidents that were reported to have attended its Passage, many were the Effect of Fancy, such as the hearing it hiss as it went along, as if it had been very near at hand…

Witnesses continued to tell of sounds from bright meteors – and also during auroral displays – but these observations gradually disappeared from the scientific literature because they were considered psychological reactions rather than physical descriptions. And indeed, any investigator looking to explain these anomalous sounds through physical causes must contend with several serious problems. First, the sounds are very rare and their occurrence is inherently unpredictable. Very few people have heard them, and the sounds themselves were not recorded until the 1990s. Second, witnesses standing next to one another sometimes disagree on whether or not a sound was produced. Third, sounds heard when the meteor is seen imply some sort of electromagnetic radiation moving at the speed of light – but no known method for producing audible noise this way could work at the distances involved. Fourth, the portion of the electromagnetic spectrum considered most likely to be responsible was the radio regime, but until recently it had not been demonstrated that meteors emitted radio waves.

That's where things stood on Apr. 7, 1978, when a fireball some 40 times brighter than a full Moon streaked above the cities of Sydney and Newcastle in New South Wales, Australia. There were numerous reports of sounds from the meteor: "like an express train or bus traveling at high speed," "an electrical crackling sound," "a loud swishing noise," "like steam hissing out of a railway engine." Colin Keay, a physicist at the University of Newcastle, heard of these reports and became convinced there was some physical explanation. He found that meteors had been observed in all but the Very Low Frequency (VLF) portion of the electromagnetic spectrum, a band that contains frequencies from 3 to 30,000 Hz. Keay showed in laboratory experiments that VLF radio waves could set ordinary objects like hair or eyeglasses into motion, vibrating them at audio frequencies and creating a noise very close to the observer. This would explain how adjacent observers could disagree on whether or not a meteor had made a sound. He argued that the source of the VLF radio waves lay in the highly ionized wake of the fireball, which briefly entraps a portion of Earth's magnetic field. As the wake cools, this "magnetic spaghetti" disentangles itself and in the process releases a burst of radio energy. Keay applies the term "electrophonic" to describe sounds generated in this way. His ideas gained ground in the 1980s with theoretical studies of the wakes of bright fireballs and with the first detection of VLF radiation from a meteor in 1990.

In his study of the 1833 Leonids, Denison Olmsted noted reports of "slight explosions, which usually resembled the noise of a child's pop-gun" but didn't know what to make of them. Fittingly, it was during the 1998 Leonid outburst that researchers in Mongolia obtained the first instrumental detection of electrophonic meteor sounds. Two fireballs of magnitude −6.5 and −12 produced short, low-frequency sounds described by observers as "deep pops," and they were simultaneously recorded by microphones in a special setup.

So while meteor sounds remain a poorly understood phenomenon, they are unquestionably real, and even casual meteor watchers would do well to keep both their eyes and ears open. This rare phenomenon produces sounds variously described as "crackling," "rushing," "popping," "vits" and "clicks" when meteoric radio waves interact with nearby electrically conducting materials.

In ancient China meteors were viewed as heavenly messengers. The messages they carry – whether a flash of light, a burst of radio waves, or even the rare rock from space – are eagerly deciphered by planetary scientists and have revealed volumes about Earth's history and the violent past of the solar system. Until 2006, when NASA's Stardust mission returned dust captured during a flyby of comet 81P/Wild, meteorites and samples returned by lunar missions during the 1960s and 1970s were the only pieces of extraterrestrial material available for direct study. That's slowly changing, most recently with the 2010 return of dust from asteroid Itokawa, a feat carried out by the Japanese probe Hayabusa. Still, the expense of these missions guarantees that they will be infrequent, which means that meteorites will remain "the poor man's space probe" for some time to come.

References

Brown P (1999) The Leonid meteor shower: Historical visual observations. Icarus 138: 287–308

Caswell D (1998) Olympus and the 1993 Perseids: Lessons for the Leonids, Nov. 6, 1998. http://sci2.estec.esa.nl/leonids/leonids98/OLYMPUS_and_the_1993_Perseids/index.htm. Accessed 15 Oct. 2010

Close S (2010) Space invaders: Shooting stars can shoot down satellites. IEEE Spectrum, April, 39–43

Cooke WJ (2010) Impressions of 2001 Leonid display. Private communication

Dodd RT (1986) Thunderstones and shooting stars: The meaning of meteorites. Harvard University Press, Cambridge, Massachusetts

Duggan P (2010) Almost-close encounter: Meteorite hits Lorton doctor's office. Wash. Post. January 21, B03

Halliday I, Blackwell AT, Griffin AA (1984) The Frequency of Meteorite Falls on the Earth. Sci. 223:1405–1407

McBeath A (2011) IMO meteor shower calendar. International Meteor Organization. http://www.imo.net/calendar/2011. Accessed 1 Dec. 2010

McBeath A (2010) IMO meteor shower calendar. International Meteor Organization. http://www.imo.net/calendar/2010. Accessed 15 Oct. 2010

Jenniskens P (2006) Meteor showers and their parent comets. Cambridge University Press, New York

Jenniskens P (2004) 2003 EH1 is the Quadrantid shower parent comet. Astron. J. 127: 3018–3022

Keay CSL (1980) Anomalous sounds from the entry of meteor fireballs. Sci. 210: 11–15

Kronk G (1988) Meteor showers: A descriptive catalog. Enslow Publishers, Hillside, New Jersey

Littman M (1988) The heavens on fire. Cambridge University Press, New York

Maslov M (2011) Future Draconid outbursts (2011–2100). WGN 39:3, 64–67

Moser D (2008) Observing prospects for the 2011 Draconid meteor shower. Private communication

Penkenier DW, Xu Z, Jiang Y (2008) Archaeoastronomy in East Asia: Historical observations of comets and meteor showers from China, Japan, and Korea. Cambria Press, Amherst, New York

Phillips T (2002) Space station meteor shower. http://science.nasa.gov/science-news/science-at-nasa/2002/17may_issmeteors. Accessed 1 Dec. 2010

Roylance F (2010) Monday's meteor fell on Lorton, Va. doctors' office. Baltimore Sun. http://weblogs.marylandweather.com/2010/01/mondays_meteor_fell_on_lorton.html. Accessed 15 Oct. 2010

Sekanina Z, Chodas PW (2005) Origin of the Marsden and Kracht groups of Sunskirting comets. I. Association with comet 96P/Machholz and its interplanetary complex. Astrophys. J. Supp. Ser. 161: 551–586

Simon SB, Grossman L, Clayton RN et al (2004) The fall, recovery, and classification of the Park Forest meteorite. Meteorit. Planet Sci. 39: 625–634

Young J (1998) Photographing a thousand meteors in 90 minutes. http://leonid.arc.nasa.gov/1966james.txt. Accessed 15 Oct. 2010

Zgrablić G, Vinković D, Gradečak S et al (2002) Instrumental recording of electrophonic sounds from Leonid fireballs. J. Geophys. Res. 107: 1124–1133. doi:10.1029/2001JA000310

Chapter 9

Unpredictable Events

Over the course of about a minute on Mar. 19, 2008, keen-eyed observers looking toward the constellation Boötes might have noticed the brief, mysterious appearance of a dim star that then quickly faded from view. At 2:13 A.M. EDT, instruments on two spacecraft detected a burst of high-energy gamma rays, a signature associated with the death of a star and the birth of a black hole. The star had reached a point in its energy-producing career where it essentially ran out of fuel. It began to collapse under its own weight and in its innermost regions formed a black hole. As gas rained toward the black hole, some of it was paradoxically diverted into a pair of oppositely directed, outwardly moving particle jets. Racing at speeds only a whisker slower than that of light itself, the jets sliced completely through the collapsing star – and one of them just happened to be pointed almost directly toward Earth. Even though most of this jet's emission was radiation too energetic to see, a small amount was in the form of visible light, and anyone looking at the spot would have seen a dim star brightening to magnitude 5.3 before fading back to invisibility about 50 s later. We know this because robotic telescopes in Chile *were* looking, and they immediately slewed to the star's position once NASA's gamma-ray-sensing Swift satellite had determined it. What's amazing about this event is that the explosion occurred not within our galaxy or even in any of its nearest neighbors. Instead, the dying star was 7.5 billion light-years away — so distant that the blast occurred when the universe was less than half its present age, long before our own Sun was born. Astronomers say that the event was so bright because the particle jet happened to be oriented almost directly into our line of sight. They expect similar alignments about once a decade, on average.

Throughout this book we've emphasized what are now the easily predictable motions and arrangements of the Moon and planets, but events like the one described

F. Reddy, *Celestial Delights: The Best Astronomical Events through 2020*,
Patrick Moore's Practical Astronomy Series, DOI 10.1007/978-1-4614-0610-5_9,
© Springer Science+Business Media, LLC 2012

above are important reminders that many interesting phenomena appear without a schedule. For those of us who look forward to surprises, this chapter presents three types of unpredictable naked-eye events in order of the likelihood of their appearance this decade.

The Northern Lights

They may first appear as a ghostly milky glow low in the north, too dim for the human eye to detect any color but bright enough to silhouette clouds near the horizon. They may then develop into steady greenish arcs, morph into writhing curtains of yellow-green light, spawn quiescent patches and veils or accelerate into rapidly moving rays. In the best displays, the basic yellows, greens, reds and blues mix into a complex palette of restless light (Figs. 9.1 and 9.2).

These are the northern lights, a regular nighttime phenomenon seen from high-latitude locations like Alaska, Canada, Iceland and northern European countries. Rarely, the northern lights appear at much lower latitudes, such as the southernmost U.S. and even the Caribbean. At these times they often take the form of a diffuse crimson glow, which inspired Galileo to dub the phenomenon the *aurora borealis*, literally meaning "northern dawn." A corresponding display occurs at southern latitudes (*aurora australis*) but the zone of most frequent occurrence lies over Antarctica and the Southern Ocean, regions that were largely uninhabited until exploratory voyages by Europeans. However, the same events that push the aurora southward to the U.S. and the Mediterranean also bring the southern aurora northward to New Zealand, Australia, South Pacific islands, and the southernmost regions of Africa and South America. The term aurora refers to the general phenomenon regardless of its location.

The aurora's lively, colorful forms naturally appear in much American and European folklore. A Scottish legend connects the lights with supernatural creatures called Merry Dancers, who fight in the sky for the favor of a beautiful woman. A Danish story explains the lights as reflections from flocks of geese trapped in the northern icepack and flapping to free themselves. Among the Sámi in northern Scandinavia, it was believed that the aurora could scorch the hair of those foolish enough to leave the house without a cap. In Alaska and eastern Norway, children were believed to be at special risk from the northern lights. Greenland Inuits thought that the aurora represented signals from dead friends who were trying to contact the living. The Fox Indians of Michigan that whistling to the light would conjure up spirits. To the Tlingit Indians of southeastern Alaska, the flickering glow was a battle between the spirits of fallen warriors and its appearance foretold catastrophes and bloodshed.

When the lights stray from their usual polar hangouts, they really make an impression. The Greek philosopher Aristotle is often credited as the first to attempt to discuss the aurora scientifically. In his book *Meteorologica*, written in the fourth century B.C., he describes auroral forms as burning flames, chasms, torches and, most imaginatively, goats. An early Chinese record describes an aurora as a "red

Fig. 9.1 The sky aglow. An aurora brings a splash of color to an Alaskan evening. The glow comes from the same source as neon signs and fluorescent light tubes — gas excited by collisions with energetic electrons. The electrons stream toward polar regions whenever solar storms impact Earth's magnetic field (Photo by the author)

Fig. 9.2 From steam to green. A bright yellow-green auroral curtain blends with a plume of steam rising above Chena Hot Springs Resort near Fairbanks, Alaska. The yellow-green glow, which is emitted by oxygen atoms, is the brightest and most commonly seen auroral light because the human eye is most sensitive to this color (Photo by the author)

cloud spreading all over the sky." Toward the end of the reign of the Roman Emperor Tiberius, a red aurora fooled watchmen, who noted a dusky glow in the direction of the important port of Ostia, thought it was ablaze, and rushed to its aid. In 1583, such "fires in the air" mobilized thousands of French pilgrims, who prayed to avert the wrath of God. At Copenhagen in 1709, a red aurora led to the deployment of drum-beating military detachments. On Sept.15, 1839, an intense aurora sent fire departments throughout London, and the same thing happened in March 1926 in Salzburg, Austria. As twilight faded on Sept. 18, 1941, the sky above the central U.S. states turned crimson and strong auroral activity persisted until dawn and was sighted in Florida and southern California.

The aurora occurs in two vast glowing ovals located near Earth's north and south magnetic poles such that their longest dimensions stretch closest to the equator on the planet's midnight side. The auroral ovals mark the places where energetic electrons and, less importantly, protons pour into the atmosphere and collide with atoms and molecules of atmospheric gases. After a collision, the atoms briefly remain in an excited state. To return to their preferred minimum-energy state, they must shed the extra energy by emitting light at specific wavelengths characteristic of each gas. This is the same process operating in a neon light or a fluorescent tube. The aurora typically glows at altitudes between 62 and 155 miles (100 to 250 km). Particles energetic enough to reach the lowest altitudes excite nitrogen molecules, which emit barely visible blue and red light in response, and oxygen atoms, which emit faint red and bright yellow-green light. In a typical aurora, this yellow-green light is the brightest single emission because this is the color to which the human eye is most sensitive. Particles able to penetrate no deeper than 124 miles (200 km) excite oxygen atoms to their lower energy levels, which produces mostly red light and is responsible for the red aurora. However, if some of the incoming particles have high-enough energy to penetrate to lower altitudes, then the aurora will be green with a transition to red at the top. If some of these particles plunge to heights of less than 62 miles (100 km), they encounter mostly nitrogen and oxygen molecules that emit mostly red light, so the resulting aurora is green with red on the bottom.

By the eighteenth century, it had become clear that auroral displays were associated with significant disturbances in Earth's magnetic field, and in the latter half of the 19th century scientists increasingly understood that the source of these disturbances has something to do with sunspots. "We must therefore conclude," wrote Yale astronomer Elias Loomis (1811–1889) in the mid-1860s, "that these three phenomena – the solar spots, the mean daily range of the magnetic needle, and the frequency of auroras – are somehow dependent the one upon the other, or all are dependent upon a common cause." It's easy to forget the Sun's enormous physical scale, which dwarfs the planets. For a reminder, turn to Fig. 9.3, which compares a solar eruption imaged by NASA's Solar Dynamics Observatory satellite with Jupiter and Earth.

The changing particle fluxes arise from a vast, complex, invisible structure that surrounds Earth. It's called the magnetosphere, and we can think of it as a kind of magnetic cocoon that traps and circulates electrically charged particles near our planet. The famous Van Allen radiation belts make up a small part of this structure. The connection to the Sun comes through a fast-moving outflow of electrons and ions called the solar wind, which carries the Sun's magnetic field out into space. Typically moving at about 1 million mph (1.6 million km/h), the solar wind strikes Earth's magnetosphere and flows around it, compressing the sunward side into a blunt nose and stretching the night side far beyond the Moon's orbit. The magnetosphere responds to changes in the speed and density of the solar wind by altering its shape – denser, faster winds compress it more strongly. Constant buffeting by the solar wind sets charged particles within the magnetosphere into motion; these currents ultimately flow into the polar atmosphere, where they light up the auroral ovals. Under certain conditions, the solar wind's magnetic field actually can merge with Earth's, creating explosive events within the magnetosphere's long tail that drive clouds of particles toward our planet. In any given year, these types of events are a bit more likely during spring and autumn, when Earth makes its minimal tilt toward the Sun and the magnetosphere becomes more favorably oriented for coupling with the solar wind.

Scientists now use the term "space weather" to refer to events and processes that can affect conditions in the space environment near Earth. Transient events on the Sun can greatly alter the speed and density of the solar wind. Regions called coronal holes generate fast, broad torrents of solar wind that can persist for months, creating recurrent activity every time the Sun's rotation turns them our way.

Fig. 9.3 An erupting prominence on the Sun dwarfs Earth and Jupiter, which are shown to scale in this montage (NASA images montage by the author)

Explosive solar flares can give rise to a fast pulse of particles and a flood of X-rays capable of blacking out high-frequency radio communications on the Earth's entire day-lit hemisphere. But the most dramatic space-weather effects accompany coronal mass ejections, or CMEs.

CMEs are enormous clouds of material that erupt from the solar atmosphere and race through interplanetary space. Somehow a portion of the Sun's magnetic field undergoes a sudden disruption, stretching and twisting like a rubber band until it snaps. When it does, the Sun launches as much as 10 billion tons of gas our way. The cloud moves so fast that it can cross the gulf of space to Earth in less than 15 h, a record set by an August 1972 CME clocked at 6.4 million mph (10.3 million km/h). For maximum terrestrial effect, a CME must be launched from near the center of the Sun onto a collision course with Earth; move at least a third as fast as the CME speed record, which gives it a large amount of kinetic energy; and contain a magnetic field with the opposite polarity of Earth's, which allows the two fields to merge.

Fig. 9.4 An aurora arcs across the U.S. at 9:45 P.M. EST on Nov. 5, 2001. Cities and the dark shapes of the Great Lakes (center) can be seen, as well as clouds in eastern Canada illuminated by the northern lights. Skywatchers as far south as Florida were treated to vivid displays. This image was acquired by the U.S. Air Force DMSP F15 satellite about an hour after the geomagnetic storm began (Meteorological Satellite Applications Branch, Air Force Weather Agency)

Because CMEs move several times faster than the ambient solar wind, a shock wave forms in front of them. Within that shock, solar wind ions bounce around and accelerate until they reach speeds too high for the shock wave to contain them – nearly 20% the speed of light. Reaching Earth within an hour of the CME's launch, the particles are channeled toward the poles and create a "radiation storm" that poses a risk for astronauts, can affect satellite systems and may interfere with high-frequency communications across the polar regions. Hours to days later, the shock wave arrives and suddenly compresses Earth's magnetosphere, which immediately brightens the auroral ovals. Shortly thereafter, the CME itself arrives, ushering in the main phase of the ongoing geomagnetic storm. The CME's magnetic field merges with Earth's, which opens the door of our planet's magnetic shield and allows a flood of new energy inside. The magnetosphere erupts explosively, hurling clouds of energetic particles toward Earth, and the enhanced currents expand the auroral ovals toward the equator (Fig. 9.4). In an extreme event, observers as far south as El Salvador and Venezuela may witness the ghostly apparition of the northern lights – the visible manifestation of these incredible energies.

We're increasingly dependent on space technologies that are extremely sensitive to the gusts and gales of space weather, but the effects aren't restricted to near-Earth space or even the upper atmosphere. The intense auroral currents can induce electricity in long conductors on the ground, such as pipelines, communications cables or power distribution networks. Predicting what to expect when a CME bears down on us is important for taking precautions and mitigating damage.

Yet in September 2003, the U.S. House Appropriations Committee recommended significantly reduced funding for the monitoring work of the National Oceanic and Atmospheric Administration's Space Environment Center (SEC) in Boulder, CO, and the Senate Appropriations Committee completely eliminated funding. On Oct. 19, as if on cue, the first of a trio of enormous and complex sunspot groups appeared on the Sun's eastern limb. The most significant, dubbed Active Region 10486, was the largest sunspot group to appear in over a decade, as wide as 13 Earths. Of the 17 major solar flares that erupted between Oct. 19 and Nov. 5, a dozen came from this enormous complex, including the most powerful ever recorded. On Oct. 28, as space weather experts headed for Washington to testify for the need to fully fund the SEC, Region 10486 unleashed a major flare followed by a fast-moving CME, which reached Earth just 19 h after launch and produced the sixth most intense geomagnetic storm since 1932. (An image of the Sun taken that day appears in Fig. 4.3.) Radiation levels were high enough that NASA officials directed astronauts in the International Space Station to take precautionary shelter in the spacecraft's most shielded location. Airlines re-routed over-the-pole flights to avoid high radiation levels and areas where radio communications were blacked out; these actions cost up to $100,000 per flight in extra fuel used for the longer flights and additional stops for crew rest. From the ground, auroral displays were sighted from California to Florida, and excellent shows were reported in Australia and Europe. On Oct. 29, the same sunspot group fired off another flare, followed by another fast Earth-directed CME that produced another extreme geomagnetic storm.

On Oct. 30, with storm conditions still raging above, representatives from NOAA, NASA, the U.S. Air Force, and the commercial satellite and aviation industries testified before the House Subcommittee on Environment, Technology and Standards. That same day, strong induced currents in Malmö, Sweden, led to a failure of the power grid and left 50,000 customers in a blackout for nearly an hour. Numerous satellites and space probes were experiencing instrument anomalies, resets, and shutdowns, and several missions simply turned off instruments as a precaution. The most ironic incident occurred aboard the Mars Odyssey spacecraft, where an instrument designed to measure the radiation environment at Mars was overwhelmed by the storms and could not be recovered.

But the Sun wasn't done yet. As a parting shot, Region 10486 erupted with a flare for the record books on Nov. 4. Fortunately by then, the Sun's rotation had carried the complex across the Sun's disk and the flare exploded on its western limb, limiting its impact. Solar flares are ranked by the flux of X-rays they produce, with X-class flares being the most extreme group. For 12 min, the Nov. 4 flare saturated the sensors on the GOES 12 satellite – the most intense solar flare seen in nearly three decades of space-based X-ray measurements. That week, Rick Jacobsen, an SEC forecaster, tried to put the solar activity into perspective. "It has broken so many records and will be studied for so long, we're trying to come up with a name for it here," he said, adding "Extraordinary." As for the SEC's fiscal woes, in the end Congress provided funding and relocated its activities under the National Weather Service. In 2008, its name was changed to the Space Weather Prediction Center to emphasize its forecasting role.

Scientists now refer to the episode as the Halloween Storms of 2003. However, as impressive as they were, these storms were by no means the most powerful the Sun can produce. Many historical storms, if they occurred today, would have strong effects on orbiting satellites and power grids on the ground. The granddaddy of solar storms occurred Sept. 2, 1859, when the only telecommunications network was thousands of miles of telegraph lines. The northern lights were sighted from regions as far south as Hawaii and Venezuela, and aurora-induced electrical interference frustrated telegraph operators. Work was backing up, and evening operators in Boston, MA, and Portland, Maine, tried a little experiment.

"Please cut off your battery entirely from the line for fifteen minutes," instructed Boston.

Portland replied: "Will do so. It is now disconnected."

"Mine is also disconnected and we are working with the auroral current. How do you receive my writing?" asked Boston.

"Better than with our batteries on. Current comes and goes gradually."

"My current is very strong at times, and we can work better without batteries, as the aurora seems to neutralize and augment our batteries alternately, making the current too strong at times for our relay magnets. Suppose we work without batteries while we are affected by this trouble?" Boston suggested.

"Very well. Shall I go ahead with business?"

"Yes. Go ahead."

What was largely a curiosity in 1859 became much more significant a century later, when a storm in February 1958 caused a power failure in Toronto and interrupted communications on telegraph cables in the North Atlantic. The most dramatic space-weather incident occurred on Mar. 13, 1989, when a strong geomagnetic storm induced currents that added to the normal flow of electricity in northeastern Canada's Hydro-Québec power grid. The currents caused protective equipment to sense overload conditions and they responded as they were designed to, "tripping" five transmission lines from James Bay off-line. The sudden loss of electrical capacity – about 44% of the load on the Hydro-Québec grid at the time – knocked the legs out from under the system, and within 90 s, the entire power grid collapsed, leaving some six million in Québec without electricity for up to 9 h. In the U.S., a transformer at the Salem Nuclear Power Plant in New Jersey failed as well, and hundreds of lesser events were recorded throughout the North American power system. It happened again on Nov. 6, 2001, when induced currents led to equipment failures and outages on the New Zealand power grid. With modern lifestyles, business and industry dependent on satellite services and the easy availability of electricity, an event like the 1859 "superstorm" could have considerable impact.

The number of spots on the Sun waxes and wanes with an approximately 11-year cycle, and solar activity follows suit. Sunspots are dark regions on the Sun's visible surface that mark places where strong magnetic fields arch up from below. The magnetic fields are so strong that they suppress the normal gas motions, so sunspots are slightly cooler than their surroundings and therefore appear dark. Because sunspot magnetic fields can store tremendous amounts of energy, coronal mass ejections, solar flares and coronal holes tend to be more frequent as the number of Sunspots increases. And as we saw with the Halloween Storms, which blasted off the Sun more than 3 years after sunspot maximum, some of the largest solar eruptions can occur in the waning years of the cycle. Sunspot numbers bottomed out in late 2008 and, according to current predictions, will be on the rise until they peak in the summer of 2013. The most recent models suggest that the coming solar maximum will be less intense overall, with fewer spots and associated activity, than the previous one, but the bottom line is that we'll see improving odds for solar storms and strong auroral displays through 2013, with the possibility of occasional impressive displays into 2016.

Overall, the chances of seeing an aurora are not all that bad – especially in Canada and the United States. Because the north magnetic pole lies within North America, the auroral oval reaches the farthest south here. This means that observers at a given latitude in North America have a better chance of seeing an aurora than those at the same latitude in Europe or Asia. For example, both Rome and Chicago lie at 42° north, but an observer at a dark-sky site in the vicinity of Rome could see one aurora per decade on average, while under similar conditions near Chicago an observer might catch ten auroral displays a year. However, because no one can precisely predict when an aurora will appear, if you really want to see one, you'll either have to travel north to meet the auroral oval on its own turf or keep abreast of what's happening on the Sun so you can make the most of strong solar storms.

Use the resources in Table 9.1 to stay informed about the potential for enhanced solar activity. One helpful measure is a magnetic quantity called the planetary K, or Kp, index. This value is updated every 3 h based on measurements from ground-based magnetometers across the globe, and the Space Weather Prediction Center provides continual Kp estimates for 3 days into the future; you can even receive email alerts. Kp values of 3 or less put the auroral oval over Canada and Alaska, and a value of 4 means that it will be roughly overhead when seen from the northernmost states of the contiguous U.S. Higher values, which indicate storm conditions, indicate that the aurora has expanded more deeply into "the lower 48" (Fig. 9.5). Because the lights occur at high altitude, they're visible far away from locations where the aurora appears directly overhead.

Perhaps the strangest question connected with the northern lights is whether or not they produce audible sounds, an issue we previously visited in Chap. 8 with meteors. Among the names the Sámi people of northern Scandinavia gave the northern lights was *guovssahas* – "audible light." Reports of rustling, hissing, and crackling noises associated with strong auroral displays have persisted for centuries.

Table 9.1 Unpredictable events on the Web

Seeing an aurora

NOAA's Space Weather Prediction Center
www.swpc.noaa.gov

Solar Terrestrial Dispatch
www.spacew.com

Spaceweather.com
www.spaceweather.com

Comets

Gary Kronk's Cometography
cometography.com

Seiichi Yoshida's home page
www.aerith.net

Active space missions

Deep Impact/EPOXI
epoxi.umd.edu

Rosetta
www.esa.int/rosetta

Stardust/NeXT
stardust.jpl.nasa.gov

Stardust@Home
stardustathome.ssl.berkeley.edu

SOHO and STEREO sungrazing comets
Sungrazer.nrl.navy.mil

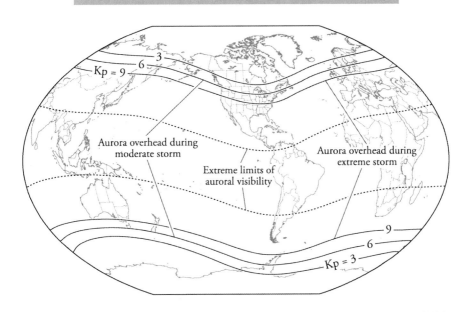

Fig. 9.5 The planetary K, or Kp, index serves as a guide to determine if an aurora can be seen from your locale. The solid curves on this map show the locations where, on average, the auroral oval will be overhead at midnight for three different values of Kp. For values greater than 3, auroral activity occurs somewhere above the contiguous U.S. Values of 5 or more indicate that a geomagnetic storm is in progress. During the most extreme storms, an aurora may be seen near the horizon from regions along the dashed curves

William Ogilvie, an explorer and surveyor in the Yukon, wrote of one incident in November 1882:

> I have often met people who said they could hear a slight rustling noise whenever the Aurora made a sudden rush. One man, a member of my party…was so positive of this that…when there was an unusually brilliant and extensive display, I took him beyond all noise of the camp, blindfolded him and told him to let me know when he heard anything, while I watched the play of the streamers. At nearly every brilliant rush of the auroral light, he exclaimed: "Don't you hear it?" All the time I was unconscious of any sound.

Other accounts also link the intensity of the sound to the intensity of the light display. Sound produced in the rarefied air where an aurora shines simply cannot reach the ground, but scientists do know that processes associated with the aurora produce Very Low Frequency (VLF) electromagnetic radiation. Under just the right conditions, this energy may cause everyday objects like hair, grass or eyeglasses to vibrate at audio frequencies, which creates sound close to the observer. So when you see a bright aurora, take a moment to listen as well. Perhaps you'll be among the fortunate few to hear it whisper.

Box 9.1 "Northern Lights at Play"

The following lines, penned by Finnish poet Ilmari Kianto (1875–1970), were set to music in 1996 by Minnesota singer-songwriter Diane Jarvi, who provided this English translation. The lyrics concisely capture the color, motion and majesty of a bright aurora.

They shimmer, they shine
They burn bold
They gleam like bronze statues
They glitter like belts of gold

They float, they fly
They gallop, they gasp
They snort like chariot horses
That heroes drive past

They flutter, they fly
They glide and make glee
They hover like angels
Or sway like the sea

They darken, they dim
They break like the ice
Then they brighten like mad
And charge into the night

Bright Comets

Previously, we discussed comets only in the context of their relationship with meteor showers, but in 2007 surprise appearances by two comets reminded millions how impressive these capricious objects can be. Most comets remain diffuse fuzz balls detectable only with telescopes or, at best, binoculars. On average, one may expect a comet to exceed 4th magnitude about every 2 years, and wait about 6 years for a comet brighter than 2nd magnitude. Because a comet's light spreads over a large angle instead of being concentrated in a starlike point, this is about the minimum brightness that could be detected from a dark suburban site. About once every 15 years, a comet exceeds Vega's brightness, and every few decades a comet becomes so bright that it can be seen near the Sun in broad daylight. Such a "great comet" can be truly impressive to the unaided eye when seen in a dark sky, with a broad dust tail spanning tens of degrees arcing away from a star-like head. Comets also have no respect for statistics: In the mid-1990s, a 20-year-long drought ended with appearances by two outstanding comets within a year of each other.

Comets are dark, solid bodies a few miles across that orbit the Sun in very eccentric paths. The solid body of a comet, called its nucleus, can be fairly described as a "dirty snowball" containing a mixture of dust and frozen gases, such as water, carbon dioxide, ammonia and methane. Some of the icy material – perhaps less than 1% – evaporates as the comet nears the Sun, creating an envelope of gas and dust that enshrouds the solid body. This envelope, called the coma, may be up to 620,000 miles (1 million km) across. Swept back by the solar wind and the pressure of sunlight, this material forms the comet's tail. Comet tails can be longer than the distance between Earth and the Sun. The record-holder is the Great March Comet of 1843 (C/1843 D1), which boasted a visible tail more than 185 million miles (299 million km or 2 AU) long, far enough to stretch from the Sun to beyond Mars. That such a small amount of material could create visible features so large has led some to describe comets as "the closest thing to nothing anything can be and still be something." To the naked eye, the coma of a bright comet looks almost starlike, a tiny ball of light set within a fainter milky glow. Sometimes this is all there is to see, but bright comets typically sport the iconic feature we associate with comets, a long tail arcing many degrees away from the coma. In fact there are usually two tails – a bright curving one made of dust, which reflects sunlight, and a straight fainter tail made of glowing gas.

The ancients gave different names to comets based on their visual appearance. To the Chinese, a comet with a prominent tail was called a "broom star" (*huixing*), while one with no obvious tail was a "bushy star" (*poxing*). The Greeks likewise recognized a comet with an extended tail as a "bearded star" (*aster pogonias*) and one without a tail was a "long-haired star" (*aster kometes*, which is where the modern word "comet" comes from). Until the mid-1400s the most detailed and complete observations of comets were made by the Chinese, who as early as 200 B.C. employed official skywatchers to record and interpret heavenly events. As in other cultures, happenings in the sky were regarded as messages from the gods to the head of state regarding the management of earthly affairs. Astute Chinese observers nevertheless concluded that comet tails always point away from the Sun, a fact not recorded by their European counterparts for another 900 years.

Oriental ideas about comets had little influence on the development of Western thought. Aristotle regarded them as a fiery atmospheric phenomenon, to be lumped together with meteors and the aurora, because their appearances far from the ecliptic ruled out the possibility that they were related to planets. Aristotle envisioned comets as being whipped up by the motion of the Sun and stars around Earth, their appearance serving as warning of coming droughts and high winds. As these ideas were extended in the Middle Ages, comets became viewed less as a portent of disaster than as a cause. They were viewed as fiery corruptions of the air, pockets of hot contaminated vapor that could bring about earthquakes, disease, and famine.

Some of these ideas were being seriously questioned when the comet of 1577 attracted the attention of Danish observer Tycho Brahe, recently installed in his deluxe observatory on the island of Ven near Copenhagen. Tycho could see no reason why comet tails should always point away from the Sun if they were products of the weather, so he measured the position of the comet with respect to the stars at different times during the night. His intent was to discover the comet's parallax – the difference in its apparent position as viewed along different lines of sight, which would indicate the object's true distance from Earth – and his observations showed that the comet was farther than the Moon but not as distant as Venus. Although Tycho's work on the comet of 1577 hardly settled the matter – Galileo, for example, dismissed his observations – it did help hasten the scientific study of comets.

When Isaac Newton published his monumental *Principia* in 1687, he showed that comets obeyed Kepler's laws of planetary motion and concluded that "comets are a sort of planet revolved in very eccentric orbits around the Sun." Any lingering doubts that comets were members of the solar system were brushed aside by further study of the bright comet of 1682.

In 1695, Newton's friend Edmond Halley began collecting accurate comet observations in order to compare the orbits of many different comets. Looking over his table, Halley quickly found that three comets seemed to follow very similar paths and shared roughly the same 76-year period, and later identified several others that also shared these properties. "Many considerations incline me to believe the comet of 1531 observed by Apianus to have been the same as that described by Kepler … in 1607 and which I again observed in 1682," Halley wrote. "Whence I would venture confidently to predict its return, namely in the year 1758. And if this occurs, there will be no further cause for doubt that the other comets ought to return also." His confidence proved well founded. The first comet ever predicted to return was spotted again on Dec. 25, 1758, and has been associated with Halley ever since.

Halley's comet – more properly referred to by its official designation, 1P/Halley – is regarded as a short-period comet, which astronomers define as a comet that completes an orbit around the Sun in less than 200 years. Halley is also the brightest and most active member of its class. A typical short-period comet follows an orbit inclined to Earth's by about 13°, circuits the Sun once every 7 years, and comes no closer to it than 1.5 AU – slightly closer than Mars' mean distance. Astronomers believe that short-period comets come from a disk-shaped region beyond Neptune called the Kuiper Belt, named for astronomer Gerard Kuiper (1905–1973), who predicted the presence of this "cold-storage zone" in the outer solar system. These icy bodies are leftovers from the solar system's formation, occasionally jostled loose by Neptune's gravitational influence and sent sunward. There, additional planetary encounters further modify comet orbits into what we see today.

The remaining population consists of long-period comets, which take more than 200 years to orbit the Sun. These comets may charge into the solar system from any direction, following orbits often steeply inclined to the ecliptic. These different properties led Dutch astronomer Jan Oort (1900–1992) to propose that long-period comets arise from a vast cloud of trillions of comets that surrounds the solar system. Today, astronomers believe that the Oort comet cloud forms a Sun-centered spherical shell extending from the outer edge of the Kuiper belt – about 55 times Earth's distance from the Sun and nearly twice Neptune's – out to a light-year or so, where the Sun's gravitational influence is so feeble that comets no longer remain bound to the solar system and drift into interstellar space. Within this zone, feeble gravitational disturbances from passing stars or massive gas clouds are all it takes to send comets on a million-year-long fall toward the Sun. Exactly how the Oort cloud formed remains an unresolved question. It's widely thought that these comets actually formed closer to the Sun than the objects in the Kuiper Belt, but were ejected to their present location by repeated encounters with Uranus and Neptune. Hal Levison at the Southwest Research Institute in Boulder, CO, thinks that this process falls far short of fully accounting for the Oort cloud's contents. Instead, he and his coworkers suggest that the overwhelming majority of the Oort cloud's mass was actually hijacked from neighboring stars that were forming in the same star cluster where the Sun was born. If true, this would mean that most of the brightest comets in history actually originated in the debris disks around other stars.

No currently known comet is expected to put on a brilliant show in the next decade, so comet fans pin their hopes to the surprise arrival of an as yet unknown long-period comet. In Aug. 7, 2006, just such a comet was found to be plunging toward the Sun. Discovered by Robert McNaught at Australian National University, an astronomer who already had 31 other comet finds to his credit, the official designation of the new object was C/2006 P1. This comet McNaught was destined to become the brightest in more than 40 years. As 2007 opened, astronomers were pleasantly surprised by the comet's sudden surge in brightness. Its breakout to a wider audience came in the second week of January, when the comet brightened to the point that it drew the attention of casual observers who weren't otherwise aware that there was a comet to see.

On the evening of Jan. 7 in Nuussuaq, a village on Greenland's west coast, Lars Christiansen was enjoying after-dinner coffee following the celebration of his son Casper's fourteenth birthday. "My son was in his room experimenting with his new digital camera," he wrote. "Suddenly he came running down from his room very excited, telling me that there was some strange light on the horizon. I told him it was probably lights from a ship." Casper ran back to his room, grabbed the camera and showed his father an image of the curious light, which clearly wasn't a ship.

Fig. 9.6 Twilight ghost. Comet McNaught (C/2006 P1) proved a welcome surprise for casual skygazers in 2007. In early January, the comet was an obvious sight immediately after sunset throughout the Northern Hemisphere. McNaught ultimately brightened enough to be visible in daylight, which placed it among the brightest and one of the most widely viewed comets in history. This photograph, taken Jan. 10, shows the comet above the village of Nuuk, Greenland (Lars T. Christiansen)

"We then spent the next two hours with our cameras and tripods photographing the phenomenon" (Fig. 9.6). They expected to hear something about it on the evening news, but there was no mention of any unusual sky event. Wondering if it might be a comet, he emailed some images to the staff at *Astronomy* magazine and asked if this was the case. Indeed, it was comet McNaught, and the magazine later published some of the images. "My son is so proud. I'm so proud," he wrote. "That was a birthday we will never forget."

To really appreciate the story of comet McNaught, it's necessary to review the factors that make some comets better visual sights than others. An unusually close approach to the Sun can heat the comet's nucleus and cause it to become more active, enhancing existing dust jets, firing up new ones, or even fragmenting the icy nucleus. An unusually close approach to Earth can turn an otherwise humdrum comet into a memorable naked-eye sight, especially if the encounter occurs after the comet has rounded the Sun and experienced strong heating. Or the comet's nucleus may be so big or react so strongly to solar heating that it becomes a fantastic sight without especially close encounters with the Sun or Earth. In Table 9.2, which lists the best comet apparitions since 1980, bold type indicates which of these factors played the most significant role in each comet's visual display.

Table 9.2 Brightest comets since 1980. Values in bold indicate the most significant element affecting the comet's brightness, either exceptional activity of the nucleus or close passes to the Sun or Earth

Comet	Nucleus size	Closest to Sun (AU)	Closest to Earth (AU)	Orbital period (Years)	Brightest magnitude	Visual ranking
C/2006 P1 (McNaught)	**15 miles** **25 km**	**0.171**	0.817	–	–6.0	Excellent
C/1995 O1 (Hale-Bopp)	**37 miles** **60 km**	0.914	1.315	2,534	–0.8	Excellent
C/1996 B2 (Hyakutake)	2.6 miles 4.2 km	0.230	**0.102**	110,187	0.0	Excellent
C/1983 H1 (IRAS-Araki-Alcock)	5.7 miles 9.2 km	0.991	**0.031**	969	1.7	Good
17P/Holmes (2007 outburst)	**2.1 miles** **3.4 km**	2.053	1.621	6.89	2.4	Good
1P/1982 U1 (Halley)	9.3 miles 14.9 km	0.586	**0.417**	75.32	2.4	Poor

For example, comet Halley was an impressive sight in 1910, but its anemic 1986 appearance was disappointing even for those who traveled far from city lights. The main difference between these two distinct appearances was the comet's distance from Earth. When Halley reached perihelion, it was on the opposite side of the Sun from Earth. Following perihelion, Halley never came closer to us than 0.417 AU – 42% of the distance between Earth and Sun or about 39 million miles (62 million km). That was three times farther away than its approach in 1910.

At the other extreme, proximity vaulted comets like IRAS-Araki-Alcock (C/1983 H1) and Hyakutake (C/1996 B2) into prominence they would otherwise never have attained because both were small and relatively inactive. IRAS-Araki-Alcock was discovered first by the Infrared Astronomical Satellite (IRAS) in late April 1983 and was originally identified as an asteroid. In early May, amateur astronomers Genichi Araki of Japan and George Alcock of England independently discovered the object, which soon became an obvious sight to the unaided eye high in the northern sky. On May 12 the comet brushed past Earth at 0.0312 AU – 3% of the Earth-Sun distance, or 2.9 million miles (4.7 million km). This was closer than any comet in more than 200 years and it remains the third closest pass on record. A typical comet might move across the sky by a degree or so a day, too slowly for the eye to pick up the motion in real time. But IRAS-Araki-Alcock was so close that its movement was clearly evident to observers, who compared it to watching the minute hand on a clock. Seen from a dark sky at its best, the coma was a faint circular cloud about as large as the Moon's apparent size. The comet resembled a star nestled within a puff of smoke and lacked any visual evidence of a tail – a fine example of a "bushy star."

Comet Hyakutake was, in the words of expert comet observer John Bortle, "One of the grandest comets of the millennium!" In late January 1996, Japanese amateur

Yuji Hyakutake discovered the inbound comet just 55 days before its Mar. 25 closest approach. By Mar. 22, observers at midnorthern latitudes could see the first-magnitude comet, which was sporting a tail at least 30° long, directly overhead before dawn. That week, it was an easy object even from cities and its motion against the stars was noticeable over the course of several minutes. On Mar. 25, Hyakutake swept past Earth at a distance of just 0.102 AU (about 9.5 million miles or 15.3 million km), making it the 20th closest comet flyby on record. On Mar. 27, when the comet passed near Polaris, Hyakutake was visible all night long and easily seen from the suburbs. From a reasonably dark sky the comet was truly something special, showing a tail spanning some 70° or more – all the more impressive because the tail contained relatively little sunlight-reflecting dust.

Sometimes a comet on an otherwise favorable trajectory simply fails to live up to expectations. The most famous of these cosmic duds was comet Kohoutek (C/1973 E1). Based on its dramatic brightening while as far from the Sun as Jupiter, Kohoutek was widely overhyped to be the "comet of the century" in early 1974. While it did reach naked-eye brightness and was well-studied scientifically, the comet fell far short of brightness predictions. The lesson learned was that some comet nuclei have veneers of highly volatile frozen gases, which respond to solar heating and vaporize long before water ice does. This creates a rapid initial brightening that isn't sustained once the coating disappears and the comet switches over to its normal activity, which is mostly powered by evaporating water ice.

And then there's a rarity like comet Hale-Bopp (C/1995 O1), which turned out to be the brightest and most active comet to pass inside Earth's orbit since the one Tycho studied back in 1577. Hale-Bopp showed unusually high activity even at great distance from the Sun and was widely expected to become a "great comet." It was discovered on Jul. 23, 1995, by Alan Hale in New Mexico and Thomas Bopp in Arizona within minutes of one another. The comet was still 7 AU from the Sun, roughly midway between the orbits of Jupiter and Saturn, which makes Hale-Bopp the most distant comet ever discovered by amateur astronomers. As an indication of the comet's unusual activity, consider that it first became bright enough to see from a dark site with the unaided eye in July 1996 – and could be viewed this way until October of the following year. Yet it was never closer to Earth than 122 million miles (197 million km) and passed no closer to the Sun than 91% of Earth's distance. At perihelion on Apr. 1, 1997, with solar heating at maximum, astronomers observing at infrared and submillimeter wavelengths showed that the comet was shedding more than 4 million pounds (2 million kilograms) of dust – equivalent to the launch mass of a space shuttle – *every second*. Following perihelion, Hale-Bopp became a striking evening object in the northwestern sky as it cruised through Cassiopeia and Perseus. For 7 weeks the comet outshone the star Vega, and it remained well placed for viewing through April. From a dark suburban site, the comet's most striking aspect was a sweeping 25°-long dust tail that arced away from the coma, but a straight, faint gas tail was also easy to see. Observers throughout the northern hemisphere could follow it with the naked eye, making it one of the most widely viewed comets in history. This extraordinary activity arose from an extraordinary nucleus, one that astronomers now estimate to be about 37 miles (60 km) across – nearly four times the longest dimension of Halley's icy core.

Fig. 9.7 Rooster tail. When comet McNaught emerged from the Sun's glare, Southern Hemisphere viewers were treated to a truly spectacular sight. The comet sported a striking dust tail that by late January extended more than 30°. This image, taken from the Radal Siete Tazas National Reserve in Chile, shows McNaught on Jan. 19, 2007. The bright head was easily visible with the unaided eye even to city observers. Portions of the broad, sweeping tail could be seen above the horizon after sunset for many northern observers. The parallel striations in the dust tail may result from the delayed breakup of larger ice-dust fragments (Miloslav Druckmüller)

With all this in mind, we can return to the early days of 2007, as comet McNaught sped toward the Sun and observers throughout the northern hemisphere were catching the comet very low in the west a few minutes after sunset. By the time Lars Christiansen and his son began imaging it, McNaught was already brighter than Hale-Bopp at its best. On Jan. 11, the day before its closest approach to the Sun, the comet exceeded magnitude −4. Despite the fact that it was positioned only a few degrees from the Sun, McNaught could be seen in broad daylight with the naked eye just by using a hand to block out the Sun's brilliance. It passed perihelion on Jan. 12 and continued to brighten for another day and a half, eventually reaching an estimated magnitude of −6. No comet had been this bright since 1965, when the remarkable sungrazing comet Ikeya-Seki (C/1965 S1) skimmed close to the Sun and possibly fragmented. The best part of McNaught's show was just beginning but, unfortunately for northern hemisphere observers, it now favored the southern hemisphere. After Jan. 14, the comet pulled farther from the Sun's glare and gradually revealed a bright, broad, curved, 30°-degree-long dust tail that was rich with parallel linear features called striae (Fig. 9.7). Even after the bright coma had set below the horizon, from some locations McNaught's glowing tail could be seen for nearly an hour. By early February it had faded to 4th magnitude, but even then visual observers in a dark sky could still make out a dim broad tail extending about 10°.

Three factors contributed to comet McNaught's exceptional display. First, the comet's nucleus was intrinsically bright and active. Estimates indicate that it was less than 15 miles (25 km) across, or somewhere between a Halley and a Hale-Bopp – perfect for lots of jet activity. Second, it approached the Sun to within about one-sixth of Earth's distance, or about 15.9 million miles (25.6 million km), so solar heating would have provided a strong boost to the activity already present. In addition, comet researcher Joseph Marcus has shown that a third factor produced the brightness surge that gave McNaught an additional two-magnitude increase needed for daytime visibility. The comet's orbit actually took it between Earth and the Sun, and by Jan. 9 McNaught was closer to Earth, which means that it was effectively backlit. This geometry has special consequences for scattering light in the presence of small particles – about a micrometer wide, roughly the size of particles in smoke. Instead of reflecting light back to the source, these particles scatter it in the forward direction, or in this case, toward Earth. A more mundane example: Forward scattering of light is why your car's windshield appears filthy when you're driving toward the setting Sun but perfectly clean at noon. So as the angle between Earth, the Sun and the comet decreased after Jan. 9, forward scattering of light by micron-size dust in the coma sent additional sunlight toward us. The comet's brightness peaked on Jan. 14, when this angle reached its minimum value, and after that the effect gradually lessened as the favorable geometry was lost.

Later in 2007, another comet reminded skywatchers just how unpredictable these objects can be. Comet 17P/Holmes orbits the Sun every 6.9 years and never comes closer than about 2 AU, well past the orbit of Mars. That's too far for dramatic solar heating and the spectacular activity driven by evaporating ice that follows it. Occasional close approaches to Jupiter slightly alter its orbital period and perihelion distance, but for the most part Holmes quietly flew around the Sun with only the weakest of cometary pulses – until Oct. 23, 2007, when the comet's nucleus underwent an enormous outburst. Within 2 days, Holmes had brightened by some 500,000 times, an unprecedented increase for any comet and one that catapulted it from obscurity to naked-eye visibility. Holmes looked like a seemingly new second-magnitude star in the constellation Perseus. As astronomers watched, a vast dusty halo steadily expanded outward from the comet's nucleus at 1,100 mph (1,800 km/h). Four days later, the dust cloud extended more than 310,000 miles (500,000 km) and the comet showed a distinct disk when viewed with binoculars. Holmes began a long, slow fadeout as November opened, but it remained visible from dark skies well into 2008. By Nov. 9, Holmes' expanding dust shroud had made it the largest object in the solar system, bigger even than the Sun (Fig. 9.8). The event is all the more impressive when you realize that this giant structure blossomed from something that happened on an icy nucleus just 2.1 miles (3.4 km) across, far too small to see.

So what created the comet Holmes "megaburst," as astronomers are calling it? According to a study by Zdenek Sekanina, a comet expert at NASA's Jet Propulsion Center, Holmes brightened after an eruption lofted several large fragments off of

Fig. 9.8 Sometimes comets do unexpected things. This is comet 17P/Holmes on Nov. 11, 2007, 19 days after an unprecedented eruption brightened it by half a million times. The comet was a naked-eye object for months (Photo by the author)

the nucleus. We think of comets as dirty snowballs, but the substance of a snowball is much more cohesive than the icy dusty mix that makes up a comet. As the fragments detached from the nucleus, they quickly underwent a fragmentation cascade: big chunks quickly broke into smaller pieces, which quickly fractured into smaller bits and eventually turned to dust. But the scale of the event is nothing short of amazing. Sekanina estimates that the dust shroud represents a loss to comet Holmes of about 100 million metric tons. Incredibly, when dust emission from the rapidly fragmenting pieces peaked late on Oct. 24, comet Holmes was shedding just as much dust as Hale-Bopp did during its closest approach to the Sun. Although the megaburst was unique in its brightness, Holmes is known for occasional outbursts; one in November 1892 actually led to the comet's discovery. The nucleus has probably spent a few thousand years in the inner solar system, basking in the sunlight and slowly evaporating. Every now and then, a pocket of gas erupts beneath the surface and launches some fragments into space, creating a visual spectacle we can all enjoy – a cry for attention from a crumbling comet.

Comets are cosmic fossils, remnants of the cloud of dust and gas out of which the Sun and planets formed nearly 5 billion years ago. They have changed relatively little since, and for this reason planetary scientists are intensely interested in them. Each comet presents a chance to glimpse the most ancient past of our solar system

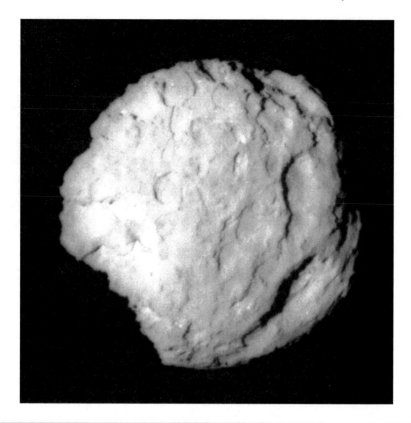

Fig. 9.9 NASA's Stardust captured this image of 81P/Wild's nucleus as it flew past the comet in January 2004. The 3-mile-long (5 km) object shows unusual relief, including deep, flat-bottom craters, terrain shaped by billions of years of impacts and gas and dust emission (NASA/JPL)

home. Comet nuclei are too small to see from Earth-based facilities, but several have been visited by spacecraft. The first, appropriately enough, was 1P/Halley, which was observed at close quarters by the European Space Agency's Giotto probe in 1986. After a long dry spell, NASA's Deep Space 1 probe flew past 19P/Borrelly in 2000. Then, in 2004, 81P/Wild ("Vilt", see Fig. 9.9) was visited by NASA's Stardust mission, which not only flew past the comet but captured dust from its coma, stored it in a reentry capsule, and returned the capsule to Earth 2 years later. The capsule landed safely and on-target southwest of Salt Lake City, Utah. Thanks to Stardust, scientists have analyzed factory-direct comet dust for the first time in history.

In 2005, NASA's Deep Impact mission attempted to look under the hood of comet 9P/Tempel. By making a big enough crater, scientists hoped to blast through the heavily modified surface and expose pristine ice below. For this purpose, Deep Impact carried a second spacecraft that, when released, stayed on a collision course while the mothership swerved out of the way to watch the show. The spacecraft

Box 9.2 How comets get their names

Comet names are complicated. Some honor astronomers who showed that multiple previous appearances were due to returns by a single object, most famously 1P/Halley but also 2P/Encke and 27P/Crommelin. Others are named for up to three independent discoverers, such as 73P/Schwassmann-Wachmann or 57P/du Toit-Neujmin-Delporte. Some are technically mistakes, bearing the names of non-discoverers due to incomplete information at the time. And increasingly, comets are named for the spacecraft or sky surveys that find them. These include SOHO, the NASA/ESA Solar and Heliospheric Observatory, which in 15 years of solar monitoring has discovered 2,000 mini-comets evaporating as they approach the Sun. These finds make SOHO the most prolific, if unintended, comet hunter in history. Sky surveys showing up in comet names include the University of Arizona's Spacewatch Project, the Lowell Observatory Near-Earth Object Search (LONEOS), the U.S. Air Force/MIT collaboration known as the Lincoln Near-Earth Asteroid Research (LINEAR) project, and the Near-Earth Asteroid Tracking (NEAT) project of NASA's Jet Propulsion Laboratory.

Because the same comet hunters often make multiple discoveries, comet names alone don't allow for unique identification and, with dozens of comets now being discovered every year, astronomers needed a more robust nomenclature. So in 1995, the International Astronomical Union adopted a new way of designating comets – that's what the alphanumeric prefix before comet names is all about. This consists of a one-letter prefix, usually a C for a comet seen at only a single apparition or a P ("periodic") for one seen in multiple apparitions. The P is also applied a one-time comet show where the object's orbital period is less than 30 years. In cases where the comet was destroyed – such as Shoemaker-Levy 9 (D/1993 F2), which broke apart and impacted Jupiter – a D (for "defunct") is used.

Periodic comets also receive a permanent number reflecting the historical order of confirmed returns, starting with Halley, the first comet so identified. So this gives us 1P/Halley as the comet's official designation.

All comets also obtain a provisional designation, and this becomes permanent for non-periodic comets. It's constructed from the discovery year plus an uppercase letter that indicates the half-month when it occurred. For example, an "A" represents Jan. 1–15, "B" is Jan. 16–31, and so on, except for "I," which isn't used to avoid confusion with an earlier scheme that used Roman numerals. Appended to this letter is a number representing the order of discovery during the half-month.

This means that comet McNaught, which was discovered in the first half of August 2006, is officially known as C/2006 P1. When referring to a specific comet apparition, such as last go-round of *Halley's comet*, astronomers also use the provisional scheme. Astronomers recovered Halley in the second half of October 1982, making the designation for this appearance 1P/1982 U1.

struck the comet perfectly, but the impact surprisingly lofted so much fine dust that the spacecraft's camera couldn't see the new crater. Thanks to some orbital ingenuity and a hardy Stardust spacecraft, scientists will have a second chance to peer into the crater in 2011, when the probe flies past the comet. Deep Impact also went on to visit another comet, 103P/Hartley, in late 2010.

By far the most ambitious mission to a comet is the European Space Agency's Rosetta, which was launched in 2004. It will reach its target, 67P/Churyumov-Gerasimenko, in 2014, but unlike earlier flyby missions, Rosetta will enter orbit around the nucleus and monitor its changing activity, providing the most detailed look at how comet's do the things they do. The spacecraft's first job after entering orbit, however, is to release a boxy three-legged lander named Philae. If all goes according to plan, Philae will carry out the first-ever landing on a comet. Immediately after touchdown, it will fire a harpoon to anchor itself to the comet, which like all small bodies has very little surface gravity. Mission planners expect Philae's instruments to provide the most comprehensive analysis of a comet's composition ever made and to return data for at least a week, and possibly months.

All this will keep comet experts busy for years. For the rest of us, a bright comet is simply one of the most spectacular sights in the celestial pageant, but all we can do is wait for the next one.

Super Stars

This chapter opened with the story of the naked-eye gamma-ray burst, a phenomenon associated with the cataclysmic death of a star. These events tend to occur in very distant galaxies, and for the bursts where distances are known, none are closer than a billion light-years away. Massive stars still explode closer to home, even in our own galaxy, but it turns out that these stellar heavyweights are relatively few and far between. By contrast, stars similar to the Sun's mass tend to be pretty common throughout the galaxy, and much smaller red dwarf stars form in even vaster numbers. It's as if there's a galactic building code that allows a couple of skyscrapers for every million single-family homes. The most massive stars, which tip the scale at more than 100 times the Sun's mass and may be the very types of stars that end their lives producing gammy-ray bursts, are extremely rare. When a molecular cloud gives birth to one of these monsters, the star's intense radiation and powerful stellar wind quickly disperse the cloud's gas before another star can form.

However, we don't need the most massive stars to produce a stellar explosion, an event called a supernova. All we need is a more common massive star with a mass of about eight solar masses. So far, astronomers have found about 5,700 supernovae in total, and within our galaxy have catalogued about 270 gaseous shells expanding so swiftly that it's clear they represent the wreckage of an exploded star. Yet for the thousands of years people have watched the sky and written down what they saw, only eight supernovae were bright enough to be seen at all by the naked eye – and of these, one went unnoticed and the two most recent actually occurred in other galaxies (Table 9.3).

Table 9.3 Visible supernovae since A.D. 1000

Year	Name	Brightness (mag.)	Type	Constellation	Distance (light-years)
1006	SN 1006	−7.5	White dwarf	Lupus	7,100
1054	SN 1054	Brighter than −4.0	Core collapse	Taurus	6,500
1181	SN 1181	−1.0	Core collapse	Cassiopeia	26,500
1572	Tycho's star	−4.0	White dwarf	Cassiopeia	7,500
1604	Kepler's star	−3.0	White dwarf	Ophiuchus	16,000
1680	Cas A	Around 6.0	Core collapse	Cassiopeia	11,000
1885	SN 1885	5.8	White dwarf	Andromeda	2.5 million
1987	SN 1987A	2.9	Core collapse	Mensa	168,000

The supernova that astronomers missed may have occurred in 1680. If so, it possibly caught the eye of only one astronomer – Britain's first Astronomer Royal, John Flamsteed – while measuring positions for the star catalog he was compiling. Later, astronomers noticed that the star Flamsteed called "3 Cassiopeiae" wasn't actually there and wrote the entry off as an error. But in 1947, one of the brightest radio sources outside the solar system was found near the very same location. Astronomers now know that this source, named Cassiopeia A, is a supernova remnant, and by backtracking the expanding gas they've determined that light from the explosion would have reached Earth about 330 years ago. The remnant is also 11,000 light-years away, and many people get hung up on the fact that the speed of light means the explosion really occurred 11,330 years ago. While this is certainly true, it's something astronomers simply take for granted. The light we see from the remnant takes just as long to reach us now as it did when the explosion occurred, so it isn't important in understanding how the object changes over time, which is what mainly interests astronomers. As the table shows, the Cas A supernova was also surprisingly dim for its distance. There is now considerable evidence in the motions of the remnant's gas that the explosion was asymmetric and unusual. In addition, a lot of dust is present throughout the region and some of it might have been in a position to attenuate the blast's light. Together, these effects may have muted the explosion as seen from Earth.

In 1885, a supernova was seen in the Andromeda Galaxy (M31). This was a time when astronomers hadn't yet invented the word "supernova" and had yet to realize that the Andromeda Galaxy *was* a galaxy. Cataloged as a variable star designated S And, the supernova brightened to magnitude 5.8 and was easily visible in a dark sky despite occurring 2.5 million light-years away. This was the first – and so far only – supernova seen in M31. At the time, astronomers cataloged all of the brightest one-time outbursts as a nova, a term derived from *nova stella* ("new star"). But after the 1920s, when they realized how far away the Andromeda Galaxy is, astronomers recognized that this event strongly hinted of a class of much more powerful stellar explosions.

Supernovae are one of nature's greatest spectacles. Each explosion releases up to 10 billion times the Sun's annual energy output, for a few weeks a supernova may even outshine its parent galaxy. This fact makes them useful probes of the very distant universe. Long before our solar system was born, most of the chemical elements heavier than helium – including the iron and carbon in our bodies and the oxygen we breathe – were forged inside the nuclear cauldron of massive stars. Their subsequent explosions dispersed these elements into the galaxy and seeded the molecular clouds that would build future stars and planets. It's even possible that an exploding star's blast wave helped trigger the collapse of the molecular cloud that gave birth to our own Sun and solar system. We owe a lot to exploding stars.

Supernovae are as rare as they are important. In spiral galaxies like the Milky Way or M31, each containing several hundred billion stars, astronomers expect an average of only two to three supernovae each century. We've only observed one in the Andromeda Galaxy since 1885, and the last observed in the Milky Way was seen by Kepler in 1604. One reason why we see far fewer than the average is that our galaxy is filled with clouds of dust and gas that obscure the view and attenuate the light. However, X-rays and radio waves can penetrate the murk and locate the remnants of unseen supernovae. In 2008, researchers established that one apparently young supernova remnant, dubbed G1.9+0.3, was only about 140 years old based on its rapid expansion captured by radio and X-ray images taken 22 years apart. Located 25,000 light-years away, in the crowded, dust-obscured center of our galaxy, the supernova's radiation swept past Earth in the 1860s – but with visible light filtered out and no radio or X-ray telescopes available, no one had a clue that anything had happened. Frustratingly, the worst place to view a supernova in our galaxy is from within our galaxy.

A typical nova is an otherwise faint star that suddenly flares up, brightening by thousands of times and then, over a period of weeks, fading back into obscurity. We now know that a nova occurs in a close binary star system containing a normal star and a white dwarf – the compact, burned-out remnant of a star like our Sun. In these close systems, hydrogen-rich gas from the normal star flows onto the white dwarf, where it accumulates and eventually explodes. Although the outburst appears quite violent, in fact only the surface material is thrown into space and the white dwarf remains intact. But by the 1920s it was clear that all novae were not equal, with astronomers using terms like "Hauptnovae" (German for "chief novae"), "giant novae" and "exceptional novae" to distinguish what were so obviously the brightest members of the group. In 1934, long before the physical details of the nova process were clear, astronomers Fritz Zwicky (1898–1974) and Walter Baade (1893–1960) argued convincingly that the exceptional novae represented a distinctly different phenomenon. "With all reserve," they wrote, "we advance the view that a supernova represents the transition of an ordinary star into a neutron star." Half a century later, this idea had become the standard view for one type of exploding star, and it would be dramatically confirmed when the next naked-eye supernova popped off much closer to home.

The Large Magellanic Cloud, the biggest of several satellite galaxies that orbit our Milky Way, is located 168,000 light-years from us. At 7:35 UT on Feb. 23, 1987, a pulse of particles called neutrinos arrived at Earth and was recorded by several underground detectors. Neutrinos, which move nearly as fast as light and are produced in vast numbers in a supernova, are also extremely reluctant to interact with other matter, which is why they can escape the incredibly dense core of a collapsing star. But this property also makes them difficult to find, requiring large facilities located underground to eliminate detections of more common particles from space. Scientists estimate that there's less than one chance in a thousand billion that a neutrino speeding through the entire Earth will interact with its matter. In fact, most of a supernova's energy is released as neutrinos, with light making up only about 1%. Over about 13 s, the Kamiokande II detector in Japan, the Baksan Neutrino Observatory in Russia, and the IMB-3 detector in Ohio saw a total of just 25 neutrino events as the burst of particles washed over Earth. But at this point, no one even realized that a supernova had occurred.

That night, at Las Campanas Observatory in Chile, telescope operator Oscar Duhalde became the first person in nearly four centuries to spot an exploding star with nothing more than his own eyes. He noticed a 5th-magnitude star that seemed out of place, but became so absorbed in his work he neglected to inform anyone at the time. Minutes later, astronomer Ian Shelton at the same observatory noticed the same star on a long-exposure photograph he had taken and began the process of informing astronomers worldwide. Before word got out, amateur astronomer Albert Jones in Nelson, New Zealand, had also seen the unexpected star. From then on, SN 1987A became the most widely observed supernova in history. Astronomers quickly identified ground zero as Sanduleak −69 202, a star estimated at about 20 times the Sun's mass. The explosion occurred because the star's core ran out of the nuclear fuel needed to support its bulk against gravity. The core collapsed under its own weight, an event that almost instantly triggered an explosion that transformed the star into a rapidly expanding ball of hot gas. The supernova's brightness peaked about 80 days later at magnitude 2.9, which made it about as bright as the stars in the Big Dipper and by far the brightest supernova seen in modern times. Dubbed SN 1987A, the letter designating it as the first supernova detected that year, the superstar shone with 280 million times the Sun's power. Since then, astronomers have watched as the ring-shaped blast wave expands into space. The ring lit up with bright X-ray-emitting knots, now numbering 30, as it plowed into dense gas puffed out by the star some 20,000 years before it exploded. In the 23 years since the explosion was detected, the blast ring has grown to 1.34 light-years across − that's 7.9 trillion miles (12.7 trillion km), or about a third of the way to the star nearest the Sun. Clearly, these are unimaginably powerful explosions. The debris will continue to expand into the interstellar environment for millennia, eventually affecting a bubble of space spanning more than 1,000 light-years across before it becomes indistinguishable from the thin matter between the stars.

Astronomers class supernovae into two broad categories called types I and II based on specific properties of their light, along with numerous subtypes. SN 1987A was a type II core-collapse supernova. All stars balance the tendency of gravity to compress them by the pressure generated through the nuclear reactions in their cores. But stars have only a limited supply of hydrogen fuel, and the more massive the star is, the faster it burns through its reserve. Every second, nuclear reactions in the Sun's core convert hundreds of millions tons of hydrogen into a slightly smaller mass of helium plus energy. It's been doing this for about 4.6 billion years and is expected to continue for billions more. Contrast this with a star ten times more massive, which will run through its hydrogen supply in just 10 million years. As their energy crisis looms, the cores of massive stars contract, heat up, and initiate nuclear reactions with the byproducts of the previous cycle. Once one set of reactions produces insufficient energy, the star's core collapses and its temperature rises until the next set of reactions ignites. So when hydrogen is depleted, the star's core reconfigures, fusing helium into carbon plus energy. This extends the star's battle against its own gravity, but every switch results in diminishing returns. Helium fusion in the core may last only a million years, and fusing heavier nuclei only postpones the inevitable a little longer. Eventually, the star's core contracts again to begin using its reserve of silicon fuel. But for a star like the one that exploded as SN 1987A, silicon fusion lasts only a week or so. The core – now full of iron nuclei, the most tightly bound nucleus – contracts again but no more energy can be extracted; the battle against gravity is lost. The core collapses for the last time and transforms into a neutron star, a remnant that packs more than the Sun's mass into a superdense ball just 12 miles (20 km) across. This event, through processes still not well understood, triggers an outward shock wave that blows the star to smithereens. Astronomers haven't seen any sign yet that SN 1987A left behind a neutron star, so perhaps in this case the star was completely destroyed.

Supernovae of types Ib and Ic are also core-collapse events and are often associated with gamma-ray bursts, but type Ia supernovae are something different. They occur in a similar physical setup as novae, in close binary systems containing a white dwarf star. Gas from the dwarf's companion floods onto the dense dwarf, but instead of flaring off periodically as in novae, the matter steadily accumulates. Eventually, this added mass triggers an explosive nuclear reaction that sweeps through the star. The reactions die out in a second, but by then the white dwarf has been transformed into a rapidly expanding fireball. Type Ia supernovae may also occur in close binary systems containing two white dwarfs. Instead of mass exchange building up to an explosion, the white dwarfs spiral closer together until they merge and explode. Recent studies suggest that such mergers may be the more common type Ia mechanism.

Despite its importance to astronomers, SN 1987A was hardly an impressive sight. The last visual supernova in our own galaxy was seen in 1604, just a few years before the invention of the telescope. No one knows when the next Milky Way supernova will blaze forth, but we can get a good idea of what to expect by examining descriptions of the handful of historical events observed.

The first supernova to attract wide attention in Europe appeared around Nov. 6, 1572, and was as bright as Venus. Among the first to begin serious study of the star was Tycho Brahe, whose candid account of his own discovery gives us a sense of how radical an event it was for astronomers at the time. The reason was a deeply held belief in the immutability of the heavens, a holdover from Aristotle that nothing beyond the planets ever changed. Poor weather kept him from noticing the new star until Nov. 11, when during a walk before dinner he noticed "directly overhead, a certain strange star ... flashing its light with a radiant gleam and it struck my eyes."

When I had satisfied myself that no star of that kind had ever shone forth before, I was led into such perplexity by the unbelievability of the thing that I began to doubt the faith of my own eyes, and so, turning to the servants who were accompanying me, I asked them whether they too could see a certain extremely bright star.... They immediately replied with one voice that they saw it completely and that it was extremely bright....

And with that, he went off to make measurements. The star remained visible for 15 months and showed no movement in the heavens, indicating that it was much farther away than the planets. "Hence this new star is located neither in the region of the Element, below the Moon, nor in the orbits of the seven wandering stars [including the Sun and Moon], but in the eighth sphere, among the other fixed stars," Tycho concluded. Based on its observed brightness and behavior, astronomers had determined that Tycho's supernova was of type Ia, a white dwarf detonated by over-generous gas transfer from a normal companion star. But what happens to the companion star when the white dwarf explodes? Astronomers assumed that the star survives and, with no partner to orbit, simply travels away from the explosion site at high speed. In 2004, astronomers announced that a survey of the central region of the supernova remnant had found a Sun-like star moving at more than three times the mean speed of other stars at that distance – the surviving companion of Tycho's supernova. His famous protégé, Johannes Kepler, would also make detailed observations of a supernova – the last one seen in our galaxy and also thought to be a type Ia – just 32 years later.

Imperial astronomers of China's Ming Dynasty (1368–1644) also noticed Tycho's star, although their measurements of its position were far less accurate. "It was seen before sunset. At the time," they reported, "the Emperor noticed it from his palace. He was alarmed and at night he prayed in the open air on the Vermillion Steps." Such an unusual star-like object was called a "guest star" (k'oxing).

A supernova of similar brightness appeared Jul. 4, 1054. In 1928, the American astronomer Edwin Hubble (1889–1953) correctly suggested that an expanding gas cloud called the Crab Nebula represented the debris of this explosion (Fig. 9.10). The supernova was first noticed in Constantinople, where it was associated with an outbreak of plague. It has long been thought that Native American groups throughout western North America were the next to record the explosion. More than a dozen rock carvings and paintings show a crescent shape Moon near a large star-like symbol, and observers in western North America would have seen the supernova 2° from the waning crescent Moon just before sunrise. However, a chemical analysis of pigment used in one of the best-known examples, at Fern Cave in the Lava Beds National Monument, CA, shows that components of the rock art critical to a supernova interpretation were put in place long after 1054.

On the other hand, a 1990 study of pottery made by the Mimbres people, who lived in New Mexico about 1,000 years ago, provided suggestive support for the notion that some North American groups recorded the event. Many earthenware bowls and plates made by the Mimbres display geometric designs of astronomical significance. Some markings appear to correspond to the number of days in the Moon's sidereal and synodic periods. But one bowl features a stylized rabbit – a common motif for the Moon – clutching a sunburst from which 23 rays emerge. Chinese records indicate that the supernova could be seen in daylight for 23 days, suggesting that it was brighter than −4.0, and it remained visible to the

Fig. 9.10 In 1054, light from a supernova explosion in Taurus reached Earth. The blast created the Crab Nebula, the most famous star wreck and one of the most studied objects in astrophysics. At its center lies the rapidly spinning core of the exploded star, a pulsar that powers the Crab's intense radio and X-ray emissions. The nebula is about 10 light-years across and can be observed with a small telescope (NASA/ESA/J. Hester and A. Loll (Arizona State Univ.))

naked-eye for 21 months. R. Robert Robbins, the astronomer at the University of Texas at Austin who first drew attention to the astronomical significance of the bowl, declared it "the most certain record of the supernova that has ever been discovered outside China and Japan." Archaeologists, however, are less convinced of this interpretation, seeing in the design an oddly distorted rabbit carrying a datura or thorn apple, a plant known to have hallucinogenic properties. The occurrence of astronomically significant numbers in other Mimbres pottery suggests that the supernova interpretation is indeed possible, but we may never really know for sure.

In terms of naked-eye visibility, the best supernova in recorded history occurred in 1006. On the night of Apr. 30, observers around the world noticed a brilliant new star in what is now the constellation of Lupus, near Scorpius. Clear historical accounts of the event have been found in Egypt, Iraq, Italy, Switzerland, China, and

Japan, with additional possible references from France and Syria. The most detailed account comes from an Egyptian physician and astrologer named Ali ibn Ridwan (c. 988–c. 1061):

> The Sun on that day was 15° in Taurus and the spectacle in the 15th degree of Scorpio. This spectacle was a large circular body, two-and-a-half to three times as large as Venus. The sky was shining because of its light. The intensity of its light was a little more than a quarter of that of moonlight. It remained where it was and it moved daily with its zodiacal sign until the Sun was in sextile with it in Virgo, when it disappeared at once.

Ibn Ridwan recorded the positions of the Sun, Moon and planets when the star first appeared, which helped modern astronomers fix the date. He noted that the supernova did not move independently of the stars and that it disappeared when the Sun came within 60° ("in sextile") of its position. That corresponds to a visibility period of about 3 months. From Ali ibn Ridwan's observing location of Al-Fustat, which now lies within modern Cairo, the star was above the horizon only during the day when it disappeared, and because Venus at its brightest can be seen during the day, his statement suggests that the star had dimmed to a magnitude of less than −4.0.

Astrologers in Europe and Asia also noticed the new star. One of the most significant observations comes from the Benedictine monastery at St. Gallen in Switzerland. "A new star of unusual size appeared, glittering in aspect, and dazzling the eyes, causing alarm…. It was seen likewise for 3 months in the inmost limits of the south, beyond all the constellations which are seen in the sky," the monastery's chronicler reported. In fact, it was the only event in 1006 they found important enough to record, placing it on a par with a famine listed in the previous year and a plague outbreak in the following year. To modern astronomers, however, the implication that the star was just barely above St. Gallen's mountainous southern horizon placed important limits on its position and helped them identify a supernova remnant discovered in 1965 with SN 1006.

In China, the strange star's appearance caused widespread alarm. There were several categories of guest stars, but only one – the yellow *Chou-po* – bore good news. No one could agree on what category SN 1006 fell into. An astute astrologer named Zhou Keming at the Imperial Astronomical Bureau rose to the occasion. "I heard that people inside and outside the court were quite disturbed about it," he reported. "I humbly suggest that the civil and military officials be permitted to celebrate in order to set the Empire's mind at rest." Not only did the emperor approve Keming's suggestion – he gave him a promotion as well. Chinese astrologers watched the star for 3 months before it became lost in the Sun's glare, in agreement with other reports, but they went even further. According to the astronomical treatise in the official history of the Song Dynasty (960–1279), imperial sky watchers recovered the fading star in December and continued to monitor it for another year and a half.

Establishing the star's brightness from this limited information is tricky, but in 2003 a team led by astronomer Frank Winkler at Middlebury College in Vermont identified the distance to the remnant and the type of supernova. With this information, Winkler estimates that the star was magnitude −7.5 – about 12 times brighter than Venus at its best and 260 times brighter than Sirius. Ali ibn Ridwan compared the supernova to both Venus and the Moon. The early evening sky of mid-May 1006 was completely dominated by these three objects, so what he's attempting to do, says Winkler, is establish a scale using Venus and the Moon as benchmarks and then describing the star's brightness relative to them. "It's taken a long time to interpret what he meant, but now I think we've finally got it right," he said, adding that the star would have been a truly dazzling sight. "In the spring of 1006, people could probably have read manuscripts at midnight by its light."

How will the next supernova in our galaxy stack up to these historical events? Only time will tell, but we can be certain that the light of the next visual supernova is already heading our way.

References

Armitage RA, Hyman M, Southon J, et al (1997) Rock art image in Fern Cave, Lava Beds National Monument, California: Not the A.D. 1054 (Crab Nebula) supernova. Antiqu. 71: 715–719

Aschwanden MJ, Poland AI, Rabin DM (2001) The new solar corona. Ann. Rev. Astron. Astrophys. 39: 175–210

Ashworth WB Jr, (1980) A probable Flamsteed observation of the Cassiopeia A supernova. J. Hist. Astron. 11: 1

Bortle JE (1998) The bright-comet chronicles. http://www.cfa.harvard.edu/icq/bortle.html. Accessed 15 Oct. 2010

Bortle JE (1998) Brightest comets seen since 1935. http://www.cfa.harvard.edu/icq/brightest. html. Accessed 15 Oct. 2010

Brekke A, Egeland A (1983) The northern light: From mythology to space research. Springer-Verlag, Berlin

Clark DH, Stephenson FR (1977) The historical supernovae. Pergamon Press, Oxford

Davis N (1992) The aurora watcher's handbook. University of Alaska Press, Fairbanks, Alaska

Davis K (2010) Supernova 1987A. http://www.aavso.org/vsots_sn1987a Accessed 15 Nov. 2010

Donsbach M (2006) The scholar's supernova. http://www.saudiaramcoworld.com/issue/200604/the.scholar.s.supernova.htm. Accessed Oct. 15, 2010

France K, McCray R, Heng K, et al (2010) Observing Supernova 1987A with the refurbished Hubble Space Telescope. Sci. 329: 1624–1627. doi: 10.1126/science.1192134

Green DA (2009) A catalogue of galactic supernova remnants. http://www.mrao.cam.ac.uk/surveys/snrs. Accessed 15 Oct. 2010

IAU Minor Planet Center: Closest approaches to the Earth by comets. http://www.cfa.harvard.edu/iau/lists/ClosestComets.html. Accessed 15 Oct. 2010

Jacobsen R (2003) Private communication

Keay CSL (1990) C. A. Chant and the mystery of auroral sounds. J. Royal Astron. Soc. Can. 84: 373–382

Krause O, Birkmann SM, Tomonori U et al (2008) The Cassiopeia A supernova was of type IIb.

Levison H (2009) Oort cloud formation — The role of the Sun's birth cluster. http://www.boulder.swri.edu/~hal/talks/oort/DPS2010/DPS.pdf. Accessed 15 Nov. 2010

Marsden BG, Williams GV (2001) Catalogue of cometary orbits, fourteenth edition. Smithsonian Astrophysical Observatory, Cambridge, Mass

National Research Council (2008) Severe space weather events — understanding societal and economic impacts workshop report. National Academies Press, Washington, D.C

O'Connor JJ, Robertson EF (2003) Tycho Brahe. http://www-history.mcs.st-and.ac.uk/Biographies/Brahe.html. Accessed Nov. 15, 2010

Odenwald S (2001) The 23 rd cycle: Learning to live with a stormy star. Columbia University Press, New York

Osterbrock DE (2001) Who really coined the word supernova? Who first predicted neutron stars? Bull. Am. Astron. Soc. 33: 1330

Racusin JL, Karpov SV, Sokolowski M. et al (2008) Broadband observations of the naked-eye burst GRB080319B. Nat. 455: 183–188. doi:10.1038/nature07270

Reynolds SP, Borkowski KJ, Green DA et al (2008) The youngest galactic supernova remnant: G1.9+0.3. Astrophys, J. Lett. 680: L4. doi: 10.1086/589570

Ruiz-Lapuente P, Comeron F, Méndez J, et al (2004) The binary progenitor of Tycho Brahe's 1572 supernova. Nat. 431: 1069–1072

Schechner SJ (1997) Comets, popular culture, and the birth of modern cosmology. Princeton University Press, Princeton, New Jersey

Sci. 320: 1195–1197. doi: 10.1126/science.1155788

Sekanina Z (2009) Comet 17P/Holmes: A megaburst survivor. Int. Comet Q. 31: 5–23

Stephenson FR (1987) Guest stars are always welcome, Nat. Hist., August, 72–76

Stephenson FR, Yau K (1984) Oriental tales of *Halley's comet*. New Sci., September 27, 31–32

Suzuki A (2008) The 20th anniversary of SN1987A. J. Phys. Conf. Ser. 120. doi:10.1088/1742-6596/120/7/072001

U.S. Dept. of Commerce (2004) Service assessment: Intense space weather storms, October 19–November 7, 2003. National Weather Service, Silver Spring, MD

Wells DA (ed.) (1860) Annual of scientific discovery. Gould and Lincoln, Boston

Wilford JN (1990) Star explosion of 1054 is seen in Indian bowl. N.Y. Times, June 15

Winkler PF (2003) Private communication

Winkler PF, Gupta D, Long KS (2003) The SN 1006 remnant: Optical proper motions, deep imaging, distance, and brightness at maximum. Astrophys, J. 585: 324–335

Yeomans DK (1991) Comets: A chronological history of observation, science, myth, and folklore. John Wiley and Sons, New York

Yeomans DK (2007) Great comets in history. http://ssd.jpl.nasa.gov/?great_comets. Accessed 15 Oct. 2010

Appendix A

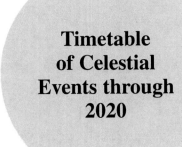

**Timetable
of Celestial
Events through
2020**

The following pages contain an astronomical almanac for the years 2011 through 2020. Think of it as a "program guide" to the sky. It lists the most observable celestial happenings – including the Moon's phases, lunar and solar eclipses, planetary gatherings and much more – in order of their occurrence. To preview what's in store for each year, read the year's highlights preceding each annual listing. To find out what's happening on a given day, check the nearest date in the almanac. Events discussed or illustrated within the book contain a reference to the page number where you'll find a diagram or more detailed information.

Some events not visible to the naked eye, such as planets in conjunction with the Sun, are included to assist you in understanding when a planet is transitioning from the evening to the morning sky or vice-versa; others, such as Mercury transits and Earth's extreme distances from the Sun, are included for their intrinsic interest.

Throughout the book, we refer to the brightness of the planets and stars by referring to the standard astronomical magnitude scale (detailed in Table 1.1 on page 8), usually by giving an object's magnitude parenthetically after the object's name – for example, Mars (1.2) or Saturn (−0.3). Keep in mind that the smaller the number, the brighter the object. Venus, the third brightest object in the sky, always has a negative magnitude around −4.0. This is so bright that, when seen in twilight or a dark sky, Venus looks artificial and much closer than it really is.

Understanding Time

The text refers to time in numerous ways. Some of these references depend on geographic location, while others do not.

F. Reddy, *Celestial Delights: The Best Astronomical Events through 2020*,
Patrick Moore's Practical Astronomy Series, DOI 10.1007/978-1-4614-0610-5,
© Springer Science+Business Media, LLC 2012

Table A.1 Convert universal time to U.S./Canadian standard time zones

Use previous calendar date

UT	Atlantic (AST) UT−4 h	Eastern (EST) UT−5	Central (CST) UT−6	Mountain (MST) UT−7	Pacific (PST) UT−8	Alaska (AKST) UT−9	Hawaii (HST) UT−10
0h	8 P.M.	7 P.M.	6 P.M.	5 P.M.	4 P.M.	3 P.M.	2 P.M.
1h	9 P.M.	8 P.M.	7 P.M.	6 P.M.	5 P.M.	4 P.M.	3 P.M.
2h	10 P.M.	9 P.M.	8 P.M.	7 P.M.	6 P.M.	5 P.M.	4 P.M.
3h	11 P.M.	10 P.M.	9 P.M.	8 P.M.	7 P.M.	6 P.M.	5 P.M.
4h	Midnight*	11 P.M.	10 P.M.	9 P.M.	8 P.M.	7 P.M.	6 P.M.
5h	1 A.M.	Midnight	11 P.M.	10 P.M.	9 P.M.	8 P.M.	7 P.M.
6h	2 A.M.	1 A.M.	Midnight	11 P.M.	10 P.M.	9 P.M.	8 P.M.
7h	3 A.M.	2 A.M.	1 A.M.	Midnight	11 P.M.	10 P.M.	9 P.M.
8h	4 A.M.	3 A.M.	2 A.M.	1 A.M.	Midnight	11 P.M.	10 P.M.
9h	5 A.M.	4 A.M.	3 A.M.	2 A.M.	1 A.M.	Midnight	11 P.M.
10h	6 A.M.	5 A.M.	4 A.M.	3 A.M.	2 A.M.	1 A.M.	Midnight
11h	7 A.M.	6 A.M.	5 A.M.	4 A.M.	3 A.M.	2 A.M.	1 A.M.
12h	8 A.M.	7 A.M.	6 A.M.	5 A.M.	4 A.M.	3 A.M.	2 A.M.
13h	9 A.M.	8 A.M.	7 A.M.	6 A.M.	5 A.M.	4 A.M.	3 A.M.
14h	10 A.M.	9 A.M.	8 A.M.	7 A.M.	6 A.M.	5 A.M.	4 A.M.
15h	11 A.M.	10 A.M.	9 A.M.	8 A.M.	7 A.M.	6 A.M.	5 A.M.
16h	Noon	11 A.M.	10 A.M.	9 A.M.	8 A.M.	7 A.M.	6 A.M.
17h	1 P.M.	Noon	11 A.M.	10 A.M.	9 A.M.	8 A.M.	7 A.M.
18h	2 P.M.	1 P.M.	Noon	11 A.M.	10 A.M.	9 A.M.	8 A.M.
19h	3 P.M.	2 P.M.	1 P.M.	Noon	11 A.M.	10 A.M.	9 A.M.
20h	4 P.M.	3 P.M.	2 P.M.	1 P.M.	Noon	11 A.M.	10 A.M.
21h	5 P.M.	4 P.M.	3 P.M.	2 P.M.	1 P.M.	Noon	11 A.M.
22h	6 P.M.	5 P.M.	4 P.M.	3 P.M.	2 P.M.	1 P.M.	Noon
23h	7 P.M.	6 P.M.	5 P.M.	4 P.M.	3 P.M.	2 P.M.	1 P.M.

*As used in this book, "midnight" refers to the *last* moment of a calendar day
Daylight Saving Time. When in effect, add one hour to the clock time indicated by this table. Note that neither Arizona (outside of
the Navajo Nation) nor Hawaii observes Daylight Saving Time. While many countries recognize some form of "summer time," they
do not necessarily change their clocks on the same dates as the U.S.

Horizon scenes. Illustrations throughout the book show the changing positions of the Moon and planets as seen by observers in the United States and southern Canada at the times and dates noted in the captions. For observers elsewhere in the Northern Hemisphere at similar latitudes, roughly between 25° and 55° N, the scenes approximate what's in the sky, although details of altitude and position will vary – especially visibility around dawn and dusk. Similarly, notes in this almanac that instruct you to look a specific time before sunrise or after sunset on a given calendar date are correct for observers in the United States and southern Canada.

Star maps. The all-sky star maps illustrate the constellations as they appear at the times on your clock – that is, your local time – and even include the annual "Spring forward/fall back" changes of Daylight Saving Time. The dates of these clock changes, which are also noted in the almanac, follow the revised U.S. rules that went into effect in 2007. If your area does not recognize daylight time, then between March and November the star charts will be one hour ahead. Just make a note to look an hour earlier than indicated.

Event listings. Some celestial events, such as conjunctions, lunar phases or eclipses, occur at specific times, whether or not North America is in darkness or pointed the right way to see them. For example, the most frequent event listings are conjunctions of the Moon with stars and planets, such as "Moon passes 1.5° south of Jupiter, 1 P.M. EST." It may at first seem strange to include a conjunction that occurs in the afternoon, when Jupiter cannot be seen. What this is really saying: When you see the Moon in a dark sky on that date, Jupiter won't be far away.

Universal time. The traditional method for communicating the time of a celestial happening is to use Universal Time (UT), which for our purposes is identical to Coordinated Universal Time (UTC), the clock time on the meridian that runs through Greenwich, England (also known as Greenwich Mean Time). With the exception of eclipses and transits, the book gives UT times that are uncorrected for irregularities in Earth's rotation, which astronomers estimate will not exceed 1.2 min by 2021. For lunar phases, oppositions, equinoxes and solstices, and Earth's closest and farthest positions from the Sun, times are rounded to the nearest minute. Times for eclipses and transits, also rounded to the nearest minute, are based on predictions by astronomer Fred Espenak at NASA's Goddard Space Flight Center.

Less time-critical events are rounded to the nearest hour. For meteor showers, unless otherwise noted, the almanac gives the hour when Earth passes closest to the orbit of the shower's parent body. The best viewing normally occurs during the hours after midnight closest to this time, although interference from the moon and specific shower behavior also come into play.

While we're most familiar with a 12-h clock, UT uses a 24-hr clock, where midnight, noon, and 6 P.M. become 0h, 12h, and 18h, respectively. To convert UT times to clock times at your location, you must add or subtract some number of hours, then switch from the 24-h UT clock to the 12-h format used by civil time. If the time shift passes midnight, the sky event may occur on a different calendar date than the one listed – and if you're not mindful of that, you may miss the event.

So to make star-gazing a little easier, the almanac lists events in both UT and Eastern Standard Time (Daylight Time, when it's in effect for the U.S.). Use Table A.1 to convert from UT to other standard U.S. and Canadian time zones. For example, say an observer in Chicago (Central Standard Time) wants to know when to look for an event occurring at 4h UT on Dec. 15. The table indicates that, for this observer, the event occurs at 10 P.M. on Dec. 14.

One final note: In everyday use, people think of midnight as being "tonight" rather than "tomorrow." To reflect this, the term as used here refers to the final instant of a calendar day.

2011 Highlights

Mercury has an excellent evening pairing with Jupiter in mid-March, is the brightest "evening star" in early July, and pops brightly into September's predawn sky near the star Regulus. And the planet closes out the year with another fine morning apparition in late December, with a waning crescent Moon and the star Antares as guides.

Venus remains a prominent predawn object through mid-May, when it pairs with Jupiter. Venus winds around the far side of its orbit, emerging from evening twilight in late November and is prominent low in the southwest as the year winds down.

Mars has one of its off years, but by mid-November the Red Planet is outshining nearby Regulus as it ramps up for the 2012 opposition.

Jupiter remains a notable evening star through mid-March, when it descends into twilight and meets Mercury going the other way. By the end of April, Jupiter emerges into morning twilight and joins a planetary gaggle (notably Venus, but also Mercury and Mars) low in the east. At opposition Oct. 28, Jupiter rises as the Sun sets, and the planet spends the rest of the year as a prominent evening object.

Saturn is prominent in early in March evenings in preparation for its April 3 opposition, and it remains a prominent evening object until September. Saturn into morning twilight in mid-October and is a predawn planet for the rest of the year.

Meteor showers. Of the year's best annual showers, only January's Quadrantids are free of lunar interference. Yet October's Draconids will be the center of attention despite less than ideal conditions. This is because astronomers predict a significant outburst from the shower as Earth encounters unusually dense streams of comet dust.

Eclipses. There are four solar eclipses, all of them partial, with the June event alone bringing the Moon's penumbra to northern Canada and Alaska. Both of the year's lunar eclipses are total; only December's brings totality to western North America, Alaska and Hawaii.

2011 Almanac		
January	1	Moon passes 2.5° north of Antares at 1 P.M. EST (18h UT)
	2	Moon near Mercury (−0.1) this morning. It will pass 3.8° south of the planet at 10 A.M. EST (15h UT)
	3	Earth at perihelion, its closest to the Sun this year – 91.4073 million miles – at 1:32 P.M. EST (18:32 UT)
		Quadrantid meteor shower peaks, 8 P.M. EST (1h UT on Jan. 4).
	4	New Moon, 4:03 A.M. EST (9:03 UT)
		Partial solar eclipse, greatest at 8:51 UT from central Sweden, where the Moon covers up to 85.8% of the Sun's disk. Visible from Europe, Africa and central Asia.
		Moon passes 2.8° north of Mars tonight, 8 P.M. EST (1h UT on Jan. 5)
	8	Venus (−4.6) at greatest western elongation (47.0°) at 11 A.M. EST (16h UT). The planet is brilliant in the predawn sky.
	10	Mercury (−0.3) at greatest western elongation (23.3°) at 9 A.M. EST (14h UT). Mercury is visible in morning twilight.
		Moon passes 7.1° north of Jupiter (−2.3) at noon EST (17h UT)
	12	First quarter Moon, 6:31 A.M. EST (11:31 UT)
	13	Look for the bright star Antares below Venus (−4.5) over the next few mornings. The planet passes 8.0° north of the star on Jan. 15, 5 P.M. EST (22h UT).
	16	Moon passes 7.5° north of the star Aldebaran at 5 A.M. EST (10h UT).
	19	Full Moon, 4:21 P.M. EST (21:21 UT)
	21	Moon passes 5.3° south of the star Regulus tonight, 9 P.M. EST (2h UT on Jan. 22).
	25	Look for Saturn (0.7) and the star Spica nearest the Moon this morning. The Moon passes 8.1° south of Saturn at 5 A.M. EST (10h UT) and 3° south of Spica eight hours later.
	26	Last quarter Moon, 7:57 A.M. EST (12:57 UT)
	28	The Moon glides 2.6° north of the star Antares at 7 P.M. EST (0h UT on Jan. 29). Together with nearby Venus, they form a striking predawn trio through month's end.
	29	The Moon passes 3.5° south of Venus (−4.4) at 11 P.M. EST (4h UT on Jan. 30).
February	1	Moon passes 3.6° north of Mercury at 1 P.M. EST (18h UT)
	2	New Moon, 9:31 P.M. EST (2:31 UT on Feb. 3)
	3	Mars in conjunction with the Sun (not visible)
	7	Moon passes 6.8° north of Jupiter at 5 A.M. EST (10h UT).
	11	First quarter Moon, 2:18 A.M. EST (7:18 UT)
	12	Moon passes 7.3° north of the star Aldebaran, 2 P.M. EST (19h UT).
	18	Full Moon, 3:36 A.M. EST (8:36 UT)
		Moon passes 5.2° south of Regulus at 8 A.M. EST (13h UT).
	21	Look for Saturn (0.5) near the rising gibbous Moon late this evening. The Moon passes 8° south of Saturn at noon EST (17h UT) and then slips 2.8° south of the star Spica at 8 P.M. EST (1h UT on Feb. 22).
	24	Last quarter Moon, 6:26 P.M. EST (23:26 UT)
	25	Moon passes 2.8° north of the star Antares at 1 A.M. EST (6h UT).
		Mercury in superior conjunction (not visible)

(continued)

2011 Almanac (continued)

28 — Look for the Moon near Venus (−4.1) in the predawn sky today and tomorrow. The Moon passes 1.6° north of Venus at 11 P.M. EST on Feb. 28 (4h UT on Mar. 1).

March 4 — New Moon, 3:46 P.M. EST (20:46 UT)

6 — Moon nearest Jupiter (−2.1) this evening. Look west in the hour after sunset. The Moon passes 6.5° north of the planet at midnight EST (5h UT on Mar. 7).

11 — Moon passes 7.1° north of the star Aldebaran at 10 P.M. EST (3h UT on Mar. 12).

12 — First quarter Moon, 6:45 P.M. EST (23:45 UT).

13 — Spring forward: U.S. Daylight Saving Time begins this morning.

14 — This is the start of Mercury's best evening appearance for the year. Use bright Jupiter (−2.1) to help you locate Mercury (−1.1) over the next few days as the innermost planet climbs into the evening sky. Look low in the west shortly after sunset; see p. 74.

16 — Jupiter passes 2.3° south of Mercury at 1 P.M. EDT (17h UT).

17 — Moon passes 5.3° south of the star Regulus, 8 P.M. EDT (0h UT on Mar. 18).

19 — Full Moon, 2:10 P.M. EDT (18:10 UT)

20 — Look for Saturn (0.4) and the star Spica near the Moon tonight. The Moon passes 8° south of Saturn at 8 P.M. EDT (0h UT on Mar. 21).

Vernal equinox, 7:21 P.M. EDT (23:21 UT)

21 — Moon slides 3° south of Spica at 7 A.M. EDT (11h UT).

22 — Mercury (−0.3) at greatest eastern elongation (18.6°), 9 P.M. EDT (1h UT on Mar. 23). It's visible low in the west about 8° above Jupiter (−2.1) in the half hour after sunset.

24 — Moon passes 3.1° north of Antares, 9 A.M. EDT (13h UT)

26 — Last quarter Moon, 8:07 A.M. EDT (12:07 UT)

31 — Moon passes 6° north of Venus (−4.0), 9 A.M. EDT (13h UT)

April 3 — New moon, 10:32 A.M. EDT (14:32 UT)

Saturn (0.4) is at opposition and nearest to Earth, 800.711 million miles away. It rises in the east at sunset and is visible all night long. Opposition occurs at 7:56 P.M. EDT (23:56 UT).

6 — Jupiter in conjunction with Sun (not visible)

7 — Look for the young crescent Moon between Aldebaran and the Pleiades star cluster this evening.

8 — Moon passes 6.8° north of the star Aldebaran, 5 A.M. EDT (9h UT)

9 — Mercury in inferior conjunction (not visible)

11 — First quarter Moon, 8:05 A.M. EDT (12:05 UT)

13 — Look for the star Regulus near the waxing gibbous Moon this evening.

14 — Moon passes 5.5° south of the star Regulus, 6 A.M. EDT (10h UT)

16 — Look for Saturn (0.4) and the star Spica near tonight's almost-full Moon.

17 — Moon passes 8.1° south of Saturn at 4 A.M. EDT (8h UT) and 2.5° south of the star Spica at 6 P.M. EDT (22h UT).

Full Moon, 10:44 P.M. EDT (2:44 UT on Apr. 18)

20 — Moon passes 3.3° north of the star Antares, 7 P.M. EDT (23h UT)

22 — Lyrid meteor shower peaks, 7 P.M. EDT (23h UT)

24 — Last quarter Moon, 10:47 P.M. EDT (2:47 UT on Apr. 25)

(continued)

2011 Almanac (continued)

| | 30 | Can you find Venus, shining at magnitude −3.8, between the Moon and the horizon shortly before dawn this morning?
Moon passes 7° north of Venus at 7 P.M. EDT (23h UT) |

May   3   New Moon, 2:51 A.M. EDT (6:51 UT)

     5   Moon passes 6.7° north of the star Aldebaran, 10 A.M. EDT (14h UT)

This is an unfavorable morning apparition for Mercury (0.5), but with Venus (−3.8), Jupiter (−2.1) and Mars (1.3) nearby, it's worth a look through binoculars in the half hour before sunrise if you have a flat eastern horizon. Venus and Jupiter, at least, will stand out. Monitor the planets' changing positions throughout the month.

Look for Eta Aquariid meteors tonight.

     6   Eta Aquariid meteor shower peaks, 9 A.M. EDT (13h UT)

     7   Mercury (0.4) at greatest western elongation (26.6°) at 3 P.M. EDT (19 UT).

     8   Look for a planetary triangle of Venus (−3.8), Jupiter (−2.1) and Mercury (0.3) low in the east in the half hour before dawn. The apex of the trio is Venus, which is only about 4° high. Jupiter is below and to the left, Mercury below and to the right. Venus and Mercury are just 1.4° apart (closest this morning at 2 A.M. EDT, 6h UT). Mars (1.3) is there, too, below and to Jupiter's left, but as the lowest and faintest of the group, it may go unnoticed. An unobstructed eastern horizon and binoculars will help.

    10   First quarter Moon, 4:33 P.M. EDT (20:33 UT)
        Jupiter passes 2.2° north of Mercury, 7 P.M. EDT (23h UT)

    11   Venus passes 0.6° south of Jupiter, 5 A.M. EDT (9h UT)
        Moon passes 5.7° south of the star Regulus, 1 P.M. EDT (17h UT)

    14   Can you find Saturn (0.6) near the Moon tonight?

    15   Moon passes 2.3° south of the star Spica, 4 A.M. EDT (8h UT)

    17   Full Moon, 7:09 A.M. EDT (11:09 UT)

    18   Moon passes 3.3° north of the star Antares, 5 A.M. EDT (9h UT)

Venus (−3.8) and Mercury (0.0) are a few arcminutes closer this morning than they were on the 8th, just under 1.4° apart (closest approach this morning at 3 A.M. EDT, 7h UT). Look very low in the east in the half hour before dawn. Jupiter (above) and Mars (below) bookend the planetary pair.

    19   Jupiter (−2.1) has broken away from the morning planetary shuffle of Venus (−3.8), Mercury (−0.2), and Mars (1.3) and stands about 4° higher than the rest; all are visible low in the east in the half hour before sunrise. Look for Mercury directly beneath Venus, with Mars off to Venus' left.

Mars passes 2.4° north of Mercury, 9 P.M. EDT (1h UT on May 20)

    22   Venus passes 1.1° south of Mars, 11 A.M. EDT (15h UT)

    24   Last quarter Moon, 2:52 P.M. EDT (18:52 UT)

    29   Watch the waning crescent Moon slide past the morning planets Jupiter (−2.1), Mars (1.3), Venus (−3.8) and Mercury (−1.0) over the next three days. Look low in the east in the half hour before sunrise. Look for Jupiter below the Moon this morning.

Moon passes 5.7° north of Jupiter, 11 A.M. EDT (15h UT)

(continued)

2011 Almanac (continued)

	30	Moon passes 3.8° north of Mars, 4 P.M. EDT (20h UT)
		Moon passes 4.4° north of Venus, midnight EDT (4h UT on May 31)
	31	Moon passes 3.7° north of Mercury, 2 P.M. EDT (18h UT)
June	1	New Moon, 5:03 P.M. EDT (21:03 UT)

Partial solar eclipse, greatest at 21:16 UT near Murmansk, Russia, where the Moon covers up to 60.1% of the Sun's disk. Visible from eastern Asia, northern Japan, central and northern Alaska, Canada, Greenland, Iceland and the Arctic Ocean.

	7	Moon passes 5.8° south of the star Regulus, 7 P.M. EDT (23h UT)
	8	First quarter Moon, 10:11 P.M. EDT (02:11 UT on June 9)
	10	Moon passes 8.2° south of Saturn, 5 P.M. EDT (21h UT)
	11	Moon passes 2.6° south of the star Spica, noon EDT (16h UT)
	12	Mercury in superior conjunction (not visible)
	14	Moon passes 3.3° north of Antares, 3 P.M. EDT (19h UT). Look for the reddish star beneath the almost-full Moon tonight.
	15	**Total lunar eclipse**, greatest at 4:13 P.M. EDT (20:13 UT). See p. 125.
		Full Moon, 4:14 P.M. EDT (20:14 UT)
	21	Summer solstice, 1:16 P.M. EDT (17:16 UT)
	23	Last quarter Moon, 7:48 A.M. EDT (11:48 UT)
	26	Jupiter (−2.2) shines brightly in the predawn sky. Look for it beneath the Moon this morning.
		Moon passes 5.5° north of Jupiter, 5 A.M. EDT (9h UT)
	27	June Boötid meteor shower peaks, 5 P.M. EDT (21h UT).
	28	Moon passes 1.8° north of Mars at 3 P.M. EDT (19h UT).
	29	Moon passes 6.7° north of Aldebaran at 1 A.M. EDT (5h UT)
July	1	New Moon, 4:54 A.M. EDT (08:54 UT)

Partial solar eclipse, greatest at 4:38 A.M. EDT (08:38 UT), off the coast of East Antarctica, where the Moon covers just 9.7 percent of the Sun's disk. Visible from a small portion of Southern Ocean and part of East Antarctica.

	2	Look for the young crescent Moon together with Mercury (−0.4) above it, low in the west-northwest in the half hour after sunset tonight.
		Moon passes 5.1° south of Mercury, 10 P.M. EDT (2h UT on July 3)
	4	Earth at aphelion, its farthest from the Sun this year – 94.5119 million miles – at 10:54 A.M. EDT (14:54 UT)
	5	Moon passes south of the star Regulus, 1 A.M. EDT (5h UT)
	6	Mars passes 5.5° north of Aldebaran, 3 A.M. EDT (7h UT)
	7	First quarter Moon, 2:29 A.M. EDT (6:29 UT)
		Look for Saturn (0.9) near the Moon tonight.
		Moon passes 8.0° south of Saturn, midnight EDT (4h UT on July 8)
	8	Moon passes 2.6° south of the star Spica, 6 P.M. EDT (22h UT)
	11	Moon passes 3.4° north of the star Antares, 10 P.M. EDT (02h UT on July 12)
	15	Full Moon, 11:07 A.M. EDT (15:07 UT)
	20	Mercury (0.4) at greatest eastern elongation (26.8°), 1 A.M. EDT (5h UT)
	23	Last quarter Moon, 1 A.M. EDT (5h UT)
		Moon passes 5.1° north of Jupiter, 9 P.M. EDT (1h UT on Jul. 24)

(continued)

2011 Almanac (continued)

	26	Moon passes 6.6° north of the star Aldebaran, 10 A.M. EDT (14h UT)
	27	Moon passes 0.5° south of Mars, 1 P.M. EDT (17h UT)
	29	Look for Southern Delta Aquariid meteors tonight.
	30	Southern Delta Aquariid meteor shower peaks, 10 A.M. EDT (14h UT)
		New Moon, 2:40 P.M. EDT (18:40 UT)
August	4	Moon passes 7.7° south of Saturn, noon EDT (16h UT)
		Moon passes 2.4° south of the star Spica, 11 P.M. EDT (3h UT on Aug. 5)
	6	First quarter Moon, 7 :08 A.M. EDT (11:08 UT)
	7	Look for the star Antares near the Moon tonight.
	8	Moon passes 3.5° north of Antares, 4 A.M. EDT (8h UT)
	12	Watch for Perseid meteors – hampered by a nearly-full Moon – tonight.
	13	Perseid meteor shower peaks this morning, 2 A.M. EDT (6h UT)
		Full moon, 2:57 P.M. EDT (18:57 UT)
	16	Venus in superior conjunction (not visible)
		Mercury in inferior conjunction (not visible)
	20	Moon passes 4.8° north of Jupiter, 8 A.M. EDT (12h UT)
	21	Last quarter Moon, 5:54 P.M. EDT (21:54 UT)
	22	Moon passes 6.4° north of the star Aldebaran, 6 P.M. EDT (22h UT)
	25	Moon passes 2.7° south of Mars, 10 A.M. EDT (14h UT)
	28	New Moon, 11:04 P.M. EDT (3:04 UT on Aug. 29)
		Begin looking for Mercury (0.7) low in the east half an hour before dawn. While you're up, check out reddish Mars (1.4) shining about 30° above it, and bright Jupiter (−2.6) high in the south.
	31	Moon passes 7.3° south of Saturn, 7 P.M. EDT (23h UT)
		Look west-southwest after sunset tonight for the young crescent Moon beneath and between Saturn (0.9, on the right) and the star Spica.
September	1	Moon passes 2.2° south of Spica, 6 A.M. EDT (10h UT)
	3	Mercury (−0.3) at greatest western elongation (18.1°) at 2 A.M. EDT (6 UT), visible in morning twilight.
	4	Moon passes 3.7° north of the star Antares, 9 A.M. EDT (13h UT)
		First quarter Moon, 1:39 P.M. EDT (17:39 UT)
	6	Watch over the next few days as descending Mercury (−0.8) meets and passes the ascending star Regulus. Look low in the east a half hour before sunrise; binoculars help. See p. 74.
	8	Mercury passes 0.7° north of Regulus, 10 P.M. EDT (2h UT on Sept. 9)
	9	Mars passes 6.0° south of the star Pollux, 10 P.M. EDT (2h UT on Sept. 10)
	12	Full Moon, 5:27 A.M. EDT (9:27 UT)
	16	Moon passes 4.7° north of Jupiter, 2 P.M. EDT (18h UT)
	19	Moon passes 6.2° north of Aldebaran, 2 A.M. EDT (6 UT)
	20	Last quarter Moon, 9:39 A.M. EDT (13:39 UT)
	23	Moon passes 4.8° south of Mars, 4 A.M. EDT (8h UT)
		Autumnal equinox, 5:05 A.M. EDT (9:05 UT)
	25	Moon passes 5.7° south of Regulus, 4 A.M. EDT (8h UT)
	27	New Moon, 7:09 A.M. EDT (11:09 UT)
	28	Mercury in superior conjunction (not visible)

(continued)

2011 Almanac (continued)

October	1	Moon passes 4.0° north of the star Antares, 4 P.M. EDT (20h UT)
	3	First quarter Moon, 11:15 P.M. EDT (3:15 UT on Oct. 4)
	8	Draconid meteor shower peaks; experts predict enhanced activity from this usually modest shower. The window for increased activity, which could briefly reach rates exceeding 600 per hour, opens at noon and closes at 7 P.M. EDT (16–23h UT). A waxing gibbous Moon makes observing a further challenge. The most favored region is the eastern Mediterranean, with the rest of Europe, northern Africa and eastern Brazil providing lesser potential.
	11	Full Moon, 10:06 P.M. EDT (2:06 UT on Oct. 12)
	13	Look low in the east as darkness falls tonight for a bright "star" near the waning Moon. That's Jupiter (−2.9), now approaching its best placement and brightness for the year.
		Moon passes 4.8° north of Jupiter, 4 P.M. EDT (20h UT)
		Saturn in conjunction with the Sun (not visible)
	16	Moon passes 5.9° north of the star Aldebaran, 8 A.M. EDT (12h UT)
	19	Last quarter Moon, midnight EDT (4h UT on Oct. 20)
	21	Orionind meteor shower peaks, 6 P.M. EDT (22h UT)
		Moon passes 6.5° south of Mars, 8 P.M. EDT (0h UT on Oct. 22)
	22	Moon passes 5.9° south of the star Regulus, 2 P.M. EDT (18h UT)
	26	Moon passes 6.7° south of Saturn, 2 A.M. EDT (6h UT)
		New Moon, 3:56 P.M. EDT (19:56 UT)
	27	Moon passes 0.2° south of Mercury, 10 P.M. EDT (2h UT on Oct. 28)
	28	Jupiter (−2.9) is at opposition and nearest to Earth, 369.013 million miles away. It rises in the east at sunset and is visible all night long. Opposition occurs at 9:42 P.M. EDT (1:42 UT on Oct. 29).
	31	Saturn passes 4.7° north of the star Spica, 1 A.M. EDT (5h UT)
		A Halloween treat: Venus (−3.8) and Mercury (−0.3) replay their double-feature from May, but in the evening sky. Look low in the west-southwest in the half hour after sunset. Venus stands above, and just under 2° separates the planets, which are closest tonight 7 P.M. EDT (23h UT on Nov. 1). A flat western horizon and binoculars help. Watch the pair slide southward and separate over the next few days.
November	2	First quarter Moon, 12:38 P.M. EDT (16:38 UT)
	6	Fall back: U.S. Daylight Saving Time ends this morning.
	9	Moon passes 5.0° north of Jupiter, 2 P.M. EST (19h UT)
		Venus passes 4° north of Antares, 4 P.M. EST (21h UT)
		Venus (−3.8) and Mercury (−0.3) remain a close (2° separation) planetary pair low in the southwest in the half hour after sunset. They don't begin to part until the 14th. The star Antares lies about 1.5° below Mercury tonight. Binoculars help.
		Mars passes 1.4° north of the star Regulus, midnight EST (5h UT on Nov. 10)
		Mercury passes 1.9° north of Antares, midnight EST (5h UT on Nov. 10)
	10	Full Moon, 3:16 P.M. EST (20:16 UT)
	12	Moon passes 5.9° north of the star Aldebaran, 1 P.M. EST (18h UT)
	13	Venus passes 1.97° from Mercury, 3 A.M. EST (8h UT)

(continued)

2011 Almanac (continued)

	14	Mercury (−0.3) at greatest eastern elongation (22.7°) at 4 A.M. EST (9h UT), visible in evening twilight
	17	Leonid meteor shower peaks, 11 P.M. EST (4h UT on Nov. 18)
	18	Moon passes 6.0° south of the star Regulus, 9 P.M. EST (2h UT on Nov. 19)
		Last quarter Moon, 10:09 A.M. EST (15:09 UT)
	19	Moon passes 7.7° south of Mars, 5 A.M. EST (10h UT)
	21	Alpha Monocerotid meteor shower peaks, 11 P.M. EST (4 UT on Nov. 22)
	22	Moon passes 2.1° south of the star Spica, 1 P.M. EST (18h UT)
		Moon passes 6.6° south of Saturn, 5 P.M. EST (22h UT)
	25	New Moon, 1:10 AM. EST (6:10 UT)
		Partial solar eclipse, greatest at 6:20 UT from near the Antarctic Peninsula, where the Moon covers up to 90.5% of the Sun's disk. Visible from the southern half of South Africa, most of Tasmania, New Zealand and nearly all of Antarctica.
		Moon passes 4.2° north of the star Antares, noon EST (17h UT)
	26	Moon passes 1.8° north of Mercury, 5 A.M. EST (10 UT)
		Look for Venus (−3.9) above and to the left of tonight's young Moon.
		Moon passes 1.8° north of Venus, 11 P.M. EST (4 UT on Nov. 27)
December	2	First quarter Moon, 4:52 A.M. EST (9:52 UT)
	4	Mercury in inferior conjunction (not visible)
	6	Moon passes 6.6° north of the star, 8 A.M. EST (13h UT)
		Moon passes 5.2° north of Jupiter, 3 P.M. EST (20h UT)
	9	Moon passes 5.9° north of the star Aldebaran, 7 P.M. EST (0h UT on Dec. 10)
	10	Full Moon, 9:36 A.M. EST (14:36 UT)
		Total lunar eclipse, greatest at 6:32 A.M. PST (14:32 UT). See p. 126.
	12	Mercury (−0.2) begins the best part of its morning show, low in the southeast in the half hour before sunrise. Look high in the south for Saturn (0.7). Ruddy Mars (0.5) rules in the southwest.
		Look for Geminid meteors tonight.
	14	Geminid meteor shower peaks, 1 P.M. EST (18h UT); look for Geminid meteors tonight, too.
	16	Moon passes 6.0° south of Regulus, 3 A.M. EST (8h UT)
	17	Moon passes 8.5° south of Mars, 8 A.M. EST (13h UT)
		Last quarter Moon, 8 P.M. EST (1h UT on Dec. 18)
	19	Moon passes 2.1° south of the star Spica, 9 P.M. EST (2h UT on Dec. 20)
	20	Moon passes 6.5° south of Saturn, 5 A.M. EST (10h UT)
	21	The moon cruises between Mercury and Antares over the next few mornings. See p. 74.
	22	Winter solstice, 12:30 A.M. EST (5:30 UT)
		Mercury passes 6.8° north of Antares, 3 P.M. EST (20h UT)
		Ursid meteor shower peaks, 9 P.M. EST (2h UT on Dec. 23)
		Mercury (−0.4) at greatest western elongation (21.8°) at 10 P.M. EST (3h UT on Dec. 23) and is visible low in the southeast in the half hour before sunrise.
		Moon passes 4.2° north of Antares, 10 P.M. EST (3h UT on Dec. 23)
		Moon passes 2.6° south of Mercury, 11 P.M. EST (4h UT on Dec. 23)

(continued)

2011 Almanac (continued)

24   New Moon, 1:06 P.M. EST (18:06 UT)
26   Look for Venus near the young Moon tonight, low in the southwest in the hour after sunset.
27   Moon passes 6.4° north of Venus, 6 A.M. EST (11h UT)

2012 Highlights

Mercury has a fine evening appearance in late February, joining Venus, Jupiter and Mars, and another good evening apparition as summer begins, with the young Moon serving as guide on Jun. 21. Mercury's best then occurs in the morning sky, where it joins Venus and Jupiter in August and Saturn and Mars in late November and December.

Venus is striking in the evening sky as the year opens and it attains its greatest prominence in early April; the apparition ends in late May as Venus approaches its inferior conjunction. When it does, on the afternoon of June 5 for the U.S., the planet will appear silhouetted on the Sun's disk – its last transit for 105 years. Venus emerges into morning twilight later that month, reaches its best in early September, and joins up with Mercury and Saturn in late November.

Mars starts the year brighter than the nearby star Regulus in Leo (see p. 179) and steadily improves through its Mar. 3 opposition, when it appears low in the east as twilight fades – a nice counterpoint to Venus, Jupiter and Mercury in the western sky. By year's end, Mars remains observable in the southwest in the hour after sunset.

Jupiter begins the year near its brightest, high in the southeast as twilight fades. By Mar. 13, Jupiter and Venus lie only 3° apart. Jupiter emerges for morning viewing in June with Venus passing nearby between Jun. 30 and Jul. 5. By the end of August, Jupiter is rising around midnight and brightening for its Dec. 2 opposition.

Saturn is rising near midnight in late January, hovering in the vicinity of the star Spica, and brightens through May 15, when it reaches opposition. Between August 12 and 16, Saturn passes less than 3° from Mars. Saturn emerges into morning twilight in mid-November.

Meteor showers. January's Quadrantids are hampered by moonlight, but it's a great year for the Perseids (August) and the Geminds (December), which have little or no lunar interference.

Eclipses. There are two solar eclipses, one annular and one total. Near the end of the annular eclipse on the afternoon of May 21, the ringed Sun appears along a strip of the western U.S., from Oregon to Texas. Of the year's two lunar eclipses, only the June event sees the Moon enter Earth's umbra; the partial eclipse is visible from the central and western U.S., Alaska, and Hawaii.

2012 Almanac		
January	1	First quarter Moon, 1:15 A.M. EST (6:15 UT)
	2	Look for Venus (–4.0) in the southwest in the hour after sunset. At the same time, look for Jupiter (–2.9) beneath the waxing gibbous Moon.
		Moon passes 5.0° north of Jupiter, 10 P.M. EST (3h UT on Jan. 3)
	3	Look for Quadrantid meteors tonight.
	4	Quadrantid meteor shower peaks, 2 A.M. EST (7h UT)
		Earth at perihelion, its closest to the Sun this year – 91.4020 million miles – at 7:32 P.M. EST (0:32 UT on Jan. 5)
	6	Moon passes 5.8° north of the star Aldebaran, 3 A.M. EST (8h UT)
	9	Full moon, 2:30 A.M. EST (7:30 UT)
	12	Moon passes 5.9° south of the star Regulus, 8 A.M. EST (13h UT)
	14	Moon passes 9.1° south of Mars, 2 A.M. EST (7h UT)
	16	Saturn (0.6) shines east of Spica and the last quarter Moon this morning.
		Moon passes 2.0° south of the star Spica, 3 A.M. EST (8h UT)
		Last quarter Moon, 4:08 A.M. EST (9:08 UT)
		Moon passes 6.4° south of Saturn, 2 P.M. EST (19h UT)
	19	Moon passes 4.2° north of the star Antares, 6 A.M. EST (11h UT)
	23	New Moon, 2:39 A.M. EST (7:39 UT)
	26	Moon passes 6.8° north of Venus, 2 P.M. EST (19h UT)
	30	Moon passes 4.5° north of Jupiter, 10 A.M. EST (15h UT)
		First quarter Moon, 11:10 P.M. EST (4:10 UT on Jan. 31)
February	2	Moon passes 5.7° north of the star Aldebaran, noon EST (17h UT)
	7	Mercury in superior conjunction (not visible)
		Full moon, 4:54 P.M. EST (21:54 UT)
	8	Moon passes 5.8° south of Regulus, 4 P.M. EST (21h UT)
	10	Moon passes 9.8° south of Mars, 7 A.M. EST (12h UT)
	12	Moon passes 1.7° south of the star Spica, 8 A.M. EST (13h UT)
		Moon passes 6.2° south of Saturn, 8 P.M. EST (1h UT on Feb. 13)
	14	Last quarter Moon, 12:04 P.M. EST (17:04 UT)
	15	Moon passes 4.5° north of the star Antares, noon EST (17h UT)
	21	New Moon, 5:35 P.M. EST (22:35 UT)
	23	This is the start of a great evening appearance for the innermost planet, and the Moon lights the way to Mercury (–1.1) tonight. Look west in the half hour after sunset and below the young Moon. Above the pair shine brilliant Venus (–4.2) and bright Jupiter (–2.2), and the Moon passes each in the days ahead. See p. 75.
	25	Moon passes 3.3° north of Venus, 5 P.M. EST (22h UT)
	26	Moon passes 3.8° north of Jupiter, 1 A.M. EST (6h UT on Feb. 27)
	29	Moon passes 5.4° north of Aldebaran, 8 P.M. EST (1h UT on March 1)
		First quarter Moon, 8:21 P.M. EST (1:21 UT on Mar. 1)
March	3	Mars (–1.2) is at opposition, rising in the east at sunset and visible all night long. Opposition occurs at 3:10 P.M. EST (20:10 UT).
		Catch four planets in one shot tonight. Start with Mercury (–0.6), low in the west in the half hour after sunset. Next up is brilliant Venus (–4.3), about 26° above Mercury. Next in line is bright Jupiter (–2.2), about 9° from Venus. Finally, turn around and cap it off with ruddy Mars, glowing low in the east and now at its best for the year.

(continued)

2012 Almanac (continued)

5	Mercury (–0.5) at greatest eastern elongation (18.2°) at 5 A.M. EST (10h UT). It's visible low in the west in the half hour after sunset.
	Mars is closest to Earth, 62.622 million miles away, noon EST (17h UT); see p. 179.
7	Moon passes 5.8° south of Regulus, 2 A.M. EST (7h UT)
8	Moon passes 9.8° south of Mars, 1 A.M. EST (6h UT)
8	Full Moon, 4:39 A.M. EST (9:39 UT)
10	Venus (–4.3) and Jupiter (–2.2) are separated by less than 4° tonight – and they'll pull even closer over the next few days. Look for the pair in the west in the hour after sunset.
	Moon passes 1.5° south of Spica, 4 P.M. EST (21h UT)
11	Spring forward: U.S. Daylight Saving Time begins this morning.
	Moon passes 6.2° south of Saturn, 3 A.M. EDT (7h UT)
13	Moon passes 4.7° north of Antares, 6 P.M. EDT (22h UT)
	Venus (–4.4) and Jupiter (–2.1) are 3° apart tonight. Look west for this conspicuous but quickly separating pair in the hour after sunset. Jupiter is headed into the twilight as Venus continues away from it.
14	Last quarter Moon, 9:25 P.M. EDT (1:25 UT on Mar. 15)
20	Vernal equinox, 1:14 A.M. EDT (5:14 UT)
21	Mercury in inferior conjunction (not visible)
22	New Moon, 10:37 A.M. EDT (14:37 UT)
25	Moon passes 3.1° north of Jupiter, 8 P.M. EDT (0h UT on Mar. 26)
26	Moon passes 1.9° south of Venus, 2 P.M. EDT (18h UT)
	Look west as twilight fades for Venus near the young Moon tonight. As the evening darkens, look for the Pleiades star cluster located about 6° above from the Moon.
27	Venus (–4.5) is at greatest eastern elongation (46.0°) at 4 A.M. EDT (8h UT). This is the planet's best evening showing through 2020. For the reason why, see p. 66 and 68.
28	Moon passes 5.2° north of Aldebaran, 4 A.M. EDT (8h UT)
30	First quarter Moon, 3:41 P.M. EDT (19:41 UT)
April 3	Moon passes 6.0° south of Regulus, 1 P.M. EDT (17h UT)
	Moon passes 8.9° south of Mars, 11 P.M. EDT (3h UT on Apr. 4)
6	Full moon, 3:19 P.M. EDT (19:19 UT)
7	Moon passes 1.5° south of Spica, 3 A.M. EDT (7h UT)
7	Moon passes 6.3° south of Saturn, 10 A.M. EDT (14h UT)
10	Moon passes 4.9° north of Antares, 2 A.M. EDT (6h UT)
13	Last quarter Moon, 6:50 A.M. EDT (10:50 UT)
16	Venus passes 10.0° north of Aldebaran, 9 P.M. EDT (1h UT on Apr. 17)
18	Mercury has switched to the morning sky, but the apparition is unfavorable. Look for the planet beneath the crescent Moon this morning, low in the east in the half-hour before dawn.
	Mercury (0.3) at greatest western elongation (27.5°) at 1 P.M. EDT (17 UT)
	Moon passes 7.6° north of Mercury, 10 P.M. EDT (2h UT on Apr. 19)
21	New Moon, 3:18 A.M. EDT (7:18 UT)
	Look for Lyrid meteors tonight.

(continued)

2012 Almanac (continued)

	22	Lyrid meteor shower peaks, 1 A.M. EDT (5h UT)
		Moon passes 2.4° north of Jupiter, 3 P.M. EDT (19h UT)
	24	Moon passes 5.0° north of Aldebaran, 11 A.M. EDT (15h UT)
		Moon passes 5.7° south of Venus, 10 P.M. EDT (2h UT on Apr. 25)
	29	First quarter Moon, 5:57 A.M. EDT (9:57 UT)
	30	Moon passes 6.1° south of Regulus, 11 P.M. EDT (3h UT on May 1)
May	1	Moon passes 7.9° south of Mars, 10 A.M. EDT (14h UT)
	4	Moon passes 1.5° south of Spica, 2 P.M. EDT (18h UT)
		Moon passes 6.4° south of Saturn, 6 P.M. EDT (22h UT)
	5	Eta Aquariid meteor shower peaks, 3 P.M. EDT (19h UT)
		Full Moon, 11:35 P.M. EDT (3:35 UT on May 6)
	7	Moon passes 5.0° north of Antares, noon ET (16h UT)
	12	Last quarter Moon, 5:47 P.M. EDT (21:47 UT)
	13	Jupiter in conjunction with the Sun (not visible)
	20	New moon, 7:47 P.M. EDT (23:47 UT)
	21	**Annular solar eclipse**, greatest at 23:53 UT on May 20, south of the Aleutian Islands. Annularity comes to portions of the western U.S. this afternoon; see p. 127.
	22	Look for Venus (−4.4) near the Moon tonight. The planet will rapidly descend into evening twilight over the next 10 days, so catch Venus while you can.
		Moon passes 4.7° south of Venus, 5 P.M. EDT (21h UT)
	25	Moon passes 10.9° south of Pollux, 9 A.M. EDT (13h UT)
	27	Mercury in superior conjunction (not visible)
	28	Moon passes 6.2° south of Regulus, 6 A.M. EDT (10h UT)
		First quarter Moon, 4:16 P.M. EDT (20:16 UT)
	29	Moon passes 6.9° south of Mars, 7 A.M. EDT (11h UT)
	31	Moon passes 1.5° south of Spica, midnight EDT (4h UT on Jun. 1)
June	1	Moon passes 6.5° south of Saturn, 1 A.M. EDT (5h UT)
	3	Moon passes 5.0° north of Antares, 11 P.M. EDT (3h UT on Jun. 4)
	4	**Partial lunar eclipse**, greatest at 4:03 A.M. PDT (11:03 UT), visible throughout the Pacific and from the western and central U.S. See p. 128.
		Full Moon, 7:12 A.M. EDT (11:12 UT)
	5	Venus at inferior conjunction. Normally this would mean that Venus cannot be seen, but starting after 6:09 P.M. EDT (22:09 UT), the planet's disk will begin tracking in silhouette across the Sun's disk – that is, **Venus transits the Sun.** This is the only portion of the event that can be seen from eastern North America because sunset occurs after the transit begins. Points west of the Great Lakes, where mid-transit occurs at about 8:29 P.M. CDT (1:29 UT Jun. 6), see this part of the event before sunset. Eye protection is required to view this event. See pp. 89 through 91 for more.
	11	Last quarter Moon, 6:41 A.M. EDT (10:41 UT)
	17	Venus (−4.2) has emerged in the morning sky and again can be near Jupiter (−2.0), low in the east-northeast in the half hour before sunrise. The waning crescent Moon joins Jupiter this morning.
		Moon passes 1.1° north of Jupiter, 4 A.M. EDT (8h UT)
		Moon passes 2.1° north of Venus, 9 P.M. EDT (1h UT on Jun. 18)

(continued)

2012 Almanac (continued)

19	New Moon, 11:02 A.M. EDT (15:02 UT)
20	Summer solstice, 7:09 P.M. EDT (23:09 UT)
21	The young Moon lies beneath Mercury (–0.1) tonight. Look low in the west in the half hour after sunset. Watch the Moon slide from Mercury to Regulus over the next few days. See p. 75.
	Mercury passes 5.2° south of the star Pollux, 2 P.M. EDT (18h UT)
	Moon passes 5.7° south of Mercury, 3 P.M. EDT (19h UT)
24	Moon passes 6.1° south of Regulus, noon EDT (16h UT)
26	Look for orange-hued Mars (0.9) near the Moon tonight.
	Moon passes 5.8° south of Mars, 11 A.M. EDT (15h UT)
	June Boötid meteor shower peaks, 11 P.M. EDT (3 UT on Jun. 27)
	First quarter Moon, 11:30 P.M. EDT (3:30 UT on Jun. 27)
28	Moon passes 1.4° south of Spica, 7 A.M. EDT (11h UT)
	Moon passes 6.3° south of Saturn, 8 A.M. EDT (12h UT)

July	1	Keep an eye on Venus (–4.6) and Jupiter (–2.0), now prominent low in the east more than an hour before sunrise.
		Moon passes 5.0° north of Antares, 9 p.m. EDT (13h UT)
	3	Full Moon, 2:52 P.M. EDT (18:52 UT)
	4	Earth at aphelion, its farthest from the Sun this year – 94.5059 million miles – at 11:32 P.M. EDT (3:32 UT on Jul. 5)
	5	This morning, Venus (–4.2) and Jupiter (–2.0) line up between the bright star Aldebaran and the prominent Pleiades star cluster. Look low in the east an hour or so before dawn. The waning crescent Moon joins the group 10 days later.
	9	Venus passes 0.9° north of Aldebaran, 3 P.M. EDT (19h UT)
	10	Last quarter Moon, 9:48 P.M. EDT (1:48 UT on Jul. 11)
	14	Moon passes 0.5° north of Jupiter, 11 P.M. EDT (3h UT on Jul. 15)
	15	Moon passes 4.9° north of Aldebaran, 6 A.M. EDT (10h UT)
		Moon passes 3.8° north of Venus, 11 A.M. EDT (15h UT)
	19	New Moon, 12:24 A.M. EDT (4:24 UT)
	21	Moon passes 5.9° south of Regulus, 5 P.M. EDT (21h UT)
	24	Moon passes 4.2° south of Mars, 6 P.M. EDT (22h UT)
		Look for the waxing crescent Moon below Mars (1.0) in the southwest tonight. Saturn (0.8), shining north of the star Spica nearby, is slightly brighter.
	25	Moon passes 1.2° south of Spica, 1 P.M. EDT (17h UT)
		Moon passes 6.0° south of Saturn, 3 P.M. EDT (19h UT)
	26	First quarter Moon, 4:56 EDT (8:56 UT)
	28	Mercury in inferior conjunction (not visible)
		Moon passes 5.2° north of Antares, 5 P.M. EDT (21h UT)
	29	Southern Delta Aquariid meteor shower peaks, 6 P.M. EDT (22 UT)

August	1	Full Moon, 11:27 P.M. EDT (3:27 UT on Aug. 2)
	3	Jupiter passes 4.7° north of Aldebaran, 1 A.M. EDT (5h UT)
	9	Last quarter Moon, 2:55 P.M. EDT (18:55 UT)
	11	Moon passes 4.7° north of Aldebaran, 2 P.M. EDT (18h UT)
		Look for Jupiter (–2.2) near the waning crescent Moon this morning and tomorrow morning. Start looking for Mercury (1.2), low in the east-northeast in the half hour before sunrise.

(continued)

2012 Almanac (continued)

		Moon passes 0.1° south of Jupiter, 5 P.M. EDT (21h UT)
		Look for Perseid meteors tonight.
	12	Perseid meteor shower peaks, 8 A.M. EDT (12h UT)
		Mars passes 1.9° north of Spica, 8 P.M. EDT (0h UT on Aug. 13)
	13	Moon passes 0.6° north of Venus, 4 P.M. EDT (20h UT)
		Look for Saturn (0.8), Mars (1.1) and the star Spica low in the west-southwest in the hour after sundown. Saturn is the brightest of the trio, ruddy Mars is in the middle. See p. 178.
	15	Venus (–4.5) at greatest western elongation (45.8°) at 5 A.M. EDT (9h UT), well placed for observation more than 2 h before sunup.
	16	Moon passes 3.6° south of Mercury, 1 A.M. EDT (5h UT)
		Mercury (–0.1) at greatest western elongation (18.7°) at 8 A.M. EDT (12h UT), visible low in the east-northeast in the half hour before sunrise. See p. 75.
	17	Mars passes 2.9° south of Saturn, 5 A.M. EDT (9h UT)
		New Moon, 11:54 A.M. EDT (15:54 UT)
		Moon passes 5.9° south of Regulus, midnight EDT (4h UT on Aug. 18)
	21	Moon passes 1.0° south of Spica, 6 P.M. EDT (22h UT)
		Look for Mars (1.2) and Saturn (0.8) above the young Moon in the hour after sunset tonight; see p. 178.
		Moon passes 5.5° south of Saturn,11 P.M. EDT (3h UT on Aug. 22)
	22	Moon passes 2.3° south of Mars, 4 A.M. EDT (8h UT)
	24	Moon passes 5.4° north of Antares, 10 P.M. EDT (2h UT on Aug. 25)
	31	Full Moon, 9:58 A.M. EDT (13:58 UT)
		Catch Venus (–4.3) near the star Pollux for the next couple of mornings. Look east in the hours before dawn. Jupiter (–2.3) is higher, traveling between the horns of the constellation Taurus. The Moon joins these planets next week.
September	1	Venus passes 8.8° south of Pollux, 6 P.M. EDT (22h UT)
		This morning, in the hour before dawn, look for Venus shining brilliantly in the east. The planet has temporarily joined the "Heavenly G," a huge asterism formed by the brightest stars of numerous constellations. Can you find them all? They are, in order: Capella in Auriga; Castor and Pollux in Gemini; Venus; Procyon in Canis Minor; Sirius in Canis Major – the brightest star in the sky, yet far outshone by Venus; Rigel in Orion; Aldebaran in Taurus; and Betelgeuse in Orion.
	8	Moon passes 0.6° south of Jupiter, 7 A.M. EDT (11h UT)
		Last quarter Moon, 9:15 A.M. EDT (13:15 UT)
	10	Mercury in superior conjunction (not visible)
	12	Venus (–4.2) is near the waning crescent Moon this morning. Look east in the hours before dawn.
		Moon passes 3.7° south of Venus, 1 P.M. EDT (17h UT)
	14	Moon passes 5.9° south of Regulus, 9 A.M. EDT (13h UT)
	15	New moon, 10:11 P.M. EDT (2:11 UT on Sept. 16)
	18	Moon passes 5.0° south of Saturn, 10 A.M. EDT (14h UT)
	19	Moon passes 0.2° south of Mars, 5 P.M. EDT (21h UT)

(continued)

2012 Almanac (continued)

	21	Moon passes 5.6° north of Antares, 4 A.M. EDT (8h UT)
	22	Autumnal equinox, 10:49 A.M. EDT (14:49 UT)
		First quarter Moon, 3:41 P.M. EDT (19:41 UT)
	29	Full Moon, 11:19 P.M. EDT (3:19 UT on Sept. 30)
October	1	Watch Venus (–4.1) approach and pass the bright star Regulus during the next few mornings. Look east in the hours before dawn. Check out the pair with binoculars.
	3	Venus passes 0.1° south of Regulus, 4 A.M. EDT (8h UT)
	5	Moon passes 4.3° north of Aldebaran, 5 A.M. EDT (9h UT)
		Moon passes 0.9° south of Jupiter, 5 P.M. EDT (21h UT)
	8	Last quarter Moon, 3:33 A.M. EDT (7:33 UT)
		Draconid meteor shower activity expected between 15h and 17h UT, which is during daylight in the Americas.
	11	Moon passes 6.0° south of Regulus, 7 P.M. EDT (23h UT)
	12	Look for Venus (–4.4) near the waning crescent Moon this morning.
		Moon passes 6.3° south of Venus, 3 P.M. EDT (19h UT)
	15	New Moon, 8:02 A.M. EDT (12:02 UT)
	16	Look for Mercury (–0.2) above the young Moon very low in the west-southwest 20 min after sunset tonight.
		Moon passes 1.3° north of Mercury, 10 P.M. EDT (2h UT on Oct. 17)
	18	Catch the young Moon above Mars (1.2) and its stellar rival, Antares, tonight in the hour after sunset. Look low in the southwest.
		Moon passes 2.0° north of Mars, 9 A.M. EDT (13h UT)
		Moon passes 5.8° north of Antares, 11 A.M. EDT (15h UT)
	20	Mars passes 3.6° north of Antares, 2 A.M. EDT (6h UT)
		Look for Orionid meteors tonight.
	21	Orionid meteor shower peaks, midnight EDT (4h UT)
		First quarter Moon, 11:32 P.M. EDT (3:32 UT on Oct. 22)
	25	Saturn in conjunction with the Sun (not visible)
	26	Mercury (–0.2) at greatest eastern elongation (24.1°) at 6 P.M. EDT (22h UT). This is not a favorable evening appearance, so look for Mercury low in the southwestern sky 20 minutes after sunset. As the sky darkens, look for Mars (1.2) about 16° above and to Mercury's left. The star Antares is roughly between them.
	29	Full Moon, 3:49 P.M. EDT (19:49 UT)
November	1	The Moon rises with bright Jupiter (–2.7) and the star Aldebaran this evening. Look east-northeast in the hours after sunset.
		Moon passes 4.2° north of Aldebaran, noon EDT (16h UT)
		Moon passes 0.9° south of Jupiter, 9 P.M. EDT (1h UT on Nov. 2)
	4	Fall back: U.S. Daylight Saving Time ends this morning.
	6	Last quarter Moon, 7:36 P.M. EST (0:36 UT on Nov. 7)
	8	Moon passes 6.1° south of Regulus, 4 A.M. EST (9h UT)
	11	Look for Venus (–4.0) near the waning crescent Moon this morning. It visits Saturn (0.6) the next day. Look low in the east-southeast in the hour before dawn.
		Moon passes 5.3° south of Venus, 1 P.M. EST (18h UT)
		Moon passes 0.8° south of Spica, 9 P.M. EST (2h UT on Nov. 12)
	12	Moon passes 4.3° south of Saturn, 4 P.M. EST (21h UT)

(continued)

2012 Almanac (continued)

	13	New Moon, 5:08 P.M. EST (22:08 UT)

Total solar eclipse, greatest at 22:12 UT in the South Pacific off New Zealand. See p. 129.

15 — Watch Venus (–3.9) slide past the star Spica this morning (look east-southeast an hour or so before dawn). Check over the next week as Venus gears up for a meeting with Saturn (0.6). See p. 209.

Venus passes 4.1° north of Spica, 6 P.M. EST (23h UT)

16 — Moon passes 4.0° north of Mars, 5 A.M. EST (10h UT)

17 — Mercury in inferior conjunction (not visible)

Leonid meteor shower peaks, 5 A.M. EST (10h UT)

20 — First quarter Moon, 9:31 A.M. EST (14:31 UT)

21 — Alpha Monocerotid meteor shower peaks, 5 A.M. EST (10h UT)

24 — Lots of action in the morning sky. Venus and Saturn rise together and close the gap between them over the next few mornings. Look low in the east-southeast an hour or so before dawn. Look out for Mercury (1.1), now returning to the morning sky for a good morning appearance; see p. 209. At the same time, look west for Jupiter (–2.8) beaming brightly between the horns of Taurus, the starry bull.

26 — Venus passes 0.6° south of Saturn, midnight EST (5h UT on Nov. 27); see p. 209.

28 — Full Moon, 9:46 A.M. EST (14:46 UT)

Look for Jupiter (–2.8) rising with tonight's full Moon. Look east-northeast in the hour after sunset. Jupiter is above and to the Moon's left; the star Aldebaran is below and to the right.

Moon passes 4.2° north of Aldebaran, 5 P.M. EST (22h UT)

Moon passes 0.6° south of Jupiter, 8 P.M. EST (1h UT on Nov. 29)

Penumbral lunar eclipse, greatest at 6:33 A.M. PST (14:33 UT), visible from the central and western continental U.S. before moonset.

December 1 — Venus (–3.9) forms a nice morning lineup with Saturn (0.6, above) and Mercury (–0.4, below) through Dec. 12. Look east-southeast in the hour before dawn. See p. 75.

2 — Jupiter (–2.8) is at opposition and nearest to Earth, 378.298 million miles away. It rises in the east at sunset and is visible all night long. Opposition occurs at 8:45 P.M. EST (1:45 UT on Dec. 3).

4 — Mercury (–0.5) at greatest western elongation (20.6°) at 6 P.M. EST (23h UT), visible in morning twilight.

5 — Moon passes 6.1° south of Regulus, 11 A.M. EST (16h UT)

6 — Last quarter Moon, 10:31 A.M. EST (15:31 UT)

7 — Jupiter passes 4.7° north of Aldebaran, 3 P.M. EST (20h UT)

9 — Follow the Moon over the next few mornings to locate the star Spica, and the planets Saturn, Venus, and Mercury. Look southeast in the hour before dawn.

Moon passes 0.8° south of Spica, 7 A.M. EST (12h UT)

10 — Moon passes 4.0° south of Saturn, 7 A.M. EST (12h UT)

11 — Moon passes 1.6° south of Venus, 9 A.M. EST (14h UT)

Moon passes 1.1° south of Mercury, 8 P.M. EST (1h UT on Dec. 12)

12 — Moon passes 5.8° north of Antares, 8 A.M. EST (13h UT)

Geminid meteor shower peaks, 7 P.M. EST (0h UT on Dec. 13)

13 — New Moon, 3:42 A.M. EST (8:42 UT)

(continued)

2012 Almanac (continued)	
15	Moon passes 5.6° north of Mars, 5 A.M. EST (10h UT)
17	Mercury passes 5.5° north of Antares, 10 A.M. EST (15h UT)
20	First quarter Moon, 12:19 A.M. EST (5:19 UT)
21	Winter solstice, 6:12 A.M. EST (11:12 UT)
22	Ursid meteor shower peaks, 3 A.M. EST (8h UT)
23	Venus passes 5.7° north of Antares, 6 A.M. EST (11h UT)
25	Moon passes 0.4° south of Jupiter, 7 P.M. EST (0h UT on Dec. 26)
	Moon passes 4.2° north of Aldebaran, midnight EST (5h UT on Dec. 26)
28	Full Moon, 5:21 A.M. EST (10:21 UT)

2013 Highlights

Mercury enjoys great evening apparitions in early February, with the crescent Moon as a guide on the 11th. In late May, the planet joins in a nice evening arrangement with Venus and Jupiter and, in mid-June, again pairs up with Venus. Mercury is a morning object in late July/early August and flies formation with Jupiter and (very faint) Mars. In late November, Mercury passes less than a degree from Saturn.

Venus hovers low in the predawn sky as the year opens. It crosses to the evening sky in March, forms up with Jupiter and Mercury in late May, and with Saturn in mid-September. But the planet never stands prominently in the dusk, preferring to hug the southwestern sky in the hours after sunset for the rest of the year. Venus attains its best altitudes in late November, when it sets nearly 3 h after the Sun.

Mars coasts through the year, emerging as a predawn object in late July, when it meets up with Jupiter. By mid-October the Red Planet is rising more than 4 hours before the Sun and is located near Regulus; by year's end, it's rising near midnight and is located a few degrees from the star Spica. See the map on p. 180.

Jupiter, just a month after opposition, is a dazzling evening sight as the year begins, glowing in the southeast between Aldebaran and the Pleiades cluster in Taurus as darkness falls. In late May, a faded Jupiter joins Venus and Mercury in a planetary triangle and then gradually slips into the Sun's glare. The planet quickly emerges into morning twilight and, in late July and early August, pairs with Mars and Mercury. By year's end – with its next opposition just days away – Jupiter ascends shortly after the Sun goes down.

Saturn brightens steadily throughout the early months of the year as it approaches its Apr. 28 opposition. In mid-September, Saturn stands above Venus in evening twilight. By the end of November, Saturn emerges into predawn twilight and takes a morning meeting with Mercury.

Meteor showers. The year's best but chilliest showers – January's Quadrantids and December's Geminids – both experience lunar interference, with the Geminids, near a full Moon, receiving the worst. However, the nearly first quarter Moon sets before midnight, leaving the best part of August's Perseids in a dark sky.

Eclipses. The year's two solar eclipses bring an annular Sun to Australia, New Zealand and the Pacific in May. In November, the decade's only hybrid eclips delivers a brief glimpse of totality to central Africa. Of the year's three lunar eclipses, just one brings the Moon into Earth's umbra, and only very briefly at that. The April partial eclipse is visible throughout Europe, Africa, Asia and Australia.

2013 Almanac		
January	1	Moon passes 5.9° south of Regulus, 5 P.M. EST (22h UT)
		Earth at perihelion, its closest to the Sun this year – 91.4026 million miles – at 11:38 P.M. EST (4:38 UT on Jan. 2)
	2	Look for Quadrantid meteors tonight.
	3	Quadrantid meteor shower peaks, 8 A.M. EST (13h UT)
	4	Last quarter Moon, 10:58 P.M. EST (3:58 UT on Jan. 5)
	5	Moon passes 0.6° south of Spica, 3 P.M. EST (20h UT)
	6	Moon passes 3.8° south of Saturn, 8 P.M. EST (1h UT on Jan. 7)
	8	Moon passes 5.9° north of Antares, 6 P.M. EST (23h UT)
	10	Catch Venus (−3.9) beneath the waning crescent Moon this morning. Look low in the southeast in the half hour before dawn.
		Moon passes 2.8° north of Venus, 7 A.M. EST (12h UT)
	11	Moon passes 5.9° north of Mercury, 9 A.M. EST (14h UT)
		New Moon, 2:44 P.M. EST (19:44 UT)
	18	Mercury in superior conjunction (not visible)
		First quarter Moon, 6:45 P.M. EST (23:45 UT)
	21	Look for the bright Jupiter (−2.6) near the Moon high in the southeast as darkness falls tonight. Look for the star Aldebaran – the eye of Taurus, the celestial bull – below them and the Pleiades star cluster above.
		Moon passes 0.5° south of Jupiter, 10 P.M. EST (3h UT on Jan. 22)
	22	Moon passes 4.0° north of Aldebaran, 7 A.M. EST (12h UT)
	26	Full Moon, 11:38 P.M. EST (4:38 UT on Jan. 27)
	28	Moon passes 5.8° south of Regulus, 10 P.M. EST (3h UT on Jan. 29)
February	1	Moon passes 0.3° south of Spica, 9 P.M. EST (2h UT on Feb. 2)
	3	Last quarter Moon, 8:56 A.M. EST (13:56 UT)
		Moon passes 3.5° south of Saturn, 5 A.M. EST (10h UT)
	5	Moon passes 6.1° north of Antares, 2 A.M. EST (7h UT)
	7	Mercury (−1.0) starts off a fine evening apparition with a little help from Mars (1.2). Watch them over the next few days. Look very low in the west-southwest in the half hour after sunset. Binoculars help. See p. 76.
	8	Mercury passes 0.3° north of Mars, 4 P.M. EST (21h UT)
	10	New Moon, 2:20 A.M. EST (7:20 UT)

(continued)

2013 Almanac (continued)

	11	Look for Mercury (−0.9) and Mars (1.2) beneath the Moon this evening. Look low in the west-southwest in the half hour after sunset.
		Moon passes 6.2° north of Mars, 9 A.M. EST (14h UT)
		Moon passes 5.2° north of Mercury, 1 P.M. EST (18h UT)
	16	Mercury (−0.6) at greatest eastern elongation (18.1°) at 5 P.M. EST (22h UT), visible in evening twilight.
	17	First quarter Moon, 3:31 P.M. EST (20:31 UT)
March	1	Moon passes 0.1° south of Spica, 2 A.M. EST (7h UT)
	2	Moon passes 3.3° south of Saturn, 10 A.M. EST (15h UT)
	4	Moon passes 6.4° north of Antares, 8 A.M. EST (13h UT)
		Mercury in inferior conjunction (not visible)
		Last quarter Moon, 4:53 P.M. EST (21:53 UT)
	10	Spring forward: U.S. Daylight Saving Time begins this morning.
	11	New Moon, 3:51 P.M. EDT (19:51 UT)
	17	Look for Jupiter (−2.2), the brightest "star" near the Moon tonight, high in the southwest as darkness falls.
		Moon passes 1.5° south of Jupiter, 9 P.M. EDT (1h UT on Mar. 18)
		Moon passes 3.5° north of Aldebaran, 11 P.M. EDT (3h UT on Mar. 18)
	19	First quarter Moon, 1:27 P.M. EDT (17:27 UT)
	20	Vernal equinox, 7:02 A.M. EDT (11:02 UT)
	24	Jupiter passes 5.1° north of Aldebaran, 2 P.M. EDT (18h UT)
		Moon passes 5.8° south of Regulus, 4 P.M. EDT (20h UT)
	27	Full Moon, 5:27 A.M. EDT (9:27 UT)
	28	Moon passes 0.0° north of Spica, 11 A.M. EDT (15h UT)
		Venus in superior conjunction (not visible)
	29	Moon passes 3.4° south of Saturn, 4 P.M. EDT (20h UT)
	31	Moon passes 6.6° north of Antares, 2 P.M. EDT (18h UT)
		Mercury (0.2) at greatest western elongation (27.8°) at 5 P.M. EDT (22h UT), visible very low in the east-southeast in the half hour before dawn.
April	3	Last quarter Moon, 12:37 A.M. EDT (4:37 UT)
	8	Moon passes 7.1° north of Mercury, 6 A.M. EDT (10h UT)
	10	New Moon, 5:35 A.M. EDT (9:35 UT)
	14	Moon passes 3.4° north of Aldebaran, 7 A.M. EDT (11h UT)
		Moon passes 2.1° south of Jupiter, 2 P.M. EDT (18h UT)
	18	Mars in conjunction with the Sun (not visible)
		First quarter Moon, 8 A.M. EDT (12:31 UT)
	21	Moon passes 5.9° south of Regulus, 1 A.M. EDT (5h UT)
	22	Lyrid meteor shower peaks, 8 A.M. EDT (12h UT)
	24	Moon passes 0.0° north of Spica, 8 P.M. EDT (0h UT on Apr. 25)
	25	Full Moon, 3:57 P.M. EDT (19:57 UT)
		Moon passes 3.5° south of Saturn, 10 P.M. EDT (2h UT on Apr. 26)
		Partial lunar eclipse, greatest at 20:07 UT; see p. 130.
	27	Moon passes 6.6° north of Antares, 10 P.M. EDT (2h UT on Apr. 28)
	28	Saturn (0.1) is at opposition and nearest to Earth, 809.051 million miles away. It rises in the east at sunset and is visible all night long. Opposition occurs at 4:27 A.M. EDT (8:27 UT).

(continued)

2013 Almanac (continued)

May	2	Last quarter Moon, 7:14 A.M. EDT (11:14 UT)
	5	Eta Aquariid meteor shower peaks 9 P.M. EDT (1h UT on May 6)
	8	Venus (−3.9) begins to emerge from evening twilight. Look west-northwest in the half hour after sunset. Jupiter (−2.0) stands about 21° above.
	9	New Moon, 8:28 P.M. EDT (0:28 UT on May 10)
	10	Moon passes 1.4° south of Venus, 8 P.M. EDT (0h UT on May 11)
		Annular solar eclipse, greatest at 0:25 UT; see p. 131.
	11	Moon passes 3.4° north of Aldebaran, 2 P.M. EDT (18h UT)
		Mercury in superior conjunction (not visible)
	12	Moon passes 2.6° south of Jupiter, 9 A.M. EDT (13h UT)
	18	First quarter Moon, 12:35 A.M. EDT (4:35 UT)
		Moon passes 5.9° south of Regulus, 10 A.M. EDT (14h UT)
	22	Moon passes 0.0° north of Spica, 7 A.M. EDT (11h UT)
		Catch Saturn (0.3) near the Moon tonight. Look southeast as darkness falls.
	23	Moon passes 3.7° south of Saturn, 6 A.M. EDT (10h UT)
		Mercury (−1.1) joins Venus (−3.9) and Jupiter (−1.9) in evening twilight. Look low in the west-northwest in the half hour after sunset. Watch them shift positions over the next few days.
	24	Mercury passes 1.4° north of Venus, midnight EDT (4h UT on May 25)
		Penumbral lunar eclipse, greatest at 11:10 P.M. EDT (4:10 UT on May 25) visible throughout the continental U.S., all be extreme eastern Canada, Central and South America, and the western half of Africa.
	25	Full Moon, 12:25 A.M. EDT (4:25 UT)
		Moon passes 6.6° north of Antares, 8 A.M. EDT (12h UT)
	27	Mercury passes 2.4° north of Jupiter, 6 A.M. EDT (10h UT)
	28	Venus passes 1.0° north of Jupiter, 5 P.M. EDT (21h UT)
	31	Last quarter Moon, 2:58 P.M. EDT (18:58 UT)
June	8	New Moon, 11:56 A.M. EDT (15:56 UT)
	10	Look for a pair of planets – Mercury (0.3) and Venus (−3.8) – and a pair of stars – Castor and Pollux in Gemini – low in the west-northwest in the hour after sunset. Together with the young Moon, the planets form a triangle with Mercury at the apex. About 10° above Mercury you'll find Castor and Pollux. See p. 76.
		Moon passes 5.3° south of Venus, 7 A.M. EDT (11h UT)
		Moon passes 5.9° south of Mercury, 7 P.M. EDT (23h UT)
	12	Mercury (0.9) at greatest eastern elongation (24.3°) at 1 P.M. EDT (17h UT), visible in evening twilight about 4° above Venus tonight. Look low in the west-northwest in the hour after sunset.
	14	Moon passes 5.8° south of Regulus, 4 P.M. EDT (20h UT)
	16	First quarter Moon, 1:24 P.M. EDT (17:24 UT)
	18	Mercury (1.0) and Venus (−3.8) sit side by side tonight, almost reflecting the positions of the stars Castor and Pollux 6° above them. Binoculars help. Look low in the west-northwest in the hour after sunset.
		Moon passes 0.1° north of Spica, 4 P.M. EDT (20h UT)

(continued)

2013 Almanac (continued)

	19	Jupiter in conjunction with the Sun (not visible)
		Moon passes 3.7° south of Saturn, 1 P.M. EDT (17h UT)
	20	Venus passes 1.9° north of Mercury, 2 P.M. EDT (18h UT)
	27	June Boötid meteor shower peaks, 5 A.M. EDT (9h UT)
July	5	Moon passes 3.3° north of Aldebaran, 3 A.M. EDT (7h UT)
		Earth at aphelion, its farthest from the Sun this year – 94.5090 million miles – at 10:44 A.M. EDT (14:44 UT)
	6	Moon passes 3.7° south of Mars, 8 A.M. EDT (12h UT)
	8	New Moon, 3:14 A.M. EDT (7:14 UT)
	9	Mercury in inferior conjunction (not visible)
	10	See Venus (–3.9) above the young Moon tonight. Look west in the hour after sunset.
		Moon passes 7.0° south of Venus, 7 P.M. EDT (23h UT)
	11	Moon passes 5.6° south of Regulus, 10 P.M. EDT (2h UT on Jul. 12)
	15	First quarter Moon, 11:18 P.M. EDT (3:18 UT on Jul. 16)
		Moon passes 0.3° north of Spica, midnight EDT (4h UT on Jul. 16)
	16	Catch Saturn (0.6) above the Moon tonight. Look south-southwest an hour or so after sunset. In the morning sky, Jupiter (–1.9) rises to meet Mars (1.6) low in the east-northeast in the hour before dawn. They're less than 3° apart this morning.
		Moon passes 3.3° south of Saturn, 9 P.M. EDT (1h UT on Jul. 17)
	19	Moon passes 6.8° north of Antares, 5 A.M. EDT (9h UT)
	22	Can you spot Regulus near Venus (–3.9) tonight? Look low in the west-northwest in the hour after sunset.
		Venus passes 1.2° north of Regulus, 1 A.M. EDT (5h UT)
		Mars and Jupiter are separated by less than 1° this morning. Look for the planet pair low in the east-northeast in the hour before dawn. Find bright Jupiter first, then look for fainter, redder Mars above and to the left. Watch the two separate in the following mornings.
		Mars passes 0.8° north of Jupiter, 2 A.M. EDT (6h UT)
		Full Moon, 2:16 P.M. EDT (18:16 UT)
	29	Last quarter Moon, 1:43 P.M. EDT (17:43 UT)
		Southern Delta Aquariid meteor shower peaks, midnight EDT (5h UT on Jul. 30)
	30	Mercury (0.8) at greatest western elongation (19.6°) at 5 A.M. EDT (9h UT), visible in morning twilight low in the east-northeast in the half hour before dawn. Look for it 8° below bright Jupiter (–1.9) and faint Mars (1.6), which are still less than 2° apart.
August	1	Moon passes 3.2° north of Aldebaran, 9 A.M. EDT (13h UT)
	3	This morning, the waning crescent Moon joins the arc of predawn planets. Look low in the east-northeast in the hour before dawn. The closest and brightest planet to the Moon, Jupiter (–1.9), is about 6° below and to its left. Mars (1.6), fainter and redder, is a similar distance below and to Jupiter's left. Finally, Mercury (–0.4) puts in a solid appearance about 8° below Mars. The Moon will lie nearest Mars tomorrow. See p. 76.
		Moon passes 4.0° south of Jupiter, 6 P.M. EDT (22h UT)
	4	Moon passes 5.2° south of Mars, 7 A.M. EDT (11h UT)
		Mercury passes 7.3° south of Pollux, 11 P.M. EDT (3h UT on Aug. 5)

(continued)

2013 Almanac (continued)

5	Moon passes 4.4° south of Mercury, 5 A.M. EDT (9h UT)
6	New Moon, 5:51 P.M. EDT (21:51 UT)
9	Moon passes 5.1° south of Venus, 10 P.M. EDT (2h UT on Aug. 10)
12	Moon passes 0.6° north of Spica, 5 A.M. EDT (9h UT)
	Perseid meteor shower peaks, 2 P.M. EDT (18h UT); look for meteors tonight.
13	Moon passes 2.9° south of Saturn, 4 A.M. EDT (8h UT)
14	First quarter Moon, 6:56 A.M. EDT (10:56 UT)
15	Moon passes 7.0° north of Antares, noon EDT (16h UT)
19	Mars passes 5.9° south of Pollux, 8 A.M. EDT (12h UT)
20	Full Moon, 9:45 P.M. EDT (1:45 UT on Aug. 21)
24	Mercury in superior conjunction (not visible)
28	Last quarter Moon, 5:35 A.M. EDT (9:35 UT)
	Moon passes 2.9° north of Aldebaran, 4 P.M. EDT (20h UT)
31	Moon passes 4.5° south of Jupiter, 1 P.M. EDT (17h UT)
September 2	Moon passes 6.2° south of Mars, 6 A.M. EDT (10h UT)
5	New Moon, 7:36 A.M. EDT (11:36 UT)
	Venus passes 1.8° north of Spica, 9 A.M. EDT (13h UT)
	Look the star Spica below Venus (−4.1) tonight, low in the west-southwest in the hour after sunset.
8	Moon passes 0.8° north of Spica, 11 A.M. EDT (15h UT)
	Moon passes 0.4° south of Venus, 5 P.M. EDT (21h UT)
9	Moon passes 2.4° south of Saturn, 1 P.M. EDT (17h UT)
11	Moon passes 7.2° north of Antares, 6 P.M. EDT (22h UT)
12	First quarter Moon, 1:08 P.M. EDT (17:08 UT)
16	Venus (−4.1) has an evening rendezvous with Saturn (0.4). Look for the pair low in the west-southwest in the hour after sunset. Saturn remains above and about 4° from Venus though Jun. 20.
19	Full Moon, 7:13 A.M. EDT (11:13 UT)
	Venus passes 3.8° south of Saturn, 8 P.M. EDT (0h UT on Sept. 20)
22	Autumnal equinox, 4:44 P.M. EDT (20:44 UT)
24	Mercury passes 0.8° north of Spica, 3 P.M. EDT (19h UT)
	Moon passes 2.7° north of Aldebaran, midnight EDT (4h UT on Sept. 25)
26	Last quarter Moon, 11:55 P.M. EDT (3:55 UT on Sept. 27)
28	Moon passes 4.9° south of Jupiter, 5 A.M. EDT (9h UT)
October 1	Moon passes 6.7° south of Mars, 2 A.M. EDT (6h UT)
	Moon passes 5.6° south of Regulus, 7 P.M. EDT (23h UT)
4	New Moon, 8:35 P.M. EDT (0:35 UT on Oct. 5)
6	Look for the young Moon low in the west-southwest in the half hour after sunset. About 2° below it is Mercury (−0.1), and this is your best chance to catch the planet in this unfavorable apparition. Saturn lies 4° above the moon. In the southwest, left of the Moon and its planetary attendants, Venus (−4.3) gleams brightly.
	Moon passes 2.8° north of Mercury, 6 P.M. EDT (22h UT)
	Moon passes 1.9° south of Saturn, midnight EDT (4h UT on Oct. 7)
8	Moon passes 4.7° north of Venus, 8 A.M. EDT (12h UT)
	Look for Venus (−4.3) beneath the young Moon tonight, low in the southwest in the hour after sunset.

(continued)

2013 Almanac (continued)

		Moon passes 7.4° north of Antares, 11 P.M. EDT (3h UT on Oct. 9)
	9	Mercury (−0.1) at greatest eastern elongation (25.3°) at 6 A.M. EDT (10h UT)
	10	Saturn passes 5.4° north of Mercury, 3 P.M. EDT (19h UT)
	11	First quarter Moon, 7:02 P.M. EDT (23:02 UT)
	13	Mars (1.6) gives Leo, the celestial lion, a red heart over the next few mornings. Look for the planet high in the east in the hours before dawn near the constellation's brightest star, Regulus.
	14	Mars passes 1.0° north of Regulus, 6 P.M. EDT (22h UT)
	16	Venus passes 1.6° north of Antares, noon EDT (16h UT)
	18	Full Moon, 7:38 P.M. EDT (23:38 UT)
		Penumbral lunar eclipse, greatest at 7:50 P.M. EDT (23:50 UT), is visible throughout the continental U.S. – all but the northeastern states see the eclipse in progress as the Moon rises – South America, Africa, Europe and Asia.
	21	Orionid meteor shower peaks, 7 A.M. EDT (11h UT)
	22	Moon passes 2.6° north of Aldebaran, 8 A.M. EDT (12h UT)
	25	Moon passes 5.1° south of Jupiter, 6 P.M. EDT (22h UT)
	26	Last quarter Moon, 7:40 P.M. EDT (23:40 UT)
	29	Moon passes 5.7° south of Regulus, 4 A.M. EDT (8h UT)
		Moon passes 6.4° south of Mars, 9 P.M. EDT (1h UT on Oct. 30)
November	1	Venus (−4.5) at greatest eastern elongation (47.1°) at 4 A.M. EDT (8h UT). Although this is a poor evening apparition, the planet is prominent low in the southwest for more than 2 h after sunset.
		Mercury in inferior conjunction (not visible)
	2	Moon passes 0.8° north of Spica, 3 A.M. EDT (7h UT)
	3	Fall back: U.S. Daylight Saving Time ends this morning.
		Hybrid solar eclipse, greatest at 12:46 UT; see p. 132.
		New Moon, 7:50 A.M. EST (12:50 UT)
		Moon passes 1.6° south of Saturn, 1 P.M. EST (18h UT)
	5	Moon passes 7.4° north of Antares, 6 A.M. EST (11h UT)
	6	Saturn in conjunction with the Sun (not visible)
		Moon passes 8.0° north of Venus, 8 P.M. EST (1h UT on Nov. 7)
	10	First quarter Moon, 12:57 A.M. EST (5:57 UT)
	17	Full Moon, 10:16 A.M. EST (15:16 UT)
		Mercury (−0.2) at greatest western elongation (19.5°) at 9 P.M. EST (2h UT on Nov. 18), visible in morning twilight.
		Leonid meteor shower peaks, 11 A.M. EST (16h UT)
	18	Moon passes 2.7° north of Aldebaran, 3 P.M. EST (20h UT)
	21	Look for Jupiter (−2.5) near the waning gibbous Moon tonight.
		Moon passes 5.1° south of Jupiter, midnight EST (5h UT on Nov. 22)
		Alpha Monocerotid meteor shower peaks, 11 A.M. EST (16h UT)
	25	Moon passes 5.6° south of Regulus, 11 A.M. EST (16h UT)
		Last quarter Moon, 2:28 P.M. EST (19:28 UT)
	22	Over the next few mornings, watch Mercury (−0.7) cruise past Saturn (0.6) low in the east-southeast in the hour before dawn. They're less than a degree apart on the mornings of Nov. 25 and 26. See p. 76.

2013 Almanac (continued)

	25	Mercury passes 0.3° south of Saturn, 11 P.M. EST (4h UT on Nov. 26)
	27	Moon passes 5.7° south of Mars, 11 A.M. EST (16h UT)
	29	Moon passes 0.9° north of Spica, noon EST (17h UT)
December	1	The waning crescent Moon joins Saturn (0.6) and Mercury (–0.6) this morning. Look low in the east-southeast in the hour before dawn.
		Moon passes 1.3° south of Saturn, 5 A.M. EST (10h UT)
		Moon passes 0.4° north of Mercury, 5 P.M. EST (22h UT)
	2	New Moon, 7:22 P.M. EST (0:22 UT on Dec. 3)
	5	Moon passes 7.7° north of Venus, 7 P.M. EST (0h UT on Dec. 6)
	9	First quarter Moon, 10:12 A.M. EST (15:12 UT)
	14	Geminid meteor shower peaks, 1 A.M. EST (6h UT)
	15	Moon passes 2.7° north of Aldebaran, 10 P.M. EST (3h UT on Dec.16)
	17	Full Moon, 4:28 A.M. EST (9:28 UT)
	18	Look for bright Jupiter (–2.7) left of the Moon tonight, low in the east-northeast a couple of hours after sunset.
	19	Moon passes 5.0° south of Jupiter, 2 A.M. EST (7h UT)
	21	Winter solstice, 12:11 P.M. EST (17:11 UT)
	22	Moon passes 5.4° south of Regulus, 6 P.M. EST (23h UT)
		Ursid meteor shower peaks, 9 A.M. EST (14 UT)
	25	Last quarter Moon, 8:48 A.M. EST (13:48 UT)
		Moon passes 4.6° south of Mars, 10 P.M. EST (3h UT on Dec. 26)
	26	Moon passes 1.1° north of Spica, 10 P.M. EST (3h UT on Dec 27)
	28	Moon passes 0.9° south of Saturn, 8 P.M. EST (1h UT on Dec. 29)
	29	Mercury in superior conjunction (not visible)
	30	Moon passes 7.4° north of Antares, 3 A.M. EST (8h UT)

2014 Highlights

Mercury is an evening star as January closes, and it is again at the end of May. In late July, Mercury appears in the predawn sky with Venus. In all cases, the waning crescent Moon serves as a helpful guide.

Venus emerges in the predawn sky in January and hugs the southeastern horizon throughout the apparition, never rising more than two and a half hours before the Sun. In mid-August, just before the apparition ends, Jupiter and Venus shine together in morning twilight.

Mars begins this opposition year near the star Spica, which it just outshines as the year opens (see map, p. 180). The Red Planet steadily brightens until its Apr. 8 opposition, when it beams throughout the night as a tawny star almost 7° from Spica.

Jupiter opens the year at its best, with opposition occurring on Jan. 5 among the stars of Gemini. The planetary giant crosses to the morning sky in July and tags up with descending Venus in mid-August. By November, Jupiter is brightening again as it approaches the stars of Leo and its next opposition.

Saturn begins the year in the constellation Libra, sharing the predawn sky with Mars (high in the south) and Jupiter (low in the west). Still in Libra, Saturn reaches its brightest for the year at opposition on May 10. In August, Mars joins Saturn for an evening conjunction in the southwest.

Meteor showers. With January's Quadrantids peaking during the day for North Americans and the Perseids hampered by a waning gibbous Moon, the Geminids offer the year's best meteor viewing opportunity – at least until around midnight, when the last quarter Moon rises.

Eclipses. Of the four eclipses this year, two solar and two lunar, all but one includes North Americans. Total lunar eclipses in April and October bring a reddened Moon to U.S. skies, and a partial solar eclipse in late October covers more than 40% of the Sun for most of the country west of the Mississippi. Alaska and the western half of Canada see an even deeper eclipse, but it misses Hawaiians entirely.
2014 Almanac

January	1	New Moon, 6:14 A.M. EST (11:14 UT)
	2	Moon passes 2.0° north of Venus, 7 A.M. EST (12h UT)
		Look for Venus (−4.4) beneath the young Moon tonight, low in the west-southwest in the half hour after sunset.
		Look for Quadrantid meteors tonight.
	3	Quadrantid meteor shower peaks, 3 P.M. EST (20h UT); look for them tonight.
	4	Earth at perihelion, its closest to the Sun this year – 91.4067 million miles – at 6:59 A.M. EST (11:59 UT)
	5	Jupiter (−2.7) is at opposition and nearest to Earth, 391.404 million miles away. It rises in the east-northeast at sunset and is visible all night long. Opposition occurs at 4:11 P.M. EST (21:11 UT)
	7	First quarter Moon, 10:39 P.M. EST (3:39 UT on Jan. 8)
	11	Venus in inferior conjunction (not visible)
	12	Moon passes 2.6° north of Aldebaran, 4 A.M. EST (9h UT)
	14	The brightest "star" near the Moon tonight is the planet Jupiter (−2.7), now at its best.
	15	Moon passes 4.9° south of Jupiter, 1 A.M. EST (6h UT)
		Full Moon, 11:52 P.M. EST (4:52 UT on Jan. 16)
	18	Moon passes 5.2° south of Regulus, midnight EST (5h UT on Jan. 19)
	22	Start looking for Mercury (−0.9), low in the west-southwest in the half hour after sunset.
	23	Moon passes 3.7° south of Mars, 1 A.M. EST (6h UT)
		Moon passes 1.3° north of Spica, 5 A.M. EST (10h UT)
		Over the coming mornings, the Moon guides you to Mars (0.4), Saturn (0.6), the star Antares (Mars' rival in color), and Venus (−4.5), now becoming prominent in the predawn sky. This morning, you'll find Mars above the Moon high in the south in the hour before sunrise.

(continued)

2014 Almanac (continued)

24	Last quarter Moon, 12:19 A.M. EST (5:19 UT)
25	Saturn hovers above the moon this morning. Look high in the south in the hour before sunrise.
	Moon passes 0.6° south of Saturn, 9 A.M. EST (14h UT)
26	Moon passes 7.6° north of Antares, 1 PM. EST (18h UT)
28	Mars passes 4.9° north of Spica, 3 PM. EST (20h UT)
	Moon passes 2.3° south of Venus, 10 P.M. EST (3h UT on Jan. 29)
29	Look for the Moon below brilliant Venus (−4.7) this morning, low in the southeast in the half hour before sunrise.
30	New Moon, 4:39 P.M. EST (21:39 UT)
31	Mercury (−0.6) at greatest eastern elongation (18.4°) at 5 A.M. EST (10h UT), visible in evening twilight. Tonight, look for the planet above the young crescent Moon, low in the west-southwest shortly after sunset (see p. 77). Then turn around and look high in the east for Jupiter (−2.6), the only other planet visible in the early evening − and the brightest star-like object you can now see; see p. 77.
February 1	Moon passes 4.1° north of Mercury, 2 A.M. EST (7h UT)
6	First quarter Moon, 2:22 P.M. EST (19:22 UT)
8	Moon passes 2.3° north of Aldebaran, 10 A.M. EST (15h UT)
11	Moon passes 5.0° south of Jupiter, 1 A.M. EST (6h UT)
14	Full Moon, 6:53 P.M. EST (23:53 UT)
15	Moon passes 5.1° south of Regulus, 6 A.M. EST (11h UT)
	Mercury in inferior conjunction (not visible)
19	Look for Mars (−0.2) above the Moon this morning. Look high in the southwest in the hours before dawn. The Moon will guide you to Saturn, Venus and Mercury in the coming mornings.
	Moon passes 1.6° north of Spica, 10 A.M. EST (15h UT)
19	Moon passes 3.1° south of Mars, 7 P.M. EST (0h UT on Feb. 20)
21	Look for Saturn (0.5) near the Moon this morning, high in the south-southwest in the hours before dawn.
	Moon passes 0.3° south of Saturn, 5 P.M. EST (22h UT)
22	Last quarter Moon, 12:15 P.M. EST (17:15 UT)
	Moon passes 7.9° north of Antares, 9 P.M. EST (2h UT on Feb. 23)
25	Moon passes 0.4° north of Venus, midnight EST (5h UT on Feb. 26)
26	Look for Venus (−4.8, its brightest) near the waning crescent Moon this morning, low in the southeast in the hour before sunrise.
27	Look for Mercury (1.0) near the Moon this morning, low in the south-southwest in the hours before dawn.
	Moon passes 2.9° north of Mercury, 4 P.M. EST (21h UT)
March 1	New Moon, 3:00 A.M. EST (8:00 UT)
7	Moon passes 2.1° north of Aldebaran, 6 P.M. EST (23h UT)
8	First quarter Moon, 8:27 A.M. EST (13:27 UT)
9	Spring ahead: U.S. Daylight Saving Time begins this morning.
	As the sky darkens tonight, look nearly overhead for the Moon. The brightest star near it is the planet Jupiter, shining at magnitude −2.4, the only planet visible in the early evening. It isn't alone for long. Mars (−0.7) rises in the east about 3 h after sunset and is rapidly brightening as its April opposition approaches.

(continued)

2014 Almanac (continued)

10	Moon passes 5.2° south of Jupiter, 7 A.M. EDT (11h UT)
14	Mercury (0.1) at greatest western elongation (27.6°) at 2 A.M. EDT (6h UT), visible in morning twilight low in the east-southeast in the half hour before sunrise.
	Moon passes 5.1° south of Regulus, 2 P.M. EDT (18h UT)
16	Full Moon, 1:08 P.M. EDT (17:08 UT)
18	Moon passes 1.7° north of Spica, 5 P.M. EDT (21h UT)
	Mars (−1.0) rises with the Moon tonight. Look east-southeast a couple of hours after sunset. The Red Planet looks more peach-colored than the crimson. Mars will brighten until Apr. 8, when it reaches opposition.
	Moon passes 3.2° south of Mars, 11 P.M. EDT (3h UT on Mar. 19)
20	Vernal equinox, 12:57 P.M. EDT (16:57 UT)
	Look for Saturn (0.3) near the Moon tonight, rising in the east-southeast about five hours after sunset.
	Moon passes 0.3° south of Saturn, 11 P.M. EDT (3h UT on Mar. 21)
22	Moon passes 8.1° north of Antares, 3 A.M. EDT (7h UT)
	Venus (−4.5) at greatest western elongation (46.6°) at 4 P.M. EDT (20h UT), brilliant in morning twilight. Look east-southeast in the hour before sunrise. While you're up, try to catch Mercury (0.0), far below and to the left of Venus, in the half hour before dawn. The Moon joins the pair on Mar. 27.
23	Last quarter Moon, 9:46 P.M. EDT (1:46 UT on Mar. 24)
27	Moon passes 3.6° north of Venus, 6 A.M. EDT (10h UT)
29	Moon passes 6.2° north of Mercury, 1 A.M. EDT (5h UT)
30	New Moon, 2:45 P.M. EDT (18:45 UT)
	Mars passes 5.1° north of Spica, midnight EDT (4h UT on Mar. 31)
April 3	Tonight, in the hours after sunset, look west for the young Moon and, to its left, the star Aldebaran. The Pleiades star cluster is below and to the Moon's right. High above, Jupiter (−2.2) beams brightly. Then turn around and look for Mars (−1.4) low in the east-southeast.
4	Moon passes 2.0° north of Aldebaran, 3 A.M. EDT (7h UT)
6	Moon passes 5.4° south of Jupiter, 7 P.M. EDT (23h UT)
7	First quarter Moon, 4:31 A.M. EDT (8:31 UT)
8	Mars (−1.5) is at opposition. It rises in the east at sunset and is visible all night long. Opposition occurs at 5:03 P.M. EDT (21:03 UT).
10	Moon passes 5.2° south of Regulus, 10 P.M. EDT (2h UT on Apr. 11)
14	Mars is closest to Earth, 57.406 million miles away, 9 A.M. EDT (13h UT)
	Moon passes 3.5° south of Mars, 2 P.M. EDT (18h UT)
	Moon passes 1.7° north of Spica, midnight EDT (4h UT on Apr. 15)
15	Full Moon, 3:42 A.M. EDT (7:42 UT)
	Total lunar eclipse, greatest at 3:46 A.M. EDT (7:46 UT), all of totality visible throughout the Americas. See p. 133.
17	Moon passes 0.4° south of Saturn, 3 A.M. EDT (7h UT)
18	Moon passes 8.1° north of Antares, 9 A.M. EDT (13h UT)
22	Last quarter Moon, 3:52 A.M. EDT (7:52 UT)
	Lyrid meteor shower peaks, 2 P.M. EDT (18h UT)

(continued)

2014 Almanac (continued)

	25	Look for Venus (−4.2) near the waning crescent Moon this morning, low in the east-southeast in the hour before dawn.
		Moon passes 4.4° north of Venus, 7 P.M. EDT (23h UT)
		Mercury in superior conjunction (not visible)
	29	**Annular solar eclipse**, greatest at 6:03 UT; annularity visible only in Antarctica.
		New Moon, 2:14 A.M. EDT (6:14 UT)
		Moon passes 1.6° south of Mercury, 9 A.M. EDT (13h UT)
May	1	Moon passes 2.0° north of Aldebaran, noon EDT (16h UT)
	4	Moon passes 5.5° south of Jupiter, 10 A.M. EDT (14h UT)
	6	First quarter Moon, 11:15 P.M. EDT (3:15 UT on May 7)
		Eta Aquariid meteor shower peaks, 3 A.M. EDT (7h UT)
	7	Tonight, start looking for Mercury (−1.1) very low in the west-northwest in the half hour after sunset. Then take a moment to consider the rest of the celestial scene: Jupiter (−2.0) shines high in the west. The bright star nearest the Moon tonight is Regulus. Mars (−1.0) is the bright ochre star high in the southeast, and Saturn (0.1) is rising low in the east-southeast. See p. 77.
	8	Moon passes 5.2° south of Regulus, 6 A.M. EDT (10h UT)
	10	Saturn (0.1) is at opposition and nearest to Earth, 827.277 million miles away. It rises in the east at sunset and is visible all night long. Opposition occurs at 2:28 P.M. EDT (18:28 UT).
	11	Moon passes 3.0° south of Mars, 10 A.M. EDT (14h UT)
	12	Moon passes 1.7° north of Spica, 9 A.M. EDT (13h UT)
	13	Mercury passes 7.8° north of Aldebaran, noon EDT (16h UT)
	14	Moon passes 0.6° south of Saturn, 8 A.M. EDT (12h UT)
		Full Moon, 3:16 P.M. EDT (19:16 UT)
	15	Moon passes 8.1° north of Antares, 5 P.M. EDT (21h UT)
	21	Last quarter Moon, 8:59 A.M. EDT (12:59 UT)
	25	Mercury (0.4) at greatest eastern elongation (22.7°) at 3 A.M. EDT (7h UT), visible low in the west-northwest in the hour after sunset. The Moon helps you find Mercury on May 30. Look for bright Jupiter (−1.9) above and to Mercury's left. See p. 77.
	25	The waning crescent Moon visits Venus this morning, low in the east in the hour before dawn.
		Moon passes 2.3° north of Venus, noon EDT (16h UT)
	28	New Moon, 2:40 P.M. EDT (18:40 UT)
		Moon passes 2.0° north of Aldebaran, 8 P.M. EDT (0h UT on May 29)
	30	Moon passes 5.9° south of Mercury, noon EDT (16h UT)
	31	Look west in the hours after sunset to see Jupiter (−1.9) above the Moon.
June	1	Moon passes 5.5° south of Jupiter, 4 A.M. EDT (8h UT)
	4	Moon passes 5.0° south of Regulus, 2 P.M. EDT (18h UT)
	5	First quarter moon, 4:39 P.M. EDT (20:39 UT)
	7	Moon passes 1.6° south of Mars, 9 P.M. EDT (1h UT on Jun. 8)
	8	Moon passes 1.8° north of Spica, 6 P.M. EDT (22h UT)
	10	Moon passes 0.6° south of Saturn, 3 P.M. EDT (19h UT)
	12	Moon passes 8.1° north of Antares, 2 A.M. EDT (6h UT)
	13	Full Moon, 12:11 A.M. EDT (4:11 UT)

(continued)

2014 Almanac (continued)

	19	Last quarter Moon, 2:39 P.M. EDT (18:39 UT)
		Mercury in inferior conjunction (not visible)
	21	Summer solstice, 6:51 A.M. EDT (10:51 UT)
		Jupiter passes 6.4° south of Pollux, 8 A.M. EDT (12h UT)
	24	Look for Venus (−3.9) beside the waning crescent Moon this morning, low in the east in the hour before dawn.
		Moon passes 1.3° south of Venus, 9 A.M. EDT (13h UT)
	25	Moon passes 2.0° north of Aldebaran, 3 A.M. EDT (7h UT)
	27	New Moon, 4:08 A.M. EDT (8:08 UT)
		June Boötid meteor shower peaks, 11 A.M. EDT (15h UT)
	28	Moon passes 5.5° south of Jupiter, 11 P.M. EDT (3h UT on Jun. 29)
July	1	Moon passes 4.8° south of Regulus, 9 P.M. EDT (1h UT on Jul. 2)
	3	Earth at aphelion, its farthest from the Sun this year – 94.5065 million miles – at 8:13 P.M. EDT (0:13 UT on Jul. 4)
	5	First quarter Moon, 7:59. A.M. EDT (11:59 UT)
		Look for Mars (0.1) above the Moon tonight, high in the southwest as darkness falls.
		Moon passes 0.2° north of Mars, 9 P.M. EDT (1h UT on Jul. 6)
	6	Moon passes 2.1° north of Spica, 3 A.M. EDT (7h UT)
	7	Moon passes 0.4° south of Saturn, 10 P.M. EDT (2h UT on Jul. 8)
	9	Moon passes 8.2° north of Antares, noon EDT (16h UT)
	12	Venus (−3.8) serves is your guide for finding Mercury (0.3). Both shine low in the east-northeast in the half hour before dawn. This morning, Mercury lies about 7° beneath Venus; see p. 77.
		Full Moon, 7:25 A.M. EDT (11:25 UT)
		Mercury at greatest western elongation (20.9°) at 2 P.M. EDT (18h UT)
		Mars passes 1.4° north of Spica, 7 P.M. EDT (23h UT)
		You'll find Mars (0.2) near the star Spica this evening; look southwest in the hours after sunset. Look for Saturn about 20° left of Mars. Both of these planets are easy to distinguish – they're they brightest star-like objects in this part of the sky. Compare their subtle colors.
	18	Last quarter Moon, 10:08 P.M. EDT (2:08 UT on Jul. 19)
	22	Moon passes 1.8° north of Aldebaran, 8 A.M. EDT (12h UT)
	24	This morning, Venus (−3.8) is the brightest "star" near the waning crescent Moon, low in the east-northeast in the half hour before sunrise.
		Moon passes 4.4° south of Venus, 2 P.M. EDT (18h UT)
		Jupiter in conjunction with the Sun (not visible)
	25	This morning, look for Mercury (−0.9) is near the waning crescent Moon, low in the east-northeast in the half hour before sunrise; see p. 77.
		Moon passes 5.1° south of Mercury, 11 A.M. EDT (15h UT)
	26	New Moon, 6:42 P.M. EDT (22:42 UT)
	30	Southern Delta Aquariid meteor shower peaks, 6 A.M. EDT (10h UT)
August	2	Moon passes 2.3° north of Spica, 10 A.M. EDT (14h UT)
	3	Moon passes 2.2° north of Mars, 6 A.M. EDT (10h UT)
		The Moon lies roughly between Mars (0.4) and Saturn (0.5) tonight. Look in the southwest as darkness falls. The Moon is closer to Saturn (above and to the left) than Mars (below right), the brighter member of the planet pair. Keep an eye on these two: The planets drift closer together in the coming weeks.

(continued)

2014 Almanac (continued)

		First quarter Moon, 8:50 P.M. EDT (0:50 UT on Aug. 4)
	4	Moon passes 0.1° south of Saturn, 7 A.M. EDT (11h UT)
	5	Moon passes 8.4° north of Antares, 10 P.M. EDT (2h UT on Aug. 6)
	7	Venus passes 6.6° south of Pollux, 5 P.M. EDT (21h UT)
	8	Mercury in superior conjunction (not visible)
	10	Full Moon, 2:09 P.M. EDT (18:09 UT)
	12	Perseid meteor shower peaks, 8 P.M. EDT (0h on Aug.13)
	13	Venus (−3.8) and Jupiter (−1.8) are heading for a meeting in the predawn sky. Look low in the east-northeast half an hour before dawn. Watch how their positions change over the next week; see p. 209.
	17	Last quarter Moon, 8:26 A.M. EDT (12:26 UT)
		Venus passes 0.2° north of Jupiter, midnight EDT (4h UT on Aug. 18)
	18	Look for Venus and Jupiter near their closest approach this morning, low in the east-northeast in the hour before sunrise. Venus is the brighter of the two. Jupiter is ascending out of morning twilight, while Venus is heading into it; see p. 209.
		Moon passes 1.6° north of Aldebaran, 2 P.M. EDT (18h UT)
	23	The waning crescent Moon joins Jupiter and Venus this morning, low in the east-northeast in the hour before sunrise. See p. 209.
		Moon passes 5.5° south of Jupiter, 1 P.M. EDT (17h UT)
	24	Moon passes 5.7° south of Venus, 2 A.M. EDT (6h UT)
	25	New Moon, 10:13 A.M. EDT (14:13 UT)
		Mars and Saturn – both at magnitude 0.6 – make their closest approach tonight. Look for the pair in the southwest as darkness falls. Both planets are the same apparent brightness, yet Saturn is 7.6 times farther from you than Mars.
	29	Moon passes 2.5° north of Spica, 4 P.M. EDT (20h UT)
	31	Moon passes 0.4° north of Saturn, 3 P.M. EDT (19h UT)
		Moon passes 4.1° north of Mars, 8 P.M. EDT (0h UT on Sept. 1)
September	2	Moon passes 8.6° north of Antares, 5 A.M. EDT (9h UT)
		First quarter Moon, 7:11 A.M. EDT (11:11 UT)
	5	This morning, Venus (−3.9) is closest to the celestial Lion's heart – it's brightest star, Regulus. Look low in the east half an hour before dawn. Check out Jupiter (−1.8), which hovers higher in the twilight, as well.
		Venus passes 0.8° north of Regulus, 8 A.M. EDT (12h UT)
	8	Full Moon, 9:38 P.M. EDT (1:38 UT on Sept. 9)
	14	Moon passes 1.4° north of Aldebaran, 9 P.M. EDT (1h UT on Sept. 15)
	15	Last quarter Moon, 10:05 P.M. EDT (2:05 UT on Sept. 16)
	20	Moon passes 5.4° south of Jupiter, 7 A.M. EDT (11h UT)
		Mercury passes 0.6° south of Spica, 10 P.M. EDT (2h UT on Sept. 21)
	21	Moon passes 4.7° south of Regulus, 3 P.M. EDT (19h UT)
		Mercury (0.0) at greatest eastern elongation (26.4°), 6 P.M. EDT (22h UT). Throughout this unfavorable apparition, the planet remains very low in the west-southwest in the half hour after sunset.
	22	Autumnal equinox, 10:29 P.M. EDT (2:29 UT on Sept. 23)
	23	Moon passes 4.0° south of Venus, noon EDT (16h UT)
	24	New Moon, 2:14 A.M. EDT (6:14 UT)

(continued)

2014 Almanac (continued)

	25	Moon passes 2.6° north of Spica, 9 P.M. EDT (1h UT on Sept. 26)
	26	Moon passes 4.2° north of Mercury, 6 A.M. EDT (10h UT)
	27	Mars passes 3.1° north of Antares, 5 P.M. EDT (21h UT)
		Look for Saturn (0.6) near the Moon tonight, low in the southwest as darkness falls.
		Moon passes 0.7° north of Saturn, midnight EDT (4h UT on Sept. 28)
	29	Moon passes 8.7° north of Antares, 11 A.M. EDT (15h UT)
		Moon passes 5.7° north of Mars, 1 P.M. EDT (17h UT)
		This evening, look southwest for the young Moon as darkness falls. Mars (0.8) lies beneath it – and directly above its stellar rival, Antares. They are nearly the same brightness, with Mars the brighter of the two.
October	1	First quarter Moon, 3:33 P.M. EDT (19:33 UT)
	6	Draconid meteor shower activity is possible around 20h UT, which is during daylight for the Americas
	8	Full moon, 6:51 A.M. EDT (10:51 UT)
		Total lunar eclipse, greatest at 5:55 A.M. EDT (10:55 UT). All of totality is visible from Australia to the U.S. Great Lakes; the Moon sets during totality as seen from the East Coast. See p. 134.
	12	Moon passes 1.4° north of Aldebaran, 6 A.M. EDT (10h UT)
	15	Last quarter Moon, 3:12 P.M. EDT (19:12 UT)
	16	Mercury in inferior conjunction (not visible)
	17	Moon passes 5.4° south of Jupiter, midnight EDT (4h UT on Oct. 18)
	18	Moon passes 4.7° south of Regulus, 11 P.M. EDT (3h UT on Oct. 19)
	21	Orionid meteor shower peaks, 1 P.M. EDT (17h UT)
	23	New Moon, 5:57 P.M. EDT (21:57 UT)
		Partial solar eclipse, greatest at 2:44 PDT (21:44 UT) when 81 percent of the Sun is obscured near Gateshead Island, Nunavut, Canada. All of the U.S. except the Northeast states and Hawaii see at least some part of the eclipse, with areas north of central Oregon seeing the Sun more than 60% obscured.
	25	Venus in superior conjunction (not visible)
		Moon passes 1.0° north of Saturn, noon EDT (16h UT)
	26	Moon passes 8.7° north of Antares, 4 P.M. EDT (20h UT)
	28	Moon passes 6.5° north of Mars, 9 A.M. EDT (13h UT)
	30	First quarter Moon, 10:48 P.M. EDT (2:48 UT on Oct. 31)
November	1	Mercury (–0.6) at greatest western elongation (18.7°), 9 A.M. EDT (13h UT), and is visible in morning twilight. Look low in the east in the hour before dawn. See p. 77.
	2	Fall back: U.S. Daylight Saving Time ends this morning.
	3	Mercury passes 4.6° north of Spica, 1 A.M. EST (6h UT)
	6	Full Moon, 5:23 P.M. EST (22:23 UT)
	8	Moon passes 1.4° north of Aldebaran, 3 P.M. EST (20h UT)
	14	Last quarter Moon, 10:15 A.M. EST (15:15 UT)
		Moon passes 5.3° south of Jupiter, 1 P.M. EST (18h UT)
	15	Moon passes 4.6° south of Regulus, 5 A.M. EST (10h UT)
	17	Leonid meteor shower peaks, 5 P.M. EST (22h UT)
	18	Saturn in conjunction with the Sun (not visible)

(continued)

2014 Almanac (continued)

	19	Moon passes 2.6° north of Spica, 11 A.M. EST (16h UT)
	21	Moon passes 1.9° north of Mercury, noon EST (17h UT)
		Alpha Monocerotid meteor shower peaks, 5 P.M. EST (22h UT)
	22	New Moon, 7:32 A.M. EST (12:32 UT)
	26	Moon passes 6.6° north of Mars, 5 A.M. EST (10h UT)
	29	First quarter Moon, 5:06 A.M. EST (10:06 UT)
December	5	Moon passes 1.5° north of Aldebaran, midnight EST (5h UT on Dec. 6)
	6	Full Moon, 7:27 A.M. EST (12:27 UT)
	8	Mercury in superior conjunction (not visible)
	11	Moon passes 5.1° south of Jupiter, 11 P.M. EST (4h UT on Dec. 12)
	12	Moon passes 4.4° south of Regulus, 2 P.M. EST (19h UT)
	13	Look for Geminid meteors tonight.
	14	Geminid meteor shower peaks, 7 A.M. EST (12h UT)
		Last quarter Moon, 7:51 A.M. EST (12:51 UT)
	16	Moon passes 2.8° north of Spica, 8 P.M. EST (1h UT on Dec. 17)
	19	Moon passes 1.5° north of Saturn, 4 P.M. EST (21h UT)
	20	Moon passes 8.7° north of Antares, 8 A.M. EST (13h UT)
	21	Winter solstice, 6:03 P.M. EST (23:03 UT)
		New Moon, 8:36 P.M. EST (1:36 UT on Dec. 22)
	22	Ursid meteor shower peaks, 3 P.M. EST (20h UT)
		Moon passes 6.2° north of Venus, midnight EST (5h UT on Dec. 23)
	23	Look for Venus (−3.9) near the young Moon tonight, very low in the west-southwest in the half hour after sunset.
	24	The brightest "star" near the Moon tonight is the dim planet Mars (1.1). Look southwest as darkness falls.
	25	Moon passes 5.7° north of Mars, 3 A.M. EST (8h UT)
	28	First quarter Moon, 1:31 P.M. EST (18:31 UT)

2015 Highlights

Mercury opens the year with a fine conjunction with Venus that peaks with these evening planets less than a degree apart on Jan. 10. Then, as April turns to May, evening-star Mercury scoots past the Pleiades star cluster in Taurus while dazzling Venus looks on. In October, Mercury has a nice treat for early risers as it makes up the tail end of a planetary line-up that includes Venus, Jupiter, Mars and the waning Moon.

Venus emerges in evening twilight as January closes, an apparition that will peak in May and last into July. The planet meets with Mars and the Moon in late February and with Jupiter at the end of June. After crossing into morning twilight in September, Venus pairs up with Jupiter again in late October and with an even fainter Mars in early November. This is when Venus is at its best for this appearance, but it remains visible in the morning sky through year's end.

Mars, despite conjunctions with Venus and Jupiter, is experiencing an off year in preparation for its 2016 opposition.

Jupiter starts the year bright and steadily improves through Feb. 6 and its opposition. As June closes, Jupiter meets Venus in evening twilight, with the two planets a third of a degree apart on Jun. 30 – much closer than their October conjunction. It will be a great sight with binoculars or a small telescope.

Saturn opens the year near Antares in Scorpius and brightens steadily through spring; its opposition comes May 22. Saturn keeps to itself most of the year, having been left behind by the gaggle of planets that crossed to the morning sky. As December draws to a close, Saturn emerges in morning twilight and prepares to meet Venus.

Meteor showers. The Moon washes out the Quadrantids, but the Perseids have no lunar interference. The broad Geminid peak fares well with a slender and early-setting waxing crescent Moon.

Eclipses. Both of the year's lunar eclipses are total – although in April the Moon passes through the Earth's umbra in just five minutes – and each favors a different North American coast. For the March total solar eclipse, the Moon's shadow swings past Iceland and over the Faroe Islands and Svalbard, while September's partial eclipse is deepest over Antarctica.

2015 Almanac		
January	2	Moon passes 1.4° north of Aldebaran, 7 A.M. EST (12h UT)
	3	Quadrantid meteor shower peaks, 9 P.M. EST (2h UT on Jan. 4)
	4	Earth at perihelion, its closest to the Sun this year – 91.4013 million miles – at 1:36 A.M. EST (6:36 UT)
		Full Moon, 11:53 P.M. EST (4:53 UT on Jan. 5)
	8	Moon passes 5.1° south of Jupiter, 3 A.M. EST (8h UT)
		Moon passes 4.1° south of Regulus, 9 P.M. EST (2h UT on Jan. 9)
	7	Venus (–3.9) is becoming a prominent evening sight low in the southwest in the half hour after sunset. Tonight, Mercury (–0.8) is just over 1° away; see p. 78. Watch the two planets close over the next 3 days.
		As the sky darkens, scan above this bright duo for Mars (1.1). Although relatively faint, Mars is brighter than any stars within about 25° of Venus.
	13	Last quarter Moon, 4:46 A.M. EST (9:46 UT)
		Moon passes 3.1° north of Spica, 5 A.M. EST (10h UT)
	10	Venus (–3.9) passes 0.7° from Mercury (–0.8), 8 P.M. EST (1h UT on Jan. 11). Look for this mismatched pair low in the southwest in the half hour after sunset. You'll spy Venus first. Mercury lies below and to the right, with enough separation from Venus that a full Moon would fit between them with little room to spare. See p. 78.
	14	Mercury (–0.6) at greatest eastern elongation (18.9°) at 4 P.M. EST (21h UT). Mercury is visible in evening twilight, 1.3° to the right of much brighter Venus (–3.9), low in the southwest in the half hour after sunset.

(continued)

2015 Almanac (continued)

	16	The Moon guides you to Saturn (0.6) today. Look in the southeast in the hours before dawn. Saturn is the brightest "star" near the waning lunar crescent. Turn to the west to find an even brighter planet – Jupiter (–2.5), second only to the Moon in brilliance this morning.

Moon passes 1.9° north of Saturn, 7 A.M. EST (12h UT)

Moon passes 8.8° north of Antares, 6 P.M. EST (23h UT)

20 New Moon, 8:14 A.M. EST (13:14 UT)

21 Moon passes 3.0° north of Mercury, 1 P.M. EST (18h UT)

Tonight, in the half hour after sunset, look west-southwest for the Moon topping a triangle formed with Venus (–3.9) and Mercury (0.5).

Moon passes 5.6° north of Venus, midnight EST (5h UT on Jan. 22)

22 The Moon visits fading Mars (1.2) tonight.

Moon passes 3.9° north of Mars, midnight EST (5h UT on Jan. 23)

26 First quarter Moon, 11:48 P.M. EST (4:48 UT on Jan. 27)

29 Moon passes 1.2° north of Aldebaran, 1 P.M. EST (18h UT)

30 Mercury in inferior conjunction (not visible)

February 3 Jupiter (–2.6) and the nearly full Moon rise together tonight. Look for them low in the east-northeast as evening twilight fades.

Full Moon, 6:09 P.M. EST (23:09 UT)

4 Moon passes 5.2° south of Jupiter, 4 A.M. EST (9h UT)

5 Moon passes 4.0° south of Regulus, 4 A.M. EST (9h UT)

6 Jupiter (–2.6) is at opposition and nearest to Earth, 406.898 million miles away. It rises in the east at sunset and is visible all night long. Opposition occurs at 1:20 P.M. EST (18:20 UT)

9 Moon passes 3.3° north of Spica, noon EST (17h UT)

11 Mercury (0.8) appears low in the east-southeast in the half hour before dawn.

Last quarter Moon, 10:50 P.M. EST (3:50 UT on Feb. 12)

12 Moon passes 2.1° north of Saturn, 7 P.M. EST (0h UT on Feb. 13)

13 Moon passes 9.0° north of Antares, 4 A.M. EST (9h UT)

17 Moon passes 3.5° north of Mercury, 1 A.M. EST (6h UT)

Look for Mercury (0.1) near the waning crescent Moon this morning, low in the east-southeast in the half hour before dawn.

18 New Moon, 6:47 P.M. EST (23:47 UT)

20 Moon passes 2.0° north of Venus, 8 P.M. EST (1h UT on Feb. 21)

Moon passes 1.5° north of Mars, 8 P.M. EST (1h UT on Feb. 21)

Look for the trio of the young Moon, dazzling Venus, and Mars (1.3) in the west in the hour after sunset; see p. 178. Mars and Venus are separated by less than half a degree tomorrow night. Although relatively faint, Mars is the brightest "star" near Venus; binoculars help. Don't forget to turn around and catch Jupiter (–2.5) shining low in the east.

21 Venus passes 0.5° south of Mars, 3 P.M. EST (20h UT)

24 Mercury (0.0) at greatest western elongation (26.7°) at 11 A.M. EST (16h UT). The planet is dimly visible in morning twilight.

25 First quarter Moon, 12:14 P.M. EST (17:14 UT)

Moon passes 1.0° north of Aldebaran, 6 P.M. EST (23h UT)

(continued)

2015 Almanac (continued)

March	3	Moon passes 5.5° south of Jupiter, 3 A.M. EST (8h UT)
		Jupiter (−2.5) rises with tonight's not-quite-full Moon. Look east-northeast in the hour after sunset.
	4	Moon passes 4.0° south of Regulus, 10 A.M. EST (15h UT)
	5	Full Moon, 1:05 P.M. EST (18:05 UT)
	8	Spring ahead: U.S. Daylight Saving Time begins this morning.
		Moon passes 3.5° north of Spica, 7 P.M. EDT (23h UT)
	12	Saturn (0.4) is the brightest "star" near the Moon this morning. Look southeast in the hours before dawn.
		Moon passes 2.2° north of Saturn, 4 A.M. EDT (8h UT)
		Moon passes 9.2° north of Antares, noon EDT (16h UT)
	13	Last quarter Moon, 1:48 P.M. EDT (17:48 UT)
	19	Moon passes 5.2° north of Mercury, 1 A.M. EDT (5h UT)
	20	New moon, 5:36 A.M. EDT (9:36 UT)
		Total solar eclipse, greatest at 9:46 UT; the path of totality runs only over the Faroe Islands and Svalbard; see p. 135.
		Vernal equinox, 6:45 P.M. EDT (22:45 UT)
	21	In the hour after sunset, look low in the west for faint Mars (1.3) paired with the young Moon. Binoculars help. Look for Jupiter (−2.4) high in the southeast.
		Moon passes 1.0° south of Mars, 6 P.M. EDT (22h UT)
	22	Moon passes 2.8° south of Venus, 4 P.M. EDT (20h UT)
		Venus (−4.0) is next to the crescent Moon tonight. They're visible in the west for more than two hours after sunset.
	25	Moon passes 0.9° north of Aldebaran, 3 A.M. EDT (7h UT)
	27	First quarter Moon, 3:43 A.M. EDT (7:43 UT)
	30	Moon passes 5.6° south of Jupiter, 6 A.M. EDT (10h UT)
	31	Moon passes 4.0° south of Regulus, 5 P.M. EDT (21h UT)
April	4	Full Moon, 8:06 A.M. EDT (12:06 UT)
		Total lunar eclipse, greatest at 4:00 A.M. PDT (12:00 UT) and favoring the Pacific and the western half of North America; see p. 136.
	5	Moon passes 3.5° north of Spica, 1 A.M. EDT (5h UT)
	8	Moon passes 2.2° north of Saturn, 9 A.M. EDT (13h UT)
		Moon passes 9.2° north of Antares, 5 P.M. EDT (21h UT)
	10	Mercury in superior conjunction (not visible)
	11	Last quarter Moon, 11:44 P.M. EDT (3:44 UT on Apr. 12)
	18	New moon, 2:57 P.M. EDT (18:57 UT)
	19	Moon passes 3.5° south of Mercury, 7 A.M. EDT (11h UT)
		Moon passes 3.1° south of Mars, 3 P.M. EDT (19h UT)
	21	Moon passes 0.9° north of Aldebaran, 1 P.M. EDT (17h UT)
		Moon passes 6.6° south of Venus, 2 P.M. EDT (18h UT)
		Lyrid meteor shower peaks, 8 P.M. EDT (0h UT on Apr. 22)
	23	Mercury passes 1.4° north of Mars, 3 A.M. EDT (7h UT)
	25	First quarter Moon, 7:55 P.M. EDT (23:55 UT)
	26	Moon passes 5.5° south of Jupiter, 2 P.M. EDT (18h UT)
	27	Moon passes 4.0° south of Regulus, midnight EDT (4h UT on Apr. 28)

2015 Almanac (continued)

	30	Mercury (−0.4) is visible low in the west-northwest for more than an hour after sunset. Look for the Pleiades star cluster above it as twilight fades; see p. 78. Don't forget to take in the rest of the scene, too. Venus (−4.2) is brilliant about 22° above and to Mercury's left. Jupiter (−2.1) shines high in the southwest.
May	2	Moon passes 3.5° north of Spica, 8 A.M. EDT (12h UT)
	3	Full Moon, 11:42 P.M. EDT (3:42 UT on May 4)
	5	Moon passes 2.0° north of Saturn, noon EDT (16h UT)
		Moon passes 9.1° north of Antares, 11 P.M. EDT (3h UT on May 6)
	6	Eta Aquariid meteor shower peaks, 9 A.M. EDT (13h UT)
	7	Mercury (0.4) at greatest eastern elongation (21.2°) at 1 A.M. EDT (5h UT). Look for Mercury low in the west-northwest in the hour after sunset.
	11	Last quarter Moon, 6:36 A.M. EDT (10:36 UT)
		Mercury passes 7.8° north of Aldebaran, 9 P.M. EDT (1h UT on May 12)
	18	New Moon, 12:13 A.M. EDT (4:13 UT)
		Moon passes 4.7° south of Mars, noon EDT (16h UT)
		Moon passes 1.0° north of Aldebaran, 11 P.M. EDT (3h UT on May 19)
	19	Moon passes 5.7° south of Mercury, 3 A.M. EDT (7h UT)
	21	Moon passes 7.9° south of Venus, 3 P.M. EDT (19h UT)
	22	Moon passes 11.8° south of Pollux, 9 A.M. EDT (13h UT)
		Saturn (0.0) is at opposition and nearest to Earth, 833.507 million miles away. It rises in the east at sunset and is visible all night long. Opposition occurs at 9:35 P.M. EDT (1:35 UT on May 23)
	24	Moon passes 5.1° south of Jupiter, 3 A.M. EDT (7h UT)
	25	Moon passes 3.8° south of Regulus, 8 A.M. EDT (12h UT)
		First quarter Moon, 1:19 P.M. EDT (17:19 UT)
	27	Mars passes 6.0° north of Aldebaran, 10 A.M. EDT (14h UT)
		Mercury passes 1.7° south of Mars, 11 A.M. EDT (15h UT)
	29	Moon passes 3.6° north of Spica, 3 P.M. EDT (19h UT)
	30	Mercury in inferior conjunction (not visible)
		Venus passes 4.1° south of Pollux, 1 P.M. EDT (17h UT)
June	1	Moon passes 1.9° north of Saturn, 4 P.M. EDT (20h UT)
		Saturn (0.1) rises with the Moon tonight. Look southeast as the sky darkens.
	2	Moon passes 9.1° north of Antares, 6 A.M. EDT (10h UT)
		Full Moon, 12:19 P.M. EDT (16:19 UT)
	6	Venus (−4.4) at greatest eastern elongation (45.4°) at 2 P.M. EDT (18h UT). The planet is dazzling in the west and is visible for more hours after the Sun sets. Jupiter (−1.9) is the other bright "star" nearby. The two planets drift closer throughout the month.
	9	Last quarter Moon, 11:42 A.M. EDT (15:42 UT)
	14	Mars in conjunction with the Sun (not visible)
	16	New Moon, 10:05 A.M. EDT (14:05 UT)
	20	Moon passes 5.8° south of Venus, 7 A.M. EDT (11h UT)
		Moon passes 4.7° south of Jupiter, 8 P.M. EDT (0h UT on Jun. 21)
	21	Summer solstice, 12:38 P.M. EDT (16:38 UT)
		Moon passes 3.6° south of Regulus, 4 P.M. EDT (20h UT)

(continued)

2015 Almanac (continued)

	24	Mercury passes 2.1° north of Aldebaran, 4 A.M. EDT (8h UT)
		First quarter Moon, 7:03 A.M. EDT (11:03 UT)
		Mercury (0.4) at greatest western elongation (22.5°) at 1 P.M. EDT (17h UT). The planet is visible low in the east-northeast in the half hour before dawn. See p. 78.
	25	Moon passes 3.8° north of Spica, 11 P.M. EDT (3h UT on June 26)
	27	Venus (−4.6) and Jupiter (−1.9) form a striking pair in the west and remain visible for more than 2 h after sunset. The planets are less than 2° apart tonight, and they'll move closer over the coming days.
		June Boötid meteor shower peaks, 5 P.M. EDT (21h UT)
	28	Moon passes 2.0° north of Saturn, 9 P.M. EDT (1h UT on Jun. 29)
		Look for Saturn (0.2) beneath the Moon as the sky darkens tonight. Then turn west to see another celestial pairing: Venus and Jupiter.
		Moon passes 9.2° north of Antares, 3 P.M. EDT (19h UT)
	30	Venus and Jupiter are less than 0.4° (actually, about 21 arcminutes) apart tonight. Look west in the hours after sunset.
July	1	Venus passes 0.4° south of Jupiter, 10 A.M. EDT (14h UT)
		The gap between Venus (−4.6) and Jupiter (−1.9) gradually widens over the next few days. Look west in the hour after sunset to find this prominent pair of "evening stars."
		Full Moon, 10:20 P.M. EDT (2:20 UT on Jul. 2)
	6	Earth at aphelion, its farthest from the Sun this year – 94.5065 million miles – at 3:40 P.M. EDT (19:40 UT)
	8	Last quarter Moon, 4:24 P.M. EDT (20:24 UT)
	12	Moon passes 0.9° north of Aldebaran, 2 P.M. EDT (18h UT)
	15	New Moon, 9:24 P.M. EDT (1:24 UT on July 16)
	18	Moon passes 4.1° south of Jupiter, 2 P.M. EDT (18h UT)
		The young Moon visits Venus (−4.7) and Jupiter (−1.7) tonight. Look west in the half hour after sunset. The Moon is closest to Venus, while Jupiter looks on about 6° to the planet's right. As the sky darkens, can you make out the star Regulus about 3° from Venus?
		Moon passes 0.4° south of Venus, 9 P.M. EDT (1h UT on Jul. 19)
		Moon passes 3.4° south of Regulus, 11 P.M. EDT (3h UT on Jul. 19)
	23	Moon passes 4.1° north of Spica, 7 A.M. EDT (11h UT)
		Mercury in superior conjunction (not visible)
	24	First quarter Moon, 12:04 A.M. EDT (4:04 UT)
	26	Moon passes 2.2° north of Saturn, 4 A.M. EDT (8h UT)
		Moon passes 9.3° north of Antares, midnight EDT (4h UT on Jul. 27)
	30	Southern Delta Aquariid meteor shower, 1 P.M. EDT (17h UT)
	31	Full Moon, 6:43 A.M. EDT (10:43 UT)
		Mars passes 5.8° south of Pollux, 7 A.M. EDT (11h UT)
		Venus passes 6.4° south of Jupiter, 4 P.M. EDT (20h UT)
August	6	Last quarter Moon, 10:03 P.M. EDT (2:03 UT on Aug. 7)
	8	Moon passes 0.7° north of Aldebaran, 8 P.M. EDT (0h UT on Aug. 9)
	12	Look for Perseid meteors tonight.
	13	Moon passes 5.6° south of Mars, 1 A.M. EDT (5h UT)
		Perseid meteor shower peaks, 2 A.M. EDT (6h UT)
	14	New Moon, 10:53 A.M. EDT (14:53 UT)

(continued)

2015 Almanac (continued)

	15	Venus in inferior conjunction (not visible)
	16	Moon passes 2.0° south of Mercury, 11 A.M. EDT (15h UT)
	19	Moon passes 4.2° north of Spica, 2 P.M. EDT (18h UT)
	22	Moon passes 2.5° north of Saturn, 1 P.M. EDT (17h UT)
		First quarter Moon, 3:31 P.M. EDT (19:31 UT)
	23	Moon passes 9.5° north of Antares, 9 A.M. EDT (13h UT)
	26	Jupiter in conjunction with the Sun (not visible)
	29	Venus (−4.4) has moved to the morning sky and is visible low in the east in the half hour before sunrise. Faint Mars (1.8), the brightest "star" in the vicinity, is all but lost in morning twilight.
		Venus passes 9.4° south of Mars, 1 A.M. EDT (05h UT)
		Full Moon, 2:35 P.M. EDT (18:35 UT)
September	4	Mercury (0.1) at greatest eastern elongation (27.1°) at 6 A.M. EDT (10h UT). The planet is visible low in the west in the half hour after sunset.
	5	Moon passes 0.6° north of Aldebaran, 2 A.M. EDT (6h UT)
		Last quarter Moon, 5:54 A.M. EDT (9:54 UT)
	10	Moon passes 2.7° north of Venus, 2 A.M. EDT (6h UT)
		The waning Moon appears between Venus (−4.7) and Mars (1.8) this morning. Look east in the hour before dawn.
		Moon passes 4.7° south of Mars, 7 P.M. EDT (23h UT)
	11	Moon passes 3.3° south of Regulus, noon EDT (16h UT)
	12	Look for Jupiter (−1.7) above the Moon this morning. Face east half an hour before dawn.
		Moon passes 3.2° south of Jupiter, 2 A.M. EDT (6h UT)
	13	New Moon, 2:41 A.M. EDT (6:41 UT)
		Partial solar eclipse, greatest at 6:54 UT from Antarctica. The southern extremes of Africa and Madagascar see up to 40% of the Sun covered.
	15	Moon passes 5.3° north of Mercury, 2 A.M. EDT (6h UT)
		Moon passes 4.3° north of Spica, 8 P.M. EDT (0h UT on Sept. 16)
	18	Look for Saturn (0.6) near the Moon tonight, high in the southwest as twilight fades.
		Moon passes 2.8° north of Saturn, 11 P.M. EDT (3h UT on Sept. 19)
	19	Moon passes 9.6° north of Antares, 4 P.M. EDT (20h UT)
	21	First quarter Moon, 4:59 A.M. EDT (8:59 UT)
	23	Autumnal equinox, 4:21 A.M. EDT (8:21 UT)
	24	Mars passes 0.8° north of Regulus, 1 P.M. EDT (17h UT)
	27	Full Moon, 10:50 P.M. EDT (2:50 UT on Sept. 28)
		Total lunar eclipse, greatest at 10:47 P.M. EDT (2:47 UT on Sept. 28), brings totality to nearly all of North America, South America, Europe and Africa. See p. 137.
	30	Mercury in inferior conjunction (not visible)
October	2	Moon passes 0.5° north of Aldebaran, 9 A.M. EDT (13h UT)
	4	Last quarter Moon, 5:06 P.M. EDT (21:06 UT)

(continued)

2015 Almanac (continued)

8	The waning crescent Moon guides you through the chain of morning planets over the next four days. It visits brilliant Venus (−4.7) this morning; look east in the hours before dawn. Just over 2° to Venus' left is the bright star Regulus. Faint Mars (1.8) and bright Jupiter (−1.7) pair up below them. Wait until the half hour before dawn and you may catch Mercury (1.2), far below the group. See p. 78.

Moon passes 0.7° south of Venus, 5 P.M. EDT (21h UT)

Moon passes 3.3° south of Regulus, 6 P.M. EDT (22h UT)

9	The waning Moon joins Mars and Jupiter this morning. Look east in the hours before dawn. Venus dazzles above the trio. In the half hour before dawn, try locating rapidly brightening Mercury (0.8) far below the brighter planets. See p. 78.

Moon passes 3.4° south of Mars, 1 P.M. EDT (17h UT)

Venus passes 2.6° south of Regulus, 5 P.M. EDT (21h UT)

Moon passes 2.7° south of Jupiter, 8 P.M. EDT (0h UT on Oct. 10)

11	The waning Moon joins Mercury (0.2) this morning, low in the east in the hour before dawn. Above them shine Jupiter (−1.8) and, even higher, Venus (−4.6). Tucked between them is dim Mars (1.8). See p. 78.

Moon passes 1.0° south of Mercury, 8 A.M. EDT (12h UT)

12	New Moon, 8:06 P.M. EDT (0:06 UT on Oct. 13)
13	Moon passes 4.3° north of Spica, 2 A.M. EDT (6h UT)
15	Mercury (−0.6) at greatest western elongation (18.1°), 11 P.M. EDT (3h UT on Oct. 16). The planet is visible low in the east in the hour before dawn.
16	Moon passes 2.9° north of Saturn, 9 A.M. EDT (13h UT)

Tonight, look for Saturn (0.6) – now, the lone evening planet – near the young crescent Moon, low in the southwest in the hour after sunset.

Moon passes 9.5° north of Antares, 10 P.M. EDT (2h UT on Oct. 17)

17	Mars (1.8) and Jupiter (−1.8) form a close, if mismatched, pair this morning. Look low in the east in the hours before dawn. Faint Mars is the brightest "star" near Jupiter. Compare their colors. Above them, Venus (−4.6) gleams brilliantly. In the hour before sunrise, look for Mercury (−0.7) joining the scene below them. See p. 209.

Jupiter passes 0.4° south of Mars, 10 A.M. EDT (14h UT)

20	First quarter Moon, 4:31 P.M. EDT (20:31 UT)
21	Orionid meteor shower peaks, 7 P.M. EDT (23h UT)
26	Venus (−4.5) at greatest western elongation (46.4°), 3 A.M. EDT (7h UT).

Venus passes 1.1° south of Jupiter, 4 A.M. EDT (8h UT)

The brightest morning planets – Venus and Jupiter – pair up this morning. Look east in the hours before sunrise to see them in a dark sky. Look for Mars (1.7) about 3.5° below the pair. Half an hour before sunrise, look for Mercury (−0.9) low in the east-southeast. See p. 209.

27	Full Moon, 8:05 A.M. EDT (12:05 UT)
28	Mercury passes 4.2° north of Spica, 3 P.M. EDT (19h UT)
29	Moon passes 0.6° north of Aldebaran, 7 P.M. EDT (23h UT)

(continued)

2015 Almanac (continued)

November	1	Fall back: U.S. Daylight Time ends this morning.
	3	Venus (−4.4) slides closest to Mars (1.7) this morning; see p. 178. Look east-southeast a couple of hours before sunrise to see them in a dark sky. You'll first notice brilliant Venus, then Jupiter (−1.8) higher up. While much fainter than these two, Mars is close to Venus and brighter than nearby stars. Compare the colors of the three planets.
		Last quarter Moon, 7:24 A.M. EST (12:24 UT)
		Venus passes 0.7° south of Mars, 11 A.M. EST (16h UT)
	4	Moon passes 3.2° south of Regulus, midnight EST (5h UT on Nov. 5)
	6	The waning crescent Moon joins Jupiter this morning; see p. 178. Venus and Mars shine below the pair. Look east-southeast in the hours before dawn.
		Moon passes 2.3° south of Jupiter, 11 A.M. EST (16h UT)
	7	The Moon, Venus, and Mars make an early morning triangle. Look east-southeast a couple of hours before dawn; see p. 178.
		Moon passes 1.8° south of Mars, 5 A.M. EST (10h UT)
		Moon passes 1.2° south of Venus, 9 A.M. EST (14h UT)
	9	Moon passes 4.3° north of Spica, 8 A.M. EST (13h UT)
	11	Moon passes 3.2° north of Mercury, 3 A.M. EST (8h UT)
		New Moon, 12:47 P.M. EST (17:47 UT)
	12	Moon passes 3.0° north of Saturn, 8 P.M. EST (1h UT on Nov. 13)
	17	Mercury in superior conjunction (not visible)
		Leonid meteor shower peaks, 11 P.M. EST (4h UT on Nov. 18)
	19	First quarter Moon, 1:27 A.M. EST (6:27 UT)
	21	Alpha Monocerotid meteor shower peaks, 11 P.M. EST (4h UT on Nov. 22)
	25	Full Moon, 5:44 P.M. EST (22:44 UT)
	26	Moon passes 0.7° north of Aldebaran, 5 A.M. EST (10h UT)
	28	Venus passes 4.5° north of Spica, 11 A.M. EST (16h UT)
	30	Saturn in conjunction with the Sun (not visible)
December	2	Moon passes 3.0° south of Regulus, 7 A.M. EST (12h UT)
	3	Last quarter Moon, 2:40 A.M. EST (7:40 UT)
	4	Moon passes 1.8° south of Jupiter, 1 A.M. EST (6h UT)
	5	Moon passes 0.1° south of Mars, 10 P.M. EST (3h UT on Dec. 6)
	6	Moon passes 4.5° north of Spica, 3 P.M. EST (20h UT)
	7	Moon passes 0.7° north of Venus, noon EST (17h UT)
	11	New moon, 5:29 A.M. EST (10:29 UT)
	12	Moon passes 7.2° north of Mercury, 9 A.M. EST (14h UT)
	13	Saturn passes 6.3° north of Antares, 3 A.M. EST (8h UT)
		Look for Geminind meteors tonight.
	14	Geminid meteor shower peaks, 1 P.M. EST (18h UT); look for meteors tonight, too.
	18	First quarter Moon, 10:14 A.M. EST (15:14 UT)
	21	Mars passes 3.8° north of Spica, 7 A.M. EST (12h UT)
		Winter solstice, 11:48 P.M. EST (4:48 UT on Dec. 22)
	22	Moon passes 0.7° north of Aldebaran, 3 P.M. EST (20h UT)
		Ursid meteor shower peaks, 9 P.M. EST (2h UT on Dec. 23)
	25	Full Moon, 6:11 A.M. EST (11:11 UT)

(continued)

2015 Almanac (continued)	
27	Mercury (−0.6) emerges briefly from the Sun's glare as the sole evening planet. Look southwest in the half hour after sunset.
28	Mercury (−0.6) at greatest eastern elongation (19.7°), 10 P.M. EST (3h UT on Dec. 29).
29	Moon passes 2.7° south of Regulus, 4 P.M. EST (21h UT)
31	Moon passes 1.5° south of Jupiter, 1 P.M. EST (18h UT)

2016 Highlights

Mercury meets up with Venus early February's morning twilight, and enjoys a nice evening appearance in early April, and another in the predawn sky of late September; in all of these cases, the Moon serves as a guide. Mercury closes out the year with a mid-December evening apparition beneath brilliant Venus.

Venus and Saturn meet in morning twilight as the year opens. Venus eases into the Sun's glare during February and eventually crosses to the evening sky in June. Hanging low in the southwest, it skims the horizon until late October. Then it begins to climb higher in evening twilight, heading to the apex of the apparition in 2017.

Mars begins the year high in the south during morning twilight, located near Spica in Virgo and shining slightly fainter than the star; by month's end, Mars bests the star and has moved into Libra. Mars and Saturn shine together in Scorpius in late April and are closer still in late August. On May 22, Mars shines brilliantly at opposition, and by year's end the Red Planet dallies with Venus in evening twilight.

Jupiter reaches its brightest for the year at its Mar. 8 opposition. In late August, Jupiter joins Venus and Mercury low in the west after sunset before transitioning to the morning sky, where it emerges in October, again pairing with Mercury.

Saturn reaches opposition Jun. 3. It spends most of the year in the vicinity of Mars, with their closest pairing in late August evenings. In early November, Saturn shines with Venus low in the southwest twilight.

Meteor showers. The Quadrantids fare best this year, with minor interference from a waning crescent Moon. A waxing gibbous Moon washes out the Perseids, and the Geminid peak coincides with full Moon.

Eclipses. The two lunar eclipses this year are penumbral only, but the March event is visible from North America, including Alaska and Hawaii. A total solar eclipse in March brings the Moon's shadow to Indonesia, and an annular solar eclipse in September brings a ring of Sun to central Africa, Madagascar and Réunion.

2016 Almanac		
January	1	Mercury (−0.3) is the lone evening planet tonight, but it won't be visible for long. Look low in the southwest in the half hour after sunset. Mercury is brighter than any stars in the vicinity.
	2	Last quarter Moon, 12:30 A.M. EST (5:30 UT)
		Earth at perihelion, its closest to the Sun this year – 91.4038 million miles – at 5:49 P.M. EST (22:49 UT)
		Moon passes 4.7° north of Spica,11 P.M. EST (4h UT on Jan. 3)
	3	Moon passes 1.5° north of Mars, 2 P.M. EST (19h UT)
	4	Quadrantid meteor shower peaks, 3 A.M. EST (8h UT)
	6	Venus (−4.0) and Saturn (0.5) meet up with the waning crescent Moon this morning. Look southeast in the hour before dawn. Watch the two planets close over the coming days. While you're up, look for Jupiter (−2.2) high in the southwest and dim Mars (1.2) near the star Spica. See. p. 209.
		Venus passes 6.5° north of Antares, noon EST (17h UT)
		Moon passes 9.5° north of Antares, 6 P.M. EST (23h UT)
		Moon passes 3.1° north of Venus, 7 P.M. EST (0h UT on Jan. 7)
		Moon passes 3.3° north of Saturn, midnight EST (5h UT on Jan. 7)
	8	Venus passes 0.1° north of Saturn, 11 P.M. EST (4h UT on Jan. 9)
	9	Venus and Saturn appear closest this morning. The planets are just 0.3° (20 arcminutes) apart in the hour before dawn. Look southeast for brilliant Venus; Saturn is above it. See p. 209.
		New Moon, 8:31 P.M. EST (1:31 UT on Jan. 10)
	10	Moon passes 2.1° north of Mercury, 1 P.M. EST (18h UT)
	14	Mercury in inferior conjunction (not visible)
	16	First quarter Moon, 6:26 P.M. EST (23:26 UT)
	19	Moon passes 0.5° north of Aldebaran, 10 P.M. EST (3h UT on Jan. 20)
	23	Full Moon, 8:46 P.M. EST (1:46 UT on Jan. 24)
	26	Mercury (0.4) pops into view in the morning sky. It's visible about 12° below and to the left of Venus this morning, very low in the southeast nearly an hour before sunrise. Once you've spotted it, take in the rest of the scene. From Venus, look for Saturn (0.6) higher up in the south-southeast, near the star Antares, and Jupiter (−2.3) in the west-southwest, not far from the Moon. Less prominent Mars (0.9) lies in the south and stands out in a field of fainter stars. See p. 79.
		Moon passes 2.5° south of Regulus, 1 A.M. EST (6h UT)
	27	Moon passes 1.4° south of Jupiter, 8 P.M. EST (1h UT on Jan. 28)
	30	Moon passes 5.0° north of Spica, 7 A.M. EST (12h UT)
	31	Last quarter Moon, 10:28 P.M. EST (3:28 UT on Feb. 1)
February	1	Moon passes 2.7° north of Mars, 4 A.M. EST (9h UT)
		The last quarter Moon visits Mars (0.8) this morning. Look due south about an hour before sunrise. To the right, look for Jupiter (−2.4), the bright "star" in the west-southwest. To the left, Saturn (0.6), hovering about 7.5° from the star Antares in the south-southeast. Venus (−3.9) and, closer to dawn, Mercury (0.0) glow low in the southeast. See p. 79.
	3	Moon passes 9.7° north of Antares, 3 A.M. EST (8h UT)
		Saturn (0.6) is the brightest "star" near the waning crescent Moon this morning. Look south-southeast in the hour before dawn; see p. 79.
		Moon passes 3.5° north of Saturn, 2 P.M. EST (19h UT)

(continued)

2016 Almanac (continued)

	6	Moon passes 4.3° north of Venus, 3 A.M. EST (8h UT)
		Moon passes 3.8° north of Mercury, noon EST (17h UT)
		The Moon visits Venus (−3.9) and Mercury (−0.1) this morning. Look low in the southeast half an hour before sunrise; see p. 79.
		Mercury at greatest western elongation (25.6°) at 8 P.M. EST (1h UT on Feb. 7).
	8	New Moon, 9:39 A.M. EST (14:39 UT)
	15	First quarter Moon, 2:46 A.M. EST (7:46 UT)
	16	Moon passes 0.3° north of Aldebaran, 3 A.M. EST (8h UT)
	22	Moon passes 2.5° south of Regulus, 8 A.M. EST (13h UT)
		Full Moon, 1:20 P.M. EST (18:20 UT)
	23	Moon passes 1.7° south of Jupiter, 11 P.M. EST (4h UT on Feb. 24)
		Jupiter (−2.5) accompanies the Moon all night. You can look for them in the east at moonrise (a couple of hours after sunset) or low in the west in the hour before sunrise, when they share the predawn sky with Mars, Saturn, and Venus.
	26	Moon passes 5.1° north of Spica, 3 P.M. EST (20h UT)
	29	Moon passes 3.6° north of Mars, 1 P.M. EST (18h UT)
		The brightest "star" near the Moon tonight is Mars (0.3); the two rise in the east-southeast around local midnight and are high in the south in the hour before dawn.
March	1	Moon passes 9.8° north of Antares, noon EDT (17h UT)
		Last quarter Moon, 6:11 P.M. EST (23:11 UT)
	2	Moon passes 3.6° north of Saturn, 2 A.M. EST (7h UT)
		The brightest "star" near the Moon this morning is Saturn (0.5). Look for them in the south-southeast in the hour before dawn.
	7	Moon passes 3.5° north of Venus, 6 A.M. EST (11h UT)
		Moon passes 3.9° north of Mercury, midnight EST (5h UT on Mar. 8)
	8	Jupiter (−2.5) is at opposition and nearest to Earth, 412.293 million miles away. It rises in the east at sunset and is visible all night long. Opposition occurs at 5:57 A.M. EST (10:57 UT).
		New Moon, 8:54 P.M. EST (1:54 UT on Mar. 9)
	9	**Total solar eclipse**, greatest at 1:57 UT and total on Sumatra, Borneo, Sulewesi and the Maluku Islands; see p. 138.
	13	Spring ahead: U.S. Daylight Saving Time begins this morning.
	14	Moon passes 0.3° north of Aldebaran, 10 A.M. EDT (14h UT)
	15	First quarter Moon, 1:03 P.M. EDT (17:03 UT)
	20	Vernal equinox, 12:30 A.M. EDT (4:30 UT)
		Moon passes 2.5° south of Regulus, 4 P.M. EDT (20h UT)
	21	Moon passes 2.1° south of Jupiter, midnight EDT (4h UT on Mar.22)
	23	**Penumbral lunar eclipse**, greatest at 7:47 A.M. EDT (11:47 UT); a portion of the eclipse is visible before moonset throughout North America.
		Full Moon, 8:01 A.M. EDT (12:01 UT)
		Mercury in superior conjunction (not visible)
	24	Moon passes 5.1° north of Spica, 10 P.M. EDT (2h UT on Mar. 25)

(continued)

2016 Almanac (continued)

	28	Rapidly brightening Mars (–0.4) lies near the Moon this morning. Look south in the hour before dawn. Saturn (0.4) is about 10° to the left. Antares, the reddish star whose name means "rival of Mars," is below. Jupiter (–2.4) is about to set very low in the west.

Moon passes 4.2° north of Mars, 3 P.M. EDT (19h UT)

Moon passes 9.8° north of Antares, 8 P.M. EDT (0h UT on Mar. 29)

29 Moon passes 3.5° north of Saturn, 11 A.M. EDT (15h UT)

31 Last quarter Moon, 11:17 A.M. EDT (15:17 UT)

April 5 Mercury (–1.2) emerges from the Sun's glare into the evening sky. Look low in the west in the half hour after sunset.

7 New Moon, 7:24 A.M. EDT (11:24 UT)

8 Moon passes 5.2° south of Mercury, 7 A.M. EDT (11h UT)

Look for Mercury (–1.0) to the right of the young Moon tonight, low in the west in the half hour after sunset; see p. 79.

10 Moon passes 0.4° north of Aldebaran, 6 P.M. EDT (22h UT)

13 First quarter Moon, 11:59 P.M. EDT (3:59 UT on Apr. 14)

16 Moon passes 2.5° south of Regulus, 9 P.M. EDT (1h UT on Apr. 17)

17 Tonight, Jupiter (–2.3) and the Moon shine high in the southeast as the sky darkens.

18 Moon passes 2.2° south of Jupiter, 1 A.M. EDT (5h UT)

Mercury (0.1) at greatest eastern elongation (19.9°), 10 A.M. EDT (14h UT). Look low in the west-northwest in the hour after sunset.

21 Moon passes 5.1° north of Spica, 4 A.M. EDT (8h UT)

22 Full Moon, 1:24 A.M. EDT (5:24 UT)

Lyrid meteor shower peaks, 2 A.M. EDT (6h UT)

24 Moon passes 4.9° north of Mars, midnight EDT (4h UT on Apr. 25)

25 Tonight, look for the Moon near quickly brightening Mars (–1.3). They rise in the east-southeast about three hours after sunset. Nearby Saturn (0.2) completes a triangle. Below Mars, look for the star Antares, the planet's stellar rival in color.

Moon passes 9.7° north of Antares, 2 A.M. EDT (6h UT)

Moon passes 3.3° north of Saturn, 3 P.M. EDT (19h UT)

29 Last quarter Moon, 11:29 P.M. EDT (3:29 UT on Mar. 30)

May 5 Eta Aquariid meteor shower peaks, 4 P.M. EDT (20h UT)

6 New Moon, 3:29 P.M. EDT (19:29 UT)

8 Moon passes 0.5° north of Aldebaran, 5 A.M. EDT (9h UT)

9 Mercury in inferior conjunction. Normally, this would mean that the planet is not visible but, beginning at 7:12 A.M. EDT (11:12 UT), Mercury can be seen in silhouette as it transits the Sun's disk. Mid-transit occurs at 10:57 A.M. EDT (14:57 UT). The transit ends at 2:42 P.M. EDT (18:42 UT). North and South America, Africa, Europe, western and central Asia, and India see some portion of the event. *This is a telescopic event only and requires eye protection for safe viewing.*

13 First quarter Moon, 1:02 P.M. EDT (17:02 UT)

14 Moon passes 2.3° south of Regulus, 4 A.M. EDT (8h UT)

15 Moon passes 2.0° south of Jupiter, 6 A.M. EDT (10h UT)

(continued)

2016 Almanac (continued)

	Look for Jupiter (−2.2) near the Moon as the sky darkens. Mars (−1.9), now just days away from its opposition, rises in the east-southeast in the hour after sunset.
18	Moon passes 5.2° north of Spica, 11 A.M. EDT (15h UT)
21	Moon passes 6.0° north of Mars, 4 P.M. EDT (20h UT)
	Full Moon, 5:14 P.M. EDT (21:14 UT)
22	Mars (−2.1) is at opposition. It rises in the east at sunset and is visible all night long. Opposition occurs at 7:17 A.M. EDT (11:17 UT). See p. 180.
	Moon passes 9.6° north of Antares, 8 A.M. EDT (12h UT)
	Moon passes 3.2° north of Saturn, 6 P.M. EDT (22h UT)
29	Last quarter Moon, 8:12 A.M. EDT (12:12 UT)
30	Mars is closest to Earth, 46.777 million miles away, 5 P.M. EDT (21h UT)

June

1	Venus passes 5.3° north of Aldebaran, midnight EDT (4h UT on Jun. 2)
3	Saturn (0.0) is at opposition and nearest to Earth, 837.988 million miles away. It rises in the east at sunset and is visible all night long. Opposition occurs at 2:37 A.M. EDT (6:37 UT).
	Moon passes 0.7° south of Mercury, 6 A.M. EDT (10h UT)
4	Moon passes 0.5° north of Aldebaran, 3 P.M. EDT (19h UT)
	Moon passes 5.0° south of Venus, 9 P.M. EDT (1h UT on Jun. 5)
	New Moon, 11 P.M. EDT (3:00 UT on Jun. 5)
5	Mercury (0.4) at greatest western elongation (24.2°), 5 A.M. EDT (9h UT). Look for Mercury low in the east-northeast in the half hour before dawn.
6	Venus in superior conjunction (not visible)
10	Moon passes 2.0° south of Regulus, 11 A.M. EDT (15h UT)
11	Moon passes 1.5° south of Jupiter, 4 P.M. EDT (20h UT)
12	First quarter Moon, 4:10 A.M. EDT (8:10 UT)
14	Moon passes 5.4° north of Spica, 5 P.M. EDT (21h UT)
17	Moon passes 7.1° north of Mars, 6 A.M. EDT (10h UT)
	As the sky darkens tonight, look the Moon, Mars (−1.7) and Saturn (0.1) in the south-southeast. The two planets bookend the Moon, with Mars on the right and Saturn on the left. Over in the west-southwest, don't miss Jupiter (−1.9). All of these planets are much brighter than any stars around them.
18	Moon passes 9.7° north of Antares, 2 P.M. EDT (18h UT)
	Moon passes 3.3° north of Saturn, 8 P.M. EDT (0h UT on June 19)
19	Mercury passes 3.9° north of Aldebaran, 5 P.M. EDT (21h UT)
20	Full Moon, 7:02 A.M. EDT (11:02 UT)
	Summer solstice, 6:34 P.M. EDT (22:34 UT)
26	June Boötid meteor shower peaks, midnight EDT (4 UT on Jun. 27)
27	Last quarter Moon, 2:19 P.M. EDT (18:19 UT)

July

1	Moon passes 0.4° north of Aldebaran, midnight EDT (4h UT Jul. 2)
4	Earth at aphelion, its farthest from the Sun this year – 94.5129 million miles – at 12:24 P.M. EDT (16:24 UT)
	New Moon, 7:01 A.M. EDT (11:01 UT)
6	Mercury in superior conjunction (not visible)
7	Moon passes 1.8° south of Regulus, 8 P.M. EDT (0h UT on Jul. 8)
8	Catch Jupiter (−1.8) near tonight's crescent Moon. Look west in the hours after sunset.

(continued)

2016 Almanac (continued)

9	Moon passes 0.9° south of Jupiter, 6 A.M. EDT (10h UT)
11	Mercury passes 5.1° south of Pollux, 3 A.M. EDT (7h UT)
	First quarter Moon, 8:52 P.M. EDT (0:52 UT on Jul. 12)
12	Moon passes 5.6° north of Spica, 1 A.M. EDT (5h UT)
14	Moon passes 7.8° north of Mars, 2 P.M. EDT (18h UT)
	Mars (−1.1) is the brightest "star" near the Moon tonight. Look south as twilight fades.
15	Moon passes 9.8° north of Antares, 10 P.M. EDT (2h UT on Jul. 16)
	Saturn (0.3) is the brightest "star" near the Moon tonight. Look for the star Antares about 6° beneath Saturn. Compare its color to Mars, now 16° to the right.
16	Moon passes 3.4° north of Saturn, 1 A.M. EDT (5h UT)
19	Full Moon, 6:57 P.M. EDT (22:57 UT)
26	Last quarter Moon, 7 P.M. EDT (23:00 UT)
	Both Mercury (−0.4) and Venus (−3.9) now appear very low in the west-northwest in the half hour after sunset. Binoculars help.
29	Moon passes 0.3° north of Aldebaran, 7 A.M. EDT (11h UT)
	Southern Delta Aquariid meteor shower peaks, 7 P.M. EDT (23h UT)
30	Mercury passes 0.3° north of Regulus, 1 P.M. EDT (17h UT)
	Mercury (−0.2) and Venus (−3.8) gather with Regulus over the coming week, very low in the west-northwest in the half hour after sunset. Binoculars help. Look for Jupiter (−1.7), now shining higher in the west, as well.

August	2	New Moon, 4:45 P.M. EDT (20:45 UT)
	4	Moon passes 2.9° south of Venus, 2 A.M. EDT (6h UT)
		Moon passes 1.7° south of Regulus, 5 A.M. EDT (9h UT)
		Moon passes 0.6° south of Mercury, 6 P.M. EDT (22h UT)
		The crescent Moon visits Mercury (0.0) tonight, very low in the west in the half hour after sunset. Look for Venus (−3.8) about 10° below right of the moon. Binoculars help. Jupiter (−1.7) shines higher in the west about 14° above left of the Moon.
	5	Venus passes 1.1° north of Regulus, 5 A.M. EDT (9h UT)
		Moon passes 0.2° south of Jupiter, midnight EDT (4h UT on Aug. 6)
		Look for Jupiter (−1.7) above the Moon tonight, low in the west in the hour after sunset.
	8	Moon passes 5.8° north of Spica, 9 A.M. EDT (13h UT)
	10	First quarter Moon, 2:21 P.M. EDT (18:21 UT)
	11	Moon passes 8.2° north of Mars, 6 P.M. EDT (22h UT)
		Tonight, Mars (−0.6), Saturn (0.4), the Moon and the star Antares form a celestial parallelogram. The brightest "star" is Mars, the second-brightest is Saturn. Look south-southwest as the sky darkens.
		Look for Perseid meteors tonight.
	12	Moon passes 9.9° north of Antares, 6 A.M. EDT (10h UT)
		Moon passes 3.7° north of Saturn, 8 A.M. EDT (12h UT)
		Perseid meteor shower peaks, 9 A.M. EDT (13h UT)
	16	Mercury (0.2) at greatest eastern elongation (27.4°), 5 P.M. EDT (21h UT). The planet hovers very low in the west 20 min after sunset. Tonight, you'll find it between the much more visible planets Venus (−3.8) and Jupiter (−1.7). Binoculars help.

(continued)

2016 Almanac (continued)

	18	Full Moon, 5:27 A.M. EDT (9:27 UT)
	20	Jupiter (−1.7) lies above Mercury (0.4) tonight. Look very low in the west 20 min after sunset. Bright Venus (−3.8) looks on about 7° to the pair's right. Binoculars help.
	23	Mars passes 1.8° north of Antares, midnight EDT (4h UT on Aug. 24)
		Saturn (0.4), Mars (−0.4) and Antares align tonight. Look south-southwest as darkness falls. Mars takes center stage and is by far the brightest of the three. Antares is below and Saturn lies above.
	24	Last quarter Moon, 11:41 P.M. EDT (3:41 UT on Aug. 25)
	25	Moon passes 0.2° north of Aldebaran, 1 P.M. EDT (17h UT)
		Mars passes 4.4° south of Saturn, 2 P.M. EDT (18h UT)
	27	Mercury passes 5.3° south of Venus, 1 A.M. EDT (5h UT)
		Jupiter passes 0.1° south of Venus, 6 P.M. EDT (22h UT)
		Venus (−3.8) and Jupiter (−1.7) pair up very low in the west. Look 20 min after sunset from a site with an unobstructed horizon. Binoculars help. You likely won't see Mercury (0.8), which is closer to the horizon and now much fainter than the other planets.
September	1	New Moon, 5:03 A.M. EDT (9:03 UT)
		Annular solar eclipse, greatest at 9:07 UT, annular from central Africa, Madagascar and Réunion; see p. 139.
	2	Moon passes 0.4° north of Jupiter, 6 P.M. EDT (22h UT)
		Look for (−1.7) Jupiter near the young Moon this evening, very low in the west about 20 min after sunset. Venus (−3.8) shines nearby. Binoculars help.
	3	Moon passes 1.1° north of Venus, 7 A.M. EDT (11h UT)
		Look for Venus (−3.8) beneath the Moon tonight, very low in the west in the half hour after sunset.
	4	Moon passes 5.8° north of Spica, 4 P.M. EDT (20h UT)
	8	Moon passes 10.0° north of Antares, 2 P.M. EDT (18h UT)
		Moon passes 3.8° north of Saturn, 5 P.M. EDT (21h UT)
		Saturn (0.5) is the bright star directly beneath tonight's waxing Moon. Look south-southwest as darkness falls. You'll find Mars (−0.2) to the left of the pair.
	9	First quarter Moon, 7:49 A.M. EDT (11:49 UT)
		Moon passes 7.9° north of Mars, 10 A.M. EDT (14h UT)
	12	Mercury in inferior conjunction (not visible)
	16	**Penumbral lunar eclipse**, greatest at 18:54 UT, visible throughout the Indian Ocean, Asia, Australia, Africa, Indonesia, and easternmost Brazil.
		Full Moon, 3:05 P.M. EDT (19:05 UT)
	17	Venus passes 2.7° north of Spica, 7 P.M. EDT (23h UT)
	21	Moon passes 0.2° north of Aldebaran, 7 P.M. EDT (23h UT)
	22	Autumnal equinox, 10:21 A.M. EDT (14:21 UT)
	23	Last quarter Moon, 5:56 A.M. EDT (9:56 UT)
	26	Jupiter in conjunction with the Sun (not visible)
	27	Moon passes 1.7° south of Regulus, 7 P.M. EDT (23h UT)
	28	Mercury (−0.5) at greatest western elongation (17.9°), 3 P.M. EDT (19h UT); see p. 79.

(continued)

2016 Almanac (continued)

	29	Moon passes 0.7° south of Mercury, 7 A.M. EDT (11h UT)

The Moon joins Mercury (−0.6) in the predawn sky. Look low in the east in the half hour before sunrise. See p. 79.

| | 30 | New Moon, 8:11 P.M. EDT (0:11 UT on Oct. 1) |
| October | 3 | Moon passes 5.0° north of Venus, 1 P.M. EDT (17h UT) |

Look for Venus (−3.9) under the young Moon tonight, low in the west in the half hour after sunset.

| | 5 | Moon passes 9.9° north of Antares, 10 P.M. EDT (2h UT on Oct. 6) |

Saturn (0.6) is about 6° to the left of the crescent Moon tonight. Look southwest as the sky darkens.

| | 6 | Moon passes 3.8° north of Saturn, 4 A.M. EDT (8h UT) |
| | 8 | Moon passes 7.0° north of Mars, 8 A.M. EDT (12h UT) |

Mars (0.1) is below the Moon tonight. The planet hovers by the lid of the Teapot asterism in Sagittarius and is the brightest "star" in the Moon's vicinity. Look south as the sky darkens.

| | 9 | First quarter Moon, 12:33 A.M. EDT (4:33 UT) |
| | 10 | Mercury passes 0.9° north of Jupiter, midnight EDT (4h UT on Oct. 11) |

This morning, check out Mercury (−1.1) and Jupiter (−1.7), low in the east in the half hour before dawn. The two worlds are 1.7° apart. Binoculars help.

	16	Full Moon, 12:23 A.M. EDT (4:23 UT)
	19	Moon passes 0.3° north of Aldebaran, 3 A.M. EDT (7h UT)
	21	Orionid meteor shower peaks, 1 A.M. EDT (5h UT)
	22	Last quarter Moon, 3:14 P.M. EDT (19:14 UT)
	24	Moon passes 1.6° south of Regulus, midnight EDT (4h UT on Oct. 25)
	25	Watch Venus (−4.0) cruise between Saturn (0.6) and Antares over the week. Look low in the southwest in the hour after sunset. See p. 210.

Venus passes 3.1° north of Antares, midnight EDT (4h UT on Oct. 26)

| | 27 | Mercury in superior conjunction (not visible) |

Tonight, Venus (−4.0) passes between Saturn (0.6) and Antares. It's about 3.5° above the star and 3.7° below Saturn. Look low in the southwest in the hour after sunset. See p. 210.

| | 28 | Moon passes 1.4° north of Jupiter, 6 A.M. EDT (10h UT) |

This morning, look for Jupiter (−1.7) above the waning crescent Moon, low in the east in the hour before dawn.

| | 29 | Moon passes 5.8° north of Spica, 6 A.M. EDT (10h UT) |
| | 30 | Venus passes 3.0° south of Saturn, 4 A.M. EDT (8h UT) |

New Moon, 1:38 P.M. EDT (17:38 UT)

| November | 2 | Moon passes 9.7° north of Antares, 4 A.M. EDT (8h UT) |

Moon passes 3.7° north of Saturn, 3 P.M. EDT (19h UT)

Look for the young crescent Moon tonight and you'll find two bright stars beneath it. The brightest is Venus (−4.0) off to the left. The other is Saturn (0.5). Look low in the southwest in the hour after sunset. See p. 210.

Moon passes 6.8° north of Venus, midnight EDT (4h UT on Nov. 3)

| | 6 | Fall back: U.S. Daylight Saving Time ends this morning. |

Moon passes 5.3° north of Mars, 7 A.M. EST (12h UT)

(continued)

2016 Almanac (continued)

	7	First quarter Moon, 2:51 P.M. EST (19:51 UT)
	14	Moon at perigee, its closest point to Earth for the month – 221,524.415 miles – at 6:21 A.M. EST (11:21 UT). In fact, this is the closest the Moon comes to Earth in the period covered by this book.
		Full Moon, 8:52 A.M. EST (13:52 UT)
	15	Moon passes 0.5° north of Aldebaran, noon EST (17h UT)
	17	Leonid meteor shower peaks, 5 A.M. EST (10h UT)
	21	Last quarter Moon, 3:33 A.M. EST (8:33 UT)
		Moon passes 1.3° south of Regulus, 6 A.M. EST (11h UT)
		Alpha Monocerotid meteor shower peaks, 6 A.M. EST (11h UT)
	23	Mercury passes 3.5° south of Saturn, 8 P.M. EST (1h UT on Nov. 24)
		Tonight, use binoculars to scan the southwestern horizon for Mercury (−0.5) and Saturn (0.5) about 20 min after sunset. Venus (−4.0) shines higher in the southwest. As the sky darkens, look for Mars (0.6) high in the south-southwest. Now cruising through Capricornus, Mars remains the brightest "star" in this part of the sky.
	24	Moon passes 1.9° north of Jupiter, 9 P.M. EST (2h UT on Nov. 25)
		Look east-southeast in the hours before sunrise this morning to find Jupiter (−1.8) hovering above the waning crescent Moon.
	25	Moon passes 5.9° north of Spica, 11 A.M. EST (16h UT)
	29	New moon, 7:18 A.M. EST (12:18 UT)
	30	Moon passes 3.6° north of Saturn, 3 A.M. EST (8h UT)
		The young Moon guide you to Mercury (−0.5) this evening. Look very low in the southwest 20 minutes after sunset for the brightest "star" near the Moon. Binoculars help. Saturn (0.5) is there as well, but is too low and faint to pull out of twilight. Now take in the rest of the scene. Venus (−4.2), quickly brightening, shines high in the south-southwest. Mars (0.6) is the brightest "star" in the south.
		Moon passes 7.1° north of Mercury, 11 P.M. EST (4h UT on Dec. 1)
December	3	Moon passes 5.8° north of Venus, 8 A.M. EST (13h UT)
	5	Moon passes 3.0° north of Mars, 6 A.M. EST (11h UT)
	7	First quarter Moon, 4:03 EST (9:03 UT)
	10	Mercury (−0.5) at greatest eastern elongation (20.8°), midnight EST (5h UT on Dec. 11). Look for the planet low in the southwest in the half hour after sunset. Venus (−4.3) blazes high above it. As the sky darkens, look for fading Mars (0.7) about 20° above left of Venus. See p. 79.
	12	Moon passes 0.5° north of Aldebaran, midnight EST (5h UT on Dec.13)
	13	Geminid meteor shower peaks, 7 P.M. EST (0h UT on Dec. 14)
		Full Moon, 7:06 P.M. EST (0:06 UT on Dec. 14)
	18	Moon passes 1.0° south of Regulus, 2 P.M. EST (19h UT)
	20	Last quarter Moon, 8:56 P.M. EST (1:56 UT on Dec. 21)
	21	Winter solstice, 5:44 A.M. EST (10:44 UT)
	22	Moon passes 2.4° north of Jupiter, noon EST (17h UT)
		Moon passes 6.1° north of Spica, 5 P.M. EST (22h UT)
		Ursid meteor shower peaks, 4 A.M. EST (9h UT)
	26	Moon passes 9.7° north of Antares, 3 P.M. EST (20h UT)
	27	Moon passes 3.6° north of Saturn, 4 P.M. EST (21h UT)

(continued)

	This morning, look for Saturn (0.5) beneath the waning crescent Moon, low in the east-southeast in the half hour before sunrise. Jupiter (−1.9) shines high in the south, near the star Spica.
28	Mercury in inferior conjunction (not visible)
29	New Moon, 1:53 A.M. EST (6:53 UT)

2017 Highlights

Mercury opens the year in the predawn sky with Saturn and Antares, then meets up with faint Mars in evening twilight in March. It shares the evening twilight with Jupiter and Regulus in July and joins a tight morning formation with Mars and Venus in September.

Venus is the striking evening star as the year opens, at its best mid-month when it sets 4 h after the Sun. Look for first-magnitude Mars nearby. Venus pops into the morning sky in early May and skirts a much fainter Mars in October.

Mars enjoys two conjunctions with Venus but otherwise takes a year off in preparation for next year's opposition.

Jupiter hovers near Spica all year. It reaches opposition on Apr. 7, and by mid-August shares the evening sky only with its companion gas giant, Saturn. Jupiter emerges into morning twilight in November, when it enjoys a close but very low altitude conjunction with Venus. By year's end, Jupiter is just a week from a close pass with Mars.

Saturn has a quiet year. The planet steadily brightens until Jun. 15, when it reaches opposition, and shares the summer evening sky with brighter Jupiter. Saturn emerges into morning twilight just before year's end.

Meteor showers. Waxing crescent Moons give little interference to the Quadrantid and Perseid showers, and December's Geminids are in full swing before the waning crescent rises.

Eclipses. Of the year's lunar eclipses, February's penumbral event is visible in the Americas, while the partial eclipse in August is reserved for Europe, Africa, Asia and Australia. For solar eclipses, the February annular event tracks through parts of Chile, Argentina and Africa. The big event as far as American skywatchers are concerned is the total solar eclipse that sweeps across the contiguous U.S. in August.

2017 Almanac		
January	1	Venus (−4.4) dazzles near the crescent Moon tonight. Mars (0.9) is fainter, but remains noticeably brighter than the stars in the region. Look for them in the southwest an hour or two after sunset.
	2	Moon passes 1.9° north of Venus, 4 A.M. EST (9h UT)
		The crescent Moon moves between Venus (−4.5) and Mars (0.9) tonight. Mars is about 4° above the Moon; Venus is 7° below it.
		Look for Quadrantid meteors tonight.
	3	Moon passes 0.2° north of Mars, 2 A.M. EST (7h UT)
		Quadrantid meteor shower peaks, 9 A.M. EST (14h UT)
	4	Earth at perihelion, its closest to the Sun this year – 91.4043 million miles – at 9:18 A.M. EST (14:18 UT)
	5	First quarter Moon, 2:47 P.M. EST (19:47 UT)
	9	Moon passes 0.4° north of Aldebaran, 10 A.M. EST (15h UT)
	12	Full Moon, 6:34 A.M. EST (11:34 UT)
		Venus (−4.6) at greatest eastern elongation (47.1°), 8 A.M. EST (13h UT). The planet is easily visible in evening twilight; see p. 66.
		Look for Mercury (−0.1) in the morning sky, low in the southeast in the hour before sunrise. Saturn (0.5) hovers nearby.
	14	Moon passes 0.8° south of Regulus, midnight EST (5h UT on Jan. 15)
	18	Moon passes 2.7° north of Jupiter, midnight EST (5h UT on Jan. 19)
	19	Moon passes 6.4° north of Spica, 1 A.M. EST (6h UT)
		Jupiter (−2.1) is the brightest "star" near the Moon this morning. Spica, a real star, lies beneath Jupiter and the Moon. Look south in the hours before dawn.
		Mercury (−0.2) at greatest western elongation (24.1°), 5 A.M. EST (10h UT). Look for Mercury below Saturn (0.5) low in the southeast in the hour before sunrise; see p. 80.
		Last quarter Moon, 5:13 P.M. EST (22:13 UT)
	20	Jupiter passes 3.7° north of Spica, 4 P.M. EST (21h UT)
	22	Moon passes 9.9° north of Antares, 11 P.M. EST (4h UT on Jan. 23)
	24	Moon passes 3.6° north of Saturn, 5 A.M. EST (10h UT)
		Look southeast in the hour before dawn for Saturn (0.5) near this morning's waning crescent Moon. Mercury (−0.2) is visible closer to the horizon; see p. 80.
	25	The Moon is 6° above Mercury this morning. Look southeast in the hour before dawn.
		Moon passes 3.7° north of Mercury, 8 P.M. EST (1h UT on Jan. 26)
	27	New Moon, 7:07 P.M. EST (0:07 UT on Jan. 28)
	31	Moon passes 4.1° south of Venus, 10 A.M. EST (15h UT)
		The Moon joins Venus (−4.7) and Mars (1.1) tonight. Look west-southwest in the hours after sunset.
		Moon passes 2.3° south of Mars, 8 P.M. EST (1h UT on Feb. 1)
February	3	First quarter Moon, 11:19 P.M. EST (4:19 UT on Feb. 4)
	5	Moon passes 0.2° north of Aldebaran, 5 P.M. EST (22h UT)
	10	Full Moon, 7:33 P.M. EST (0:33 UT on Feb. 11)
		Penumbral lunar eclipse, greatest at 7:44 P.M. EST (0:44 UT on Feb. 11) and visible throughout the easternmost Americas. The event is in progress at moonrise for the rest of the continental U.S. and Canada, Mexico, and western South America.

2017 Almanac (continued)

	11	Moon passes 0.8° south of Regulus, 9 A.M. EST (14h UT)
	15	Moon passes 6.5° north of Spica, 9 A.M. EST (14h UT)

Moon passes 2.7° north of Jupiter, 10 A.M. EST (15h UT)

Jupiter (–2.2) is brilliant near the Moon late tonight and into tomorrow morning. Jupiter, the Moon and the star Spica are high in the south 3 h before sunrise. At that time, Saturn (0.6) is also visible low in the southeast.

18 Last quarter Moon, 2:33 P.M. EST (19:33 UT)

19 Moon passes 10.0° north of Antares, 7 A.M. EST (12h UT)

20 The Moon is near Saturn (0.5) this morning. Look southeast in the hours before dawn. Tomorrow morning, the Moon appears on the opposite side of Saturn. Jupiter (–2.3) shines brightly in the southwest.

Moon passes 3.6° north of Saturn, 6 P.M. EST (23h UT)

23 Jupiter passes 3.8° north of Spica, 11 A.M. EST (16h UT)

25 Moon passes 2.5° north of Mercury, 9 P.M. EST (2h UT on Feb. 26)

26 **Annular solar eclipse**, greatest at 14:53 UT; the track includes the South Pacific, Chile, Argentina, South Atlantic and Angola. See p. 140.

New Moon, 9:58 A.M. EST (14:58 UT)

28 Moon passes 10.3° south of Venus, 3 P.M. EST (20h UT)

Look for the young Moon, dazzling Venus (–4.8) and faint Mars (1.3) in the west in the hour after sunset. Mars tops the triangle formed by the three objects.

March 1 Moon passes 4.3° south of Mars, 2 P.M. EST (19h UT)

4 Moon passes 0.2° north of Aldebaran, 10 P.M. EST (3h UT on Mar. 5)

5 First quarter Moon, 6:32 A.M. EST (11:32 UT)

6 Mercury in superior conjunction (not visible)

10 Moon passes 0.8° south of Regulus, 6 P.M. EST (23h UT)

12 Spring ahead: U.S. Daylight Saving Time begins this morning.

Full Moon, 10:54 A.M. EDT (14:54 UT)

14 Moon passes 2.5° north of Jupiter, 4 P.M. EDT (20h UT)

Moon passes 6.4° north of Spica, 7 P.M. EDT (23h UT)

Look for Jupiter (–2.4) near the Moon late tonight and into tomorrow morning.

16 Venus passes 9.6° north of Mercury, 7 P.M. EDT (23h UT)

Look for Mercury (–1.4) very low in the west 20 min after sunset. The planet shines below and to the left of brilliant Venus (–4.3), preparing to dive into twilight and switch to the morning sky.

18 Moon passes 9.9° north of Antares, 4 P.M. EDT (20h UT)

20 The Moon visits Saturn (0.5) this morning. Look high in the south in the hour before dawn. Turn southwest to see another celestial pair: Jupiter (–2.4) and the bright star Spica.

Moon passes 3.4° north of Saturn, 6 A.M. EDT (10h UT)

Vernal equinox, 6:29 A.M. EDT (10:29 UT)

Last quarter Moon, 11:58 A.M. EDT (15:58 UT)

25 Venus in inferior conjunction (not visible)

Venus is gone from the evening sky but Mercury (–0.7) remains, low in the west in the hour after sunset. Can you spot Mars (1.5) above it? See p. 80.

(continued)

2017 Almanac (continued)

	27	New Moon, 10:57 P.M. EDT (2:57 UT on Mar. 28)
	29	Moon passes 6.6° south of Mercury, 3 A.M. EDT (7h UT)
		Tonight, the Moon is midway between Mercury (−0.5) and Mars (1.5). Mercury closer to both the Moon and horizon; see p. 80.
	30	Moon passes 5.5° south of Mars, 9 A.M. EDT (13h UT)
		Tonight, the crescent Moon hovers above and to the left of faint Mars (1.5); see p. 80.
April	1	Moon passes 0.3° north of Aldebaran, 5 A.M. EDT (9h UT)
		Mercury (−0.2) at greatest eastern elongation (19.0°), at 6 A.M. EDT (10h UT). Look for the planet low in the west in the hour after sunset.
	3	First quarter Moon, 2:39 P.M. EDT (18:39 UT)
	4	Moon passes 10.0° south of Pollux, 10 A.M. EDT (14h UT)
	6	Venus (−4.4) is now easily visible low in the east in the half hour before sunrise.
	7	Moon passes 0.7° south of Regulus, 1 A.M. EDT (5h UT)
		Jupiter (−2.5) is at opposition and nearest to Earth, 414.123 million miles away. It rises in the east at sunset and is visible all night long. Opposition occurs at 5:39 P.M. EDT (21:39 UT)
	10	Moon passes 2.2° north of Jupiter, 5 P.M. EDT (21h UT)
	11	Moon passes 6.4° north of Spica, 2 A.M. EDT (6h UT)
		Full Moon, 2:08 A.M. EDT (6:08 UT)
	14	Moon passes 9.7° north of Antares, midnight EDT (4h UT on Apr. 15)
	16	Moon passes 3.2° north of Saturn, 2 P.M. EDT (18h UT)
	18	Mars (1.6) lies within 4° of the Pleiades star cluster until Apr. 23. Look west-northwest an hour after sunset.
	19	Last quarter Moon, 5:57 A.M. EDT (9:57 UT)
	20	Mercury in inferior conjunction (not visible)
	22	Lyrid meteor shower peaks, 8 A.M. EDT (12h UT)
	23	The waning crescent Moon visits brilliant Venus (−4.7) this morning. Look low in the east in the half hour before sunrise.
		Moon passes 5.2° south of Venus, 2 P.M. EDT (18h UT)
	26	New Moon, 8:16 A.M. EDT (12:16 UT)
	27	Look for Mars (1.6) and the Pleiades star cluster above tonight's crescent Moon, low in the west-northwest in the hour after sundown. Binoculars help.
	28	Moon passes 5.8° south of Mars, 3 A.M. EDT (7h UT)
		Moon passes 0.5° north of Aldebaran, 2 P.M. EDT (18h UT)
May	1	Moon passes 9.8° south of Pollux, 4 P.M. EDT (20h UT)
	2	First quarter Moon, 10:47 P.M. EDT (2:47 UT on May 3)
	4	Moon passes 0.5° south of Regulus, 6 A.M. EDT (10h UT)
	5	Eta Aquariid meteor shower peaks, 9 P.M. EDT (2h UT on May 6)
	7	Mars passes 6.2° north of Aldebaran, 3 A.M. EDT (7h UT)
		Moon passes 2.1° north of Jupiter, 5 P.M. EDT (21h UT)
		The bright "star" near the tonight's Moon is Jupiter (−2.4), visible in the southeast in the hour after sunset. Mars (1.6), faint and low, hovers in the west-northwest at the same time.
	8	Moon passes 6.4° north of Spica, 9 A.M. EDT (13h UT)
	10	Full Moon, 5:42 P.M. EDT (21:42 UT)
	12	Moon passes 9.6° north of Antares, 6 A.M. EDT (10h UT)

(continued)

2017 Almanac (continued)

	13	Moon passes 3.1° north of Saturn, 7 P.M. EDT (23h UT)
		Look for Saturn (0.2) near the Moon late tonight and tomorrow morning.
	17	Mercury (0.4) at greatest western elongation (25.8°) at 7 P.M. EDT (23h UT). Mercury can be found very low in the east in the half hour before sunrise, below and to the left of brilliant Venus (−4.6)
	18	Last quarter Moon, 8:33 P.M. EDT (0:33 UT on May 19)
	22	Look for Venus (−4.6) near the waning crescent Moon this morning, low in the east in the hour before dawn.
		Moon passes 2.4° south of Venus, 9 A.M. EDT (13h UT)
	23	Moon passes 1.6° south of Mercury, 9 P.M. EDT (1h UT on May 24)
	25	New Moon, 3:44 P.M. EDT (19:44 UT)
		Moon passes 0.6° north of Aldebaran, midnight EDT (4h UT on May 26)
	26	Moon passes 5.4° south of Mars, 10 P.M. EDT (2h UT on May 27)
	28	Moon passes 9.6° south of Pollux, midnight EDT (4h UT on May 29)
	31	Moon passes 0.3° south of Regulus, 1 P.M. EDT (17h UT)
June	1	First quarter Moon, 8:42 A.M. EDT (12:42 UT)
	3	Venus (−4.4) at greatest western elongation (45.9°), 8 A.M. EDT (12h UT). The planet is prominent in morning twilight, low in the east in the hour before dawn.
		Moon passes 2.3° north of Jupiter, 8 P.M. EDT (0h UT on Jun. 4)
	4	Moon passes 6.6° north of Spica, 2 P.M. EDT (18h UT)
	8	Moon passes 9.6° north of Antares, noon EDT (16h UT)
	9	Full Moon, 9:10 A.M. EDT (13:10 UT)
		Moon passes 3.1° north of Saturn, 9 P.M. EDT (1h UT on Jun. 10)
		Look for Saturn (0.0) near the Moon tonight, rising in the east-southeast shortly after sunset.
	12	Mercury passes 5.1° north of Aldebaran, 7 A.M. EDT (11h UT)
	15	Saturn (0.0) is at opposition and nearest to Earth, 840.569 million miles away. It rises in the east at sunset and is visible all night long. Opposition occurs at 6:18 A.M. EDT (10:18 UT)
	17	Last quarter Moon, 7:33 A.M. EDT (11:33 UT)
	20	Venus (−4.3) is near the waning crescent Moon this morning. Look east in the hour before sunrise.
		Moon passes 2.4° south of Venus, 5 P.M. EDT (21h UT)
	21	Summer solstice, 12:24 A.M. EDT (4:24 UT)
		Mercury in superior conjunction (not visible)
	22	Moon passes 0.5° north of Aldebaran, 11 A.M. EDT (15h UT)
	23	New Moon, 10:31 P.M. EDT (2:31 UT on Jun. 24)
	25	Moon passes 9.4° south of Pollux, 10 A.M. EDT (14h UT)
	27	Moon passes 0.0° south of Regulus, 9 P.M. EDT (1h UT on Jun. 28)
		June Boötids meteor shower peaks, 6 A.M. EDT (10h UT)
	30	First quarter Moon, 8:51 P.M. EDT (0:51 UT on Jul. 1)
		Look for Jupiter (−2.1) near the Moon as the sky darkens tonight.
July	1	Moon passes 2.7° north of Jupiter, 3 A.M. EDT (7h UT)
		Moon passes 6.8° north of Spica, 8 P.M. EDT (0h UT on Jul. 2)
	2	Mercury passes 4.9° south of Pollux, 8 P.M. EDT (0h UT on July 3)
	3	Earth at aphelion, its farthest from the Sun this year – 94.5059 million miles – at 4:11 P.M. EDT (20:11 UT)

(continued)

2017 Almanac (continued)

	5	Moon passes 9.7° north of Antares, 6 P.M. EDT (22h UT)
	6	Moon passes 3.2° north of Saturn, 11 P.M. EDT (3h UT on Jul. 7)
		Look for Saturn (0.1) near the Moon tonight, low in the southeast in the hour after sunset.
	9	Full Moon, 12:07 A.M. EDT (4:07 UT)
	16	Last quarter Moon, 3:26 P.M. EDT (19:26 UT)
	19	Moon passes 0.4° north of Aldebaran, 8 P.M. EDT (0h UT on Jul. 20)
	20	The Moon visits Venus (−4.1) this morning. Look east in the hour before dawn. The Pleiades hover high above them. Can you make out the rest of the constellation Taurus?
		Moon passes 2.7° south of Venus, 7 A.M. EDT (11h UT)
	22	Moon passes 9.4° south of Pollux, 9 P.M. EDT (1h UT on Jul. 23)
	23	New Moon, 5:46 A.M. EDT (9:46 UT)
		Mercury (0.1) emerges for an evening apparition. Look for the planet low in the west in the half hour after sunset; see p. 80. Regulus, the brightest star in Leo, is less than 3° above it. Binoculars help.
	25	Moon passes 0.9° north of Mercury, 5 A.M. EDT (9h UT)
		Moon passes 0.1° north of Regulus, 7 A.M. EDT (11h UT)
		The crescent Moon guides you to Mercury (0.2) and Regulus tonight, lying nearly 8° above the pair. Look for the planet low in the west in the half hour after sunset.
	26	Mercury passes 1.1° south of Regulus, 5 A.M. EDT (9h UT)
		Mars in conjunction with the Sun (not visible)
	28	Moon passes 3.1° north of Jupiter, 4 P.M. EDT (20h UT)
		Tonight, in the hour after sundown, look west-southwest for the Moon. The bright "star" beneath it is Jupiter (−1.9). Spica, the brightest star in Virgo, is 8° to its left.
	29	Moon passes 7.0° north of Spica, 4 A.M. EDT (8h UT)
	30	Mercury (0.3) at greatest eastern elongation (27.2°) at 1 A.M. EDT (5h UT). It's visible low in the west in the half hour after sunset.
		Southern Delta Aquariid meteor shower peaks, 1 A.M. EDT (5h UT)
		First quarter Moon, 11:23 A.M. EDT (15:23 UT)
August	2	Moon passes 9.8° north of Antares, 1 A.M. EDT (5h UT)
		Saturn (0.3) is near the Moon tonight. Look south in the hour after sunset. Turn to the west-southwest to see Jupiter (−1.9). Look half an hour earlier and you'll find Mercury (0.5) low in the west.
	3	Moon passes 3.5° north of Saturn, 3 A.M. EDT (7h UT)
	7	Full Moon, 2:11 P.M. EDT (18:11 UT)
		Partial lunar eclipse, greatest at 18:20 UT and visible throughout the Indian Ocean, Australia, Asia, Africa and Europe. See p. 141.
	11	Look for Perseid meteors tonight.
	12	Perseid meteor shower peaks, 2 P.M. EDT (19h UT); look for meteors tonight, too.
	14	Last quarter Moon, 9:15 P.M. EDT (1:15 UT on Aug. 15)
	16	Moon passes 0.4° north of Aldebaran, 3 A.M. EDT (7h UT)
	19	Moon passes 2.3° south of Venus, 1 A.M. EDT (5h UT)

(continued)

2017 Almanac (continued)

	Venus (−3.9) is the brilliant "morning star" above the waning crescent Moon this morning. Look east in hour before sunrise. Look the star Pollux to Venus' left and for the star Procyon 13° below and to the right of the Moon.
	Moon passes 9.4° south of Pollux, 6 A.M. EDT (10h UT)
21	New Moon, 2:30 P.M. EDT (18:30 UT)
	Venus passes 7.3° south of Pollux, 3 P.M. EDT (19h UT)
	Moon passes 0.1° north of Regulus, 4 P.M. EDT (20h UT)
	Total solar eclipse, greatest at 1:25 P.M. CDT (18:25 UT); totality tracks from the northern Pacific, across the continental U.S. from Oregon to South Carolina, and into the South Atlantic. See p. 142.
25	Moon passes 3.5° north of Jupiter, 9 A.M. EDT (13h UT)
	Moon passes 7.0° north of Spica, 1 P.M. EDT (17h UT)
	Jupiter (−1.8) and the bright star Spica lie beneath the Moon tonight. Look west-northwest in the hour after sunset. Then scan the southern sky for Saturn (0.4); the Moon helps on the 29th and 30th.
26	Mercury in inferior conjunction (not visible)
29	First quarter Moon, 4:13 A.M. EDT (8:13 UT)
	Moon passes 9.8° north of Antares, 9 A.M. EDT (13h UT)
	Tonight and tomorrow night, look for Saturn (0.4) near the Moon. Look south in the hour after sunset. Then, look low in the west-southwest to catch Jupiter (−1.7) near the star Spica in Virgo.
30	Moon passes 3.6° north of Saturn, 10 A.M. EDT (14h UT)
September 4	Mars passes 0.8° north of Regulus, 10 P.M. EDT (2h UT on Sept. 5)
5	Jupiter passes 3.4° north of Spica, 7 A.M. EDT (11h UT)
	Look for Jupiter and the star Spica low in the west-southwest in the half hour after sunset. Saturn (0.5), the other evening planet, shines in the south.
6	Full Moon, 3:03 A.M. EDT (7:03 UT)
10	Mercury passes 0.6° south of Regulus, 8 A.M. EDT (12h UT)
	Mercury (1.3), Mars (1.8) and Regulus gather in the morning sky. Look east-northeast beneath bright Venus (−3.9) in the half hour before dawn. Binoculars help.
12	Mercury (−0.3) at greatest western elongation (17.9°), 6 A.M. EDT (10h UT). Look for Mercury low in the east – beneath Venus (−3.9) – in the half hour before sunrise. Mars (1.8) is 3° below Mercury, the star Regulus is above it. See p. 80.
	Moon passes 0.4° north of Aldebaran, 9 A.M. EDT (13h UT)
13	Last quarter Moon, 2:25 A.M. EDT (6:25 UT)
15	Moon passes 9.4° south of Pollux, 1 P.M. EDT (17h UT)
16	Mercury passes 0.1° north of Mars, 2 P.M. EDT (18h UT)
17	Moon passes 0.6° south of Venus, 9 P.M. EDT (1h UT on Sept. 18)
18	Moon passes 0.1° north of Regulus, 1 A.M. EDT (5h UT)
	Catch the predawn Venus, Regulus, moon, Mars, Mercury morning meeting. Look for bright Venus (−3.9) above the Moon an hour before sunrise. The star Regulus is 2° below Venus. Mercury is about 6° below the Moon. Mars (1.8, binoculars help) is 1.4° below Mercury.

(continued)

2017 Almanac (continued)

		Moon passes 0.1° north of Mars, 4 P.M. EDT (20h UT)
		Moon passes 0.0° south of Mercury, 7 P.M. EDT (23h UT)
	19	Venus passes 0.5° north of Regulus, 7 P.M. EDT (23h UT)
	20	New Moon, 1:30 A.M. EDT (5:30 UT)
	21	Moon passes 6.9° north of Spica, 10 P.M. EDT (2h UT on Sept. 22)
		Look for Jupiter (−1.7) near the young Moon tonight and tomorrow night, low in the west-southwest in the half hour after sunset.
	22	Moon passes 3.7° north of Jupiter, 4 A.M. EDT (8h UT)
		Autumnal equinox, 4:02 P.M. EDT (20:02 UT)
	25	Moon passes 9.7° north of Antares, 5 P.M. EDT (21h UT)
	26	Moon passes 3.5° north of Saturn, 8 P.M. EDT (0h UT on Sept. 27)
		Saturn (0.5) is beneath the Moon tonight. Look southwest in the hour after sunset.
	27	First quarter Moon, 10:53 P.M. EDT (2:53 UT on Sept. 28)
October	3	Watch Mars (1.8) and Venus (−3.9) converge over the following mornings. Look east in the hour before dawn. Binoculars will help with faint Mars. See p. 178.
	5	Venus and Mars are closest this morning. Look east in the hour before dawn.
		Venus passes 0.2° north of Mars, 9 A.M. EDT (13h UT)
		Full Moon, 2:40 P.M. EDT (18:40 UT)
	8	Mercury in superior conjunction (not visible)
	9	Moon passes 0.6° north of Aldebaran, 3 P.M. EDT (19h UT)
	12	Last quarter Moon, 8:25 A.M. EDT (12:25 UT)
		Moon passes 9.2° south of Pollux, 6 P.M. EDT (22h UT)
	15	Moon passes 0.2° north of Regulus, 7 A.M. EDT (11h UT)
	17	Moon passes 1.8° north of Mars, 6 A.M. EDT (10h UT)
		The Moon is near Mars (1.8) this morning. Look east in the hour before dawn. Although faint, Mars is brighter than other stars near the Moon. Venus (−3.9) blazes below the pair.
		Moon passes 2.0° north of Venus, 8 P.M. EDT (0h UT on Oct. 18)
	19	Moon passes 6.9° north of Spica, 6 A.M. EDT (10h UT)
		New Moon, 3:12 P.M. EDT (19:12 UT)
	21	Orionid meteor shower peaks, 7 A.M. EDT (11h UT)
	23	Moon passes 9.5° north of Antares, 1 A.M. EDT (5h UT)
	24	Moon passes 3.3° north of Saturn, 8 A.M. EDT (12h UT)
		Saturn (0.6) lies beneath the young Moon tonight. Look southwest in the hour after sunset.
	27	First quarter Moon, 6:22 P.M. EDT (22:22 UT)
November	1	Venus passes 3.8° north of Spica, 11 P.M. EDT (15h UT)
		Look for Venus (−3.9) and Spica low in the east-southeast in the half hour before dawn.
	4	Full Moon, 1:23 A.M. EDT (5:23 UT)
	5	Fall back: U.S. Daylight Saving Time ends this morning.
		Moon passes 0.8° north of Aldebaran, 10 P.M. EST (3h UT on Nov. 6)
	8	Moon passes 8.9° south of Pollux, 11 P.M. EST (4h UT on Nov. 9)
	10	Last quarter Moon, 3:36 P.M. EST (20:36 UT)

(continued)

2017 Almanac (continued)

	11	Moon passes 0.5° north of Regulus, noon EST (17h UT)
	12	Jupiter (−1.7) emerges from the Sun's glare and meets up with Venus (−3.9). This morning, the two planets are less than a degree apart. Look low in the east-southeast in the half hour before dawn.
	13	Venus passes 0.3° north of Jupiter, 1 A.M. EST (6h UT)
		Venus and Jupiter are 17 arcminutes apart this morning. Look for the pair low in the east-southeast in the half hour before sunup.
	14	Moon passes 3.2° north of Mars, 8 P.M. EST (1h UT on Nov. 15)
		The Moon is above Mars (1.8) this morning, below it tomorrow morning.
	15	Moon passes 7.0° north of Spica, 11 A.M. EST (16h UT)
	16	This morning, the Moon hovers above Jupiter (−1.7) and Venus (−3.9). Look east-southeast in the half hour before dawn.
		Moon passes 4.1° north of Jupiter, 4 P.M. EST (21h UT)
	17	Moon passes 4.0° north of Venus, 1 A.M. EST (6h UT)
		Leonid meteor shower peaks, 11 A.M. EST (16h UT)
	18	New Moon, 6:42 A.M. EST (11:42 UT)
	20	Moon passes 6.9° north of Mercury, 4 A.M. EST (9h UT)
		The Moon brings you the Saturn (0.5) and Mercury (−0.4) this evening. Look southwest in the half hour after sunset.
		Moon passes 3.0° north of Saturn, 7 P.M. EST (0h UT on Nov. 21)
	21	Alpha Monocerotid meteor shower peaks, noon EST (17h UT)
	23	Mercury (−0.4) at greatest eastern elongation (22.0°), 7 P.M. EST (0h UT on Nov. 24). Look for the planet low in the southwest in the half hour after sunset, beneath fainter Saturn (0.5)
	26	First quarter Moon,12:03 P.M. EST (17:03 UT)
	27	Look for Mars (1.7) and Spica in the east about an hour before sunrise. You'll notice Jupiter (−1.7) before Mars. At about this time, Venus (−3.9) is just coming over the horizon below Jupiter.
		Mars passes 3.4° north of Spica, 7 P.M. EST (0h UT on Nov. 28)
	28	Mercury passes 3.1° south of Saturn, 4 A.M. EST (9h UT)
		Mercury (−0.3) outshines Saturn (0.5) tonight. Look for these two "evening stars" low in the southwest in the half hour after sunset.
December	3	Moon passes 0.8° north of Aldebaran, 8 A.M. EST (13h UT)
		Full Moon, 10:47 A.M. EST (15:47 UT)
	6	Moon passes 8.7° south of Pollux, 8 A.M. EST (13h UT)
	8	Moon passes 0.7° north of Regulus, 6 P.M. EST (23h UT)
	10	Last quarter Moon, 2:51 A.M. EST (7:51 UT)
	12	Moon passes 7.2° north of Spica, 4 P.M. EST (21h UT)
		Mercury in inferior conjunction (not visible)
	13	Slowly brightening Mars (1.6) appears beneath the Moon this morning. Look southeast in the hour before sunrise. Jupiter (−1.7) shines below.
		Moon passes 4.2° north of Mars, 11 A.M. EST (16h UT)
	14	The Moon hovers by Jupiter (−1.7) this morning, while Mars looks on above them. Look southeast in the hour before sunrise.
		Geminid meteor shower peaks, 1 A.M. EST (6h UT)
		Moon passes 4.3° north of Jupiter, 9 A.M. EST (14h UT)
	15	Mercury passes 2.2° north of Venus, 11 A.M. EST (16h UT)

(continued)

2017 Almanac (continued)

16	Moon passes 9.4° north of Antares, 1 P.M. EST (18h UT)
17	Moon passes 1.8° north of Mercury, 4 A.M. EST (9h UT)
18	New Moon, 1:30 A.M. EST (6:30 UT)
20	Mercury (1.0) joins Jupiter (−1.8) and Mars (1.6) in the morning sky. Look low in the southeast in the half hour before sunrise.
21	Winter solstice, 11:28 A.M. EST (16:28 UT) Saturn in conjunction with the Sun (not visible)
22	Ursid meteor shower peaks, 10 A.M. EST (15h UT)
26	First quarter Moon, 4:20 A.M. EST (9:20 UT)
27	Mercury (−0.2) shines brightly low in the southeast in the hour before dawn. The more conspicuous "morning star" is Jupiter (−1.8), which appears higher in the southeast. Mars (1.5) is about 6° above Jupiter.
30	Moon passes 0.8° north of Aldebaran, 8 P.M. EST (1h UT on Dec. 31)

2018 Highlights

Mercury is visible before dawn as the year opens. It puts on nice evening shows with Venus in March and again in June/July, and the two planets reign in the morning sky as the year closes.

Venus emerges into evening twilight in early April but keeps a low profile, never riding high even at its best in June. It remains an evening star into September and completes the transition to the morning sky in November.

Mars is dogged by Jupiter in early January, and the two planets appear a third of a degree apart on Jan. 7. In early April, as Mars begins brightening rapidly, it makes a close pass to Saturn. Mars reaches opposition Jul. 27 and remains a prominent evening planet through year's end.

Jupiter rendezvous with Mars as the year opens and steadily brightens toward its opposition on May 8. It remains a prominent evening planet to October, then emerges into morning twilight as the year closes.

Saturn meets with Mars after Jupiter but otherwise has an uneventful year. It reaches opposition on Jun. 27, remains prominent in evening twilight until December, and emerges into the predawn sky in January 2019.

Meteor showers. A Moon just past full washes out the Quadrantids. By contrast, near-perfect viewing conditions greet the Perseid shower, which peaks near new Moon. The Geminids contend with a waxing crescent Moon, which can be avoided by viewing late on the evening before and morning of the peak.

Eclipses. There are five eclipses this year, but only the two lunar eclipses are total. January's brings totality to the western third of the U.S. Of three partial solar eclipses, the one in August will be the most widely observed, with visibility ranging from China, Russia, Iceland, Scandanavia and northern Canada.

2018 Almanac

January	1	Mercury (−0.3) at greatest western elongation (22.7°) at 3 P.M. EST (20h UT). Mercury is visible in morning twilight, low in the southeast in the hour before sunrise. This morning, look for Jupiter (−1.8) and Mars (1.5) 30° above Mercury. Watch Jupiter and Mars come together over the next week; see p. 81.
		Full Moon, 9:24 P.M. EST (2:24 UT on Jan. 2)
	2	Moon passes 8.6° south of Pollux, 6 P.M. EST (23h UT)
	3	Earth at perihelion, its closest to the Sun this year – 91.4020 million miles – at 12:35 A.M. EST (5:35 UT)
		Quadrantid meteor shower peaks, 3 P.M. EST (20h UT)
	5	Moon passes 0.9° north of Regulus, 3 A.M. EST (8h UT)
	6	Jupiter passes 0.2° north of Mars, 11 P.M. EST (4h UT on Jan. 7)
	7	Jupiter (−1.7) and Mars (1.4) are separated by just 0.3° (18 arcminutes) this morning. Look south-southeast in the hours before dawn. Within half an hour of sunrise, look for Mercury (−0.3) and Saturn (0.5) low in the southeast. See p. 81.
	8	Last quarter Moon, 5:25 P.M. EST (22:25 UT)
		Moon passes 7.4° north of Spica, 11 P.M. EST (4h UT on Jan. 9)
		Venus in superior conjunction (not visible)
	11	Moon passes 4.3° north of Jupiter, 1 A.M. EST (6h UT)
		Moon passes 4.6° north of Mars, 5 A.M. EST (10h UT)
		The waning crescent Moon guides you to Jupiter (−1.9) and Mars (1.4), which are separated by 2.1° this morning. Look south-southeast in the hour before dawn. Saturn (0.5) and Mercury (−0.3) shine very low in the southeast.
	12	Moon passes 9.5° north of Antares, 7 P.M. EST (0h UT on Jan. 13)
	13	Mercury passes 0.6° south of Saturn, 2 A.M. EST (7h UT)
		Mercury (−0.3) and Saturn (0.5) pair up this morning, low in the southeast in the half hour before sunrise. Binoculars help. Jupiter (−1.9) and Mars (1.4) shine higher in the south-southeast; see p. 81.
	14	Moon passes 2.6° north of Saturn, 9 P.M. EST (2h UT on Jan. 15)
	15	Moon passes 3.4° north of Mercury, 2 A.M. EST (7h UT)
		The waning Moon is next to Mercury (−0.3) this morning. Saturn (0.5) is 3° above Mercury. Look low in the southeast in the half hour before sunup. At the same, Jupiter (−1.9) and Mars (1.4) shine in the south.
	16	New Moon, 9:17 P.M. EST (2:17 UT on Jan. 17)
	24	First quarter Moon, 5:20 P.M. EST (22:20 UT)
	27	Moon passes 0.7° north of Aldebaran, 6 A.M. EST (11h UT)
	30	Moon passes 8.6° south of Pollux, 6 A.M. EST (11h UT)
	31	Full Moon, 8:27 A.M. EST (13:27 UT)
		Total lunar eclipse, greatest at 5:13 A.M. PST (13:30 UT), with totality in progress as the Moon sets from the central U.S.; locations farther west seeing a larger share of totality. See p. 143.
February	1	Moon passes 1.0° north of Regulus, 2 P.M. EST (19h UT)
	5	Moon passes 7.4° north of Spica, 7 A.M. EST (12h UT)
	7	Last quarter Moon, 10:54 A.M. EST (15:54 UT)
		Moon passes 4.3° north of Jupiter, 3 P.M. EST (20h UT)

(continued)

2018 Almanac (continued)

		The brightest "star" near the Moon this morning is Jupiter (−2.0). Look south in the hours before sunrise.
		Moon passes 4.4° north of Mars, midnight EST (5h UT on Feb. 9)
	9	Moon passes 9.5° north of Antares, 2 A.M. EST (7h UT)
		The Moon stands above Mars (1.1) and its stellar rival, Antares, this morning. Planet and star are of similar brightness (slight advantage to Antares). Look southeast in the hours before sunrise.
	10	Mars passes 5.2° north of Antares, 10 A.M. EST (15h UT)
	11	The waning crescent Moon is above Saturn (0.6) this morning. Look southeast in the hour before sunrise.
		Moon passes 2.5° north of Saturn, 10 A.M. EST (15h UT)
	15	New Moon, 4:05 P.M. EST (21:05 UT)
		Partial solar eclipse, greatest at 20:51 UT, when up to 60% of the Sun is obscured as seen from Antarctica. Southern Chile and Argentina see the Sun covered up to 40%.
	16	Moon passes 0.6° south of Venus, 11 A.M. EST (16h UT)
		Look for Venus (−3.9) beneath the young Moon this evening, low in the west-southwest 20 min after sunset.
	17	Mercury in superior conjunction (not visible)
	23	First quarter Moon, 3:09 A.M. EST (8:09 UT)
		Moon passes 0.7° north of Aldebaran, 1 P.M. EST (18h UT)
	26	Moon passes 8.6° south of Pollux, 4 P.M. EST (21h UT)
March	1	Moon passes 1.0° north of Regulus, 1 A.M. EST (6h UT)
		Full Moon, 7:51 P.M. EST (0:51 UT on Mar. 2)
	3	Look for Venus (−3.9) and Mercury (−1.2) very low in the west 20 min after sunset. The two planets are 1.1° apart.
	4	Moon passes 7.4° north of Spica, 5 P.M. EST (22h UT)
	5	Venus passes 1.4° south of Mercury, 1 P.M. EST (18h UT)
		Venus (−3.9) and Mercury (−1.1) appear low in the west 20 min after sunset.
	7	Jupiter (−2.2) is the brightest "star" near the Moon late tonight and tomorrow morning.
		Moon passes 4.1° north of Jupiter, 2 A.M. EST (7h UT)
	8	Moon passes 9.4° north of Antares, 10 A.M. EST (15h UT)
	9	The Moon stands near rapidly brightening Mars (0.7) this morning. Look south-southeast in the hour before dawn.
		Last quarter Moon, 6:20 A.M. EST (11:20 UT)
		Moon passes 3.8° north of Mars, 8 P.M. EST (1h UT on Mar. 10)
	10	The Moon is between Saturn (0.6, below left) and Mars (0.7) this morning. Look south-southeast in the hour before dawn.
		Moon passes 2.2° north of Saturn, 9 P.M. EST (2h UT on Mar. 11)
	11	Spring forward: U.S. Daylight Saving Time begins this morning.
	15	Mercury (−0.4) at greatest eastern elongation (18.4°) at 11 A.M. EDT (15h UT). Look for Mercury above brighter Venus (−3.9), low in the west in the hour after sunset. See p. 81.
	17	New Moon, 9:12 A.M. EDT (13:12 UT)
		Venus passes 3.9° south of Mercury, 9 P.M. EDT (1h UT on Mar. 18)

(continued)

2018 Almanac (continued)

	18	Moon passes 7.7° south of Mercury, 2 P.M. EDT (18h UT)
		Moon passes 3.8° south of Venus, 3 P.M. EDT (19h UT)
		Look for the Moon, Venus (−3.9) and Mercury (0.3) low in the west in the hour after sunset; see p. 81.
	20	Vernal equinox, 12:45 P.M. EDT (16:45 UT)
	22	Moon passes 0.9° north of Aldebaran, 7 P.M. EDT (23h UT)
	24	First quarter Moon, 11:35 A.M. EDT (15:35 UT)
	25	Moon passes 8.5° south of Pollux, 11 P.M. EDT (3h UT on Mar. 26)
	28	Moon passes 1.0° north of Regulus, 10 A.M. EDT (14h UT)
	31	Full Moon, 8:37 A.M. EDT (12:37 UT)
April	1	Moon passes 7.3° north of Spica, 3 A.M. EDT (7h UT)
		Mercury in inferior conjunction (not visible)
	2	Mars (0.3) and Saturn (0.5) are less than 1.3° apart this morning. The two planets shine above the lid of the Teapot asterism in Sagittarius. Look south-southeast in the hour before dawn.
		Mars passes 1.3° south of Saturn, 8 A.M. EDT (12h UT)
		Jupiter (−2.4) is the bright "star" accompanying the Moon late tonight and into tomorrow morning.
	3	Moon passes 3.9° north of Jupiter, 10 A.M. EDT (14h UT)
	4	Moon passes 9.3° north of Antares, 8 P.M. EDT (0h UT on Apr. 5)
	7	The Moon joins Mars (0.2) and Saturn (0.5) this morning. Look south-southeast in the hour before dawn.
		Moon passes 1.9° north of Saturn, 9 A.M. EDT (13h UT)
		Moon passes 3.1° north of Mars, 2 P.M. EDT (18h UT)
	8	Last quarter Moon, 3:18 A.M. EDT (7:18 UT)
	14	Moon passes 3.9° south of Mercury, 5 A.M. EDT (9h UT)
	15	New Moon, 9:57 P.M. EDT (1:57 UT on Apr. 16)
	17	Moon passes 5.4° south of Venus, 3 P.M. EDT (19h UT)
		The young Moon visits Venus (−3.9) tonight. Look west in the hour after sundown. Can you find the Pleiades star cluster 9° above Venus?
	19	Moon passes 1.1° north of Aldebaran, 1 A.M. EDT (5h UT)
	22	Moon passes 8.2° south of Pollux, 5 A.M. EDT (9h UT)
		First quarter Moon, 5:46 P.M. EDT (21:46 UT)
		Lyrid meteor shower peaks, 2 P.M. EDT (18h UT)
	23	Venus (−3.9) lies within 4° of the Pleiades star cluster until Apr. 25.
	24	Moon passes 1.2° north of Regulus, 4 P.M. EDT (20h UT)
	28	Moon passes 7.3° north of Spica, noon EDT (16h UT)
	29	Mercury (0.4) at greatest western elongation (27.0°), 2 P.M. EDT (18h UT). The planet is visible low in the east in the half hour before dawn.
		Full Moon, 8:58 P.M. EDT (0:58 UT on Apr. 30)
	30	Moon passes 3.8° north of Jupiter, 1 P.M. EDT (17h UT)
		Jupiter (−2.5) rises shortly before the Moon tonight. Look east-southeast in the hour after sunset. Then turn to the west to find Venus (−3.9).
May	2	Moon passes 9.1° north of Antares, 4 A.M. EDT (8h UT)
	3	Venus passes 6.5° north of Aldebaran, 1 P.M. EDT (17h UT)

(continued)

2018 Almanac (continued)

	4	The Moon stands near Saturn (0.4) this morning. Look south in the hour before sunrise. Mars (−0.5) outshines Saturn to the par's left. Turn southwest to catch Jupiter (−2.5) in the southwest.
		Moon passes 1.7° north of Saturn, 4 P.M. EDT (20h UT)
	6	Moon passes 2.7° north of Mars, 3 A.M. EDT (7h UT)
		Mars (−0.5) shine beneath the Moon this morning. Look south-southeast in the hour before sunrise. Saturn (0.3) shines to the pair's right. In the southwest, bright Jupiter (−2.5) hovers near the horizon.
		Eta Aquariid meteor shower peaks, 4 A.M. EDT (8h UT)
	7	Last quarter Moon, 10:09 P.M. EDT (2:09 UT on May 8)
	8	Jupiter (−2.5) is at opposition and nearest to Earth, 409.019 million miles away. It rises in the east at sunset and is visible all night long. Opposition occurs at 8:39 P.M. EDT (0:39 UT on May 9).
	13	Moon passes 2.4° south of Mercury, 1 P.M. EDT (17h UT)
	15	New Moon, 7:48 A.M. EDT (11:48 UT)
	16	Moon passes 1.2° north of Aldebaran, 9 A.M. EDT (13h UT)
	17	Moon passes 4.8° south of Venus, 2 P.M. EDT (18h UT)
		Look west-northwest in the half hour after sunset to locate Venus (−3.9) near the young Moon, then turn to the southeast to spot Jupiter (−2.5).
	19	Moon passes 8.0° south of Pollux, 11 A.M. EDT (15h UT)
	21	Moon passes 1.5° north of Regulus, 9 P.M. EDT (1h UT on May 22)
		First quarter Moon, 11:49 P.M. EDT (3:49 UT on May 22)
	25	Moon passes 7.4° north of Spica, 6 P.M. EDT (22h UT)
	27	Moon passes 4.0° north of Jupiter, 2 P.M. EDT (18h UT)
		Jupiter (−2.5) accompanies the Moon tonight. Look southeast as the sky darkens. Then turn to the west-northwest to find Venus (−3.9).
	29	Full Moon, 10:20 A.M. EDT (14:20 UT)
		Moon passes 9.0° north of Antares, 11 A.M. EDT (15h UT)
	31	Moon passes 1.6° north of Saturn, 9 P.M. EDT (1h UT on Jun. 1)
		The brightest "star" near the Moon late tonight and tomorrow is Saturn (0.2).
June	3	Mars (−1.2) shines near the Moon this morning. Look south in the hour before dawn. Saturn (0.2) hovers in the southwest.
		Moon passes 3.2° north of Mars, 8 A.M. EDT (12h UT)
	5	Mercury in superior conjunction (not visible)
	6	Last quarter Moon, 2:32 P.M. EDT (18:32 UT)
	8	Venus passes 4.8° south of Pollux, 9 P.M. EDT (1h UT on Jun. 9)
	12	Moon passes 1.2° north of Aldebaran, 8 P.M. EDT (0h UT on Jun. 13)
	13	New moon, 3:43 P.M. EDT (19:43 UT)
	14	Moon passes 4.6° south of Mercury, 9 A.M. EDT (13h UT)
		Can you spot Mercury (−1.3) to the right of the Moon tonight? Look west-northwest in the half hour after sunset. Binoculars help. Venus (−4.0) shines high above the pair. Jupiter (−2.4) glows high in the south-southeast.
	15	Moon passes 7.9° south of Pollux, 7 P.M. EDT (23h UT)
		Tonight, the young Moon forms a triangle with Venus (−4.0) and the bright star Pollux in Gemini. Look west-northwest in the hour after sunset.

(continued)

2018 Almanac (continued)

	16	Moon passes 2.3° south of Venus, 9 A.M. EDT (13h UT)
	18	Moon passes 1.7° north of Regulus, 4 A.M. EDT (8h UT)
	20	First quarter Moon, 6:51 A.M. EDT (10:51 UT)
	21	Summer solstice, 6:07 A.M. EDT (10:07 UT)
		Moon passes 7.6° north of Spica, 11 P.M. EDT (3h UT on Jun. 22)
	23	Moon passes 4.2° north of Jupiter, 3 P.M. EDT (19h UT)
		Look south in the hour after sunset to locate Jupiter (−2.3) near the Moon. Venus (−4.0) shines nearby in the west.
	25	Mercury passes 4.9° south of Pollux, noon EDT (16h UT)
		Tonight, look for Mercury (−0.4) low in the west-northwest in the half hour after sunset. Venus (−4.0) stands 19° above and to the left of Mercury. See p. 81.
		Moon passes 9.1° north of Antares, 5 P.M. EDT (21h UT)
	27	Saturn (0.0) is at opposition and nearest to Earth, 841.140 million miles away. It rises in the east around sunset and is visible all night long. Opposition occurs at 9:28 A.M. EDT (13:28 UT).
		In the hour after sunset tonight, look for Saturn rising with the nearly full Moon low in the southeast. Then take in the rest of the scene: Jupiter (−2.3) is the brightest "star" in the south; Venus (−4.1) dazzles in the west; Mercury (−0.3) appears near the star Pollux very low in the west-northwest (see p. 81).
		Moon passes 1.8° north of Saturn, midnight EDT (4h UT on Jun. 28)
		June Boötid meteor shower peaks, noon EDT (16h UT)
	28	Full Moon, 12:53 A.M. EDT (4:53 UT)
	30	Moon passes 4.8° north of Mars, 10 P.M. EDT (2h UT on Jul. 1)
		A little over two hours after sunset tonight, look for bright Mars (−2.2) rising with the nearly full Moon low in the east-southeast. At this time, Saturn (0.0) is visible in the southeast, Jupiter (−2.3) glows brightly in the south-southwest, and brilliant Venus (−4.1) is setting in the west-northwest.
July	6	Earth at aphelion, its farthest from the Sun this year – 94.5078 million miles – at 12:47 P.M. EDT (16:47 UT)
		Last quarter Moon, 3:51 A.M. EDT (7:51 UT)
	9	Venus passes 1.1° north of Regulus, 4 P.M. EDT (20h UT)
	10	Moon passes 1.1° north of Aldebaran, 6 A.M. EDT (10h UT)
	12	Mercury (0.4) at greatest eastern elongation (26.4°), 1 A.M. EDT (5h UT). Look for Mercury low in the west-northwest in the half hour after sunset. Venus (−4.2) shines about 16° above left of Mercury.
		New Moon, 10:48 P.M. EDT (2:48 UT on Jul. 13)
	13	Moon passes 7.8° south of Pollux, 5 A.M. EDT (9h UT)
		Partial solar eclipse, greatest at 3:01 UT, when up to 34% of the Sun is obscured as seen from the Antarctic coast. Tasmania, the southernmost portions of New South Wales and South Australia, and the southern coast of New Zealand see up to a 10% eclipse.
	14	Moon passes 2.2° north of Mercury, 6 P.M. EDT (22h UT)
		The "evening star" beneath the young Moon tonight is Mercury (0.5). Look west in the half hour after sunset.

(continued)

2018 Almanac (continued)

15	Moon passes 1.8° north of Regulus, 1 P.M. EDT (17h UT)
	Moon passes 1.6° north of Venus, midnight EDT (4h UT on Jul. 16)
	Look west for the crescent Moon half an hour after sunset tonight, and you'll find dazzling Venus (−4.2) next to it.
19	Moon passes 7.7° north of Spica, 5 A.M. EDT (9h UT)
	First quarter Moon, 3:52 P.M. EDT (19:52 UT)
20	Moon passes 4.5° north of Jupiter, 8 P.M. EDT (0h UT on Jul. 21)
	Jupiter (−2.2) is beneath the Moon tonight.
22	Moon passes 9.1° north of Antares, 11 P.M. EDT (3h UT on Jul. 23)
25	Moon passes 2.0° north of Saturn, 2 A.M. EDT (6h UT)
	Saturn (0.2) is beneath the Moon tonight.
27	Mars (−2.8) is at opposition tonight, rising in the east near sunset and visible all night long. Opposition occurs at 1:13 A.M. EDT (5:13 UT). See p. 181.
	Full Moon, 4:20 P.M. EDT (20:20 UT)
	Moon passes 6.7° north of Mars, 6 P.M. EDT (22h UT)
	Total lunar eclipse, greatest at 20:22 UT, is visible throughout the Indian Ocean, Africa, Australia, central Asia and Europe. See p. 144.
30	Southern Delta Aquariid meteor shower peaks, 7 A.M. EDT (11h UT)
31	Mars is closest to Earth, 35.785 million miles, 4 A.M. EDT (8h UT)
August 4	Last quarter Moon, 2:18 P.M. EDT (18:18 UT)
6	Moon passes 1.1° north of Aldebaran, 3 P.M. EDT (19h UT)
8	Mercury in inferior conjunction (not visible)
9	Moon passes 7.8° south of Pollux, 4 P.M. EDT (20h UT)
11	New Moon, 5:58 A.M. EDT (9:58 UT)
	Partial solar eclipse, greatest at 9:46 UT, when up to 74% of the Sun is obscured as seen from northeastern Siberia. Iceland sees a 20% eclipse, Scandanavia up to 30%.
12	Perseid meteor shower peaks, 9 P.M. EDT (1h UT on Aug. 13)
14	Moon passes 6.3° north of Venus, 10 A.M. EDT (14h UT)
	Venus (−4.2) is beneath the Moon tonight. Look west-southwest in the half hour after sunset.
15	Moon passes 7.7° north of Spica, 1 P.M. EDT (17h UT)
17	Moon passes 4.5° north of Jupiter, 7 A.M. EDT (11h UT)
	Venus (−4.5) at greatest eastern elongation (45.9°), 2 P.M. EDT (18h UT). Venus is brilliant low in the west-southwest in the half hour after sunset; see p. 68.
18	First quarter Moon, 3:48 A.M. EDT (7:48 UT)
19	Moon passes 9.1° north of Antares, 6 A.M. EDT (10h UT)
21	Moon passes 2.1° north of Saturn, 6 A.M. EDT (10h UT)
	The bright "star" near the Moon tonight is Saturn (0.3).
22	The brilliant ruddy "star" near the Moon tonight and tomorrow night is Mars (−2.3).
23	Moon passes 6.8° north of Mars, 1 P.M. EDT (17h UT)
26	Full Moon, 7:56 A.M. EDT (11:56 UT)
	Mercury (−0.2) at greatest western elongation (18.3°), 5 P.M. EDT (21h UT). Mercury is visible low in the east in the half hour before dawn.

(continued)

2018 Almanac (continued)

September	2	Venus passes 1.4° south of Spica, 5 A.M. EDT (9h UT)
		Moon passes 1.2° north of Aldebaran, 10 P.M. EDT (2h UT on Sept. 3)
		Last quarter Moon, 10:37 P.M. EDT (2:37 UT on Sept. 3)
	5	Mercury passes 1.1° north of Regulus, 7 P.M. EDT (23h UT)
	6	Moon passes 7.8° south of Pollux, 1 A.M. EDT (5h UT)
		Mercury (−1.1) is closest to Regulus this morning, very low in the east in the half hour before sunrise.
	8	Moon passes 1.8° north of Regulus, 10 A.M. EDT (14h UT)
		Moon passes 0.9° north of Mercury, 6 P.M. EDT (22h UT)
	9	New Moon, 2:01 P.M. EDT (18:01 UT)
	11	Moon passes 7.6° north of Spica, 11 P.M. EDT (3h UT on Sept. 12)
	12	Moon passes 10.5° north of Venus, noon EDT (16h UT)
	13	Moon passes 4.4° north of Jupiter, 10 P.M. EDT (2h UT on Sept. 14)
		Jupiter (−1.9) is beneath the Moon tonight. Look southwest as half an hour after sunset. Take in the rest of the planetary lineup. Venus (−4.7), the brightest star-like object in the sky, shines low in the west-southwest, beneath Jupiter and the Moon. Saturn (0.5) is due south at this time, the brightest "star" in the region. Mars (−1.8) glows brightly in the south-southeast.
	15	Moon passes 9.0° north of Antares, 1 P.M. EDT (17h UT)
	16	First quarter Moon, 7:15 P.M. EDT (23:15 UT)
	17	Moon passes 2.1° north of Saturn, 1 P.M. EDT (17h UT)
		Saturn (0.5) stands to the right of the Moon tonight, in the south as darkness falls.
	20	Moon passes 4.8° north of Mars, 3 A.M. EDT (7h UT)
		Mercury in superior conjunction (not visible)
		As twilight fades, find the Moon. The bright orange "star" beneath it is Mars (−1.6).
	22	Autumnal equinox, 9:54 P.M. EDT (1:54 UT on Sept. 23)
	24	Full Moon, 10:52 P.M. EDT (2:52 UT on Sept. 25)
	30	Moon passes 1.4° north of Aldebaran, 4 A.M. EDT (8h UT)
October	2	Last quarter Moon, 5:45 A.M. EDT (9:45 UT)
	3	Moon passes 7.6° south of Pollux, 8 A.M. EDT (12h UT)
	5	Moon passes 1.9° north of Regulus, 6 P.M. EDT (22h UT)
	8	New Moon, 11:47 P.M. EDT (3:47 UT on Oct. 9)
		Draconid meteor shower activity is expected from 7 to 8 P.M. EDT (23h to 0h UT on Oct. 9)
	11	Moon passes 4.1° north of Jupiter, 5 P.M. EDT (21h UT)
		Tonight, look southwest and find the young crescent Moon. Jupiter (−1.8) is beneath it.
	12	Moon passes 8.7° north of Antares, 10 P.M. EDT (2h UT on Oct. 13)
	14	Moon passes 1.8° north of Saturn, 11 P.M. EDT (3h UT on Oct. 15)
		Saturn (0.6) shines near the Moon tonight. Look south-southwest as darkness falls.
	16	First quarter Moon, 2:02 P.M. EDT (18:02 UT)
	18	Moon passes 2.0° north of Mars, 9 A.M. EDT (13h UT)
		Mars (−0.9) shines beside the Moon tonight. Look south-southeast as darkness falls.

(continued)

2018 Almanac (continued)

	21	Orionid meteor shower peaks, 1 P.M. EDT (17h UT)
	23	Mercury (−0.2) appears low in the west-southwest. Look below Jupiter (−1.7) 20 minutes after sunset.
	24	Full Moon, 12:45 P.M. EDT (16:45 UT)
	26	Venus in inferior conjunction (not visible)
	27	Moon passes 1.6° north of Aldebaran, 10 A.M. EDT (14h UT)
	29	Mercury passes 3.3° south of Jupiter, midnight EDT (4h UT on Oct. 30)
	30	Moon passes 7.3° south of Pollux, 2 P.M. EDT (18h UT)
	31	Last quarter Moon, 12:40 P.M. EDT (16:40 UT)
November	2	Moon passes 2.1° north of Regulus, 1 A.M. EDT (5h UT)
	4	Fall back: U.S. Daylight Saving Time ends this morning.
	5	Moon passes 7.6° north of Spica, 5 P.M. EST (22h UT)
		Moon passes 9.5° north of Venus, 9 P.M. EST (2h UT on Nov. 6)

November 6 — Venus (−4.3) has switched to the morning sky. Look it for it beside the waning crescent Moon this morning, low in the east-southeast in the half hour before sunrise.

Mercury (−0.3) at greatest eastern elongation (23.3°), 11 A.M. EST (16h UT). Look for Mercury and Jupiter (−1.7) very low in the west-southwest 20 min. after sunset. Binoculars help.

	7	New Moon, 11:02 A.M. EST (16:02 UT)
	8	Moon passes 3.8° north of Jupiter, 1 P.M. EST (18h UT)

Look for Jupiter (−1.7) and Mercury (−0.2) near the crescent Moon tonight, low in the west-southwest 20 min. after sunset.

	9	Mercury passes 1.8° north of Antares, 1 A.M. EST (6h UT)
		Moon passes 8.6° north of Antares, 6 A.M. EST (11h UT)
		Moon passes 6.7° north of Mercury, 7 A.M. EST (12h UT)
	11	Moon passes 1.5° north of Saturn, 11 A.M. EST (16h UT)

Saturn (0.6) shines beneath the crescent Moon tonight. Look southwest as the sky darkens.

	15	First quarter Moon, 9:54 A.M. EST (14:54 UT)
		Moon passes 1.0° south of Mars, 11 P.M. EST (4h UT on Nov. 16)

Mars (−0.3) glows above the Moon tonight. Look south-southeast as twilight fades.

	17	Leonid meteor shower peaks, 6 P.M. EST (23h UT)
	21	Alpha Monocerotid meteor shower peaks, 6 P.M. EST (23h UT)
	23	Full Moon, 12:39 A.M. EST (5:39 UT)
		Moon passes 1.7° north of Aldebaran, 5 P.M. EST (22h UT)
	26	Jupiter in conjunction with the Sun (not visible)
		Moon passes 7.1° south of Pollux, 7 P.M. EST (0h UT on Nov. 27)
	27	Mercury in inferior conjunction (not visible)
	29	Moon passes 2.4° north of Regulus, 5 A.M. EST (10h UT)
		Last quarter Moon, 7:19 P.M. EST (0:19 UT on Nov. 30)
December	2	Moon passes 7.7° north of Spica, 11 P.M. EST (4h UT on Dec. 3)

December 3 — This morning, in the hour before dawn, look for the waning crescent Moon in the southeast. Beneath it shines dazzling Venus (−4.9). To the right of this pair is Spica, the brightest star in the constellation Virgo. See p. 81.

Moon passes 3.6° north of Venus, 2 P.M. EST (19h UT)

(continued)

2018 Almanac (continued)

5	Mercury (0.6) is beneath the waning crescent Moon this morning. Look southeast in the half hour before sunrise. Venus (−4.9) gleams brilliantly higher up. See p. 81.
	Moon passes 1.9° north of Mercury, 4 P.M. EST (21h UT)
6	Jupiter (−1.7) has transitioned to the predawn sky. Can you see it this morning? Look beneath the Moon, very low in the east-southeast 20 min before dawn. Binoculars help.
	Moon passes 3.5° north of Jupiter, 8 A.M. EST (13h UT)
7	New Moon, 2:20 A.M. EST (7:20 UT)
8	Moon passes 1.1° north of Saturn, midnight EST (5h UT on Dec. 9)
13	Look for Geminid meteors tonight.
14	Moon passes 3.6° south of Mars, 6 P.M. EST (23h UT)
	Geminid meteor shower peaks, 7 A.M. EST (12h UT)
15	Mercury (−0.5) at greatest western elongation (21.3°), 6 A.M. EST (11h UT). Look low in the southeast in the hour before dawn; see p. 81. Jupiter (−1.7), about 6° below Mercury, become more evident in the half hour before sunrise. Venus (−4.8) remains a brilliant object higher up in the southeast.
	First quarter Moon, 6:49 A.M. EST (11:49 UT)
19	Jupiter passes 5.3° north of Antares, 9 P.M. EST (2h UT on Dec. 20)
	Watch Jupiter (−1.8) approach and pass Mercury (−0.5) over the coming days. Look low in the east-southeast in the half hour before sunrise.
21	Moon passes 1.7° north of Aldebaran, 3 A.M. EST (8h UT)
	Mercury passes 6.2° north of Antares, 3 A.M. EST (8h UT)
	Mercury passes 0.9° north of Jupiter, 10 A.M. EST (15h UT)
	Winter solstice, 5:23 P.M. EST (22:23 UT)
22	Full Moon, 12:49 P.M. EST (17:49 UT)
	Ursid meteor shower peaks, 4 P.M. EST (21h UT)
24	Moon passes 7.0° south of Pollux, 3 A.M. EST (8h UT)
26	Moon passes 2.5° north of Regulus, noon EST (17h UT)
29	Last quarter Moon, 4:34 3 A.M. EST (9:34 UT)
30	Moon passes 7.9° north of Spica, 5 A.M. EST (10h UT)

2019 Highlights

Mercury pops into evening twilight in late February and late June, and is well placed in the predawn sky in August and November.

Venus begins the year as a brilliant morning star, but another bright morning planet, Jupiter, closes rapidly. Separated by less than 2.5° on Jan. 22, the planetary duo shines near Antares in the hour before dawn. Saturn makes an even closer pass in February. Venus slips closer to the Sun's glare in March, but it won't emerge into evening twilight until late November. When it does, both Jupiter and Saturn will meet up with Venus again.

Mars remains a ruddy first-magnitude evening star into March. The Red Planet stands alone until Mercury briefly joins it in February, and later, in November, the two planets appear together in the predawn sky. However, there is no opposition for Mars this year.

Jupiter and **Saturn** begin the year separated by about 29°; by year's end, the gap close to nearly half this value. Their oppositions arrive just a month apart now: Jupiter's on Jun. 10, Saturn's on Jul. 9.

Meteor showers. The Perseid and Geminid showers are washouts, both occurring within a couple of days of full Moon, but the frosty Quadrantid shower, which peaks near new Moon this year, is ideal – at least as far as moonlight is concerned.

Eclipses. Of the year's five eclipses, one lunar (January) and one solar (July) are total. The lunar eclipse is perfectly positioned for the Americas, while solar totality comes to Chile and Argentina. December's annular solar eclipse brings a ring of Sun to the Arabian Peninsula, India, Indonesia and Guam.

2019 Almanac		
January	1	Moon passes 1.3° north of Venus, 5 P.M. EST (22h UT)
	2	Saturn in conjunction with the Sun (not visible)
		Moon passes 8.6° north of Antares, 9 P.M. EST (2h UT on Jan. 3)
	3	Earth at perihelion, its closest to the Sun this year – 91.4036 million miles – at 12:20 A.M. EST (5:20 UT)
		Quadrantid meteor shower peaks, 9 P.M. EST (2h UT on Jan. 4)
		Moon passes 3.1° north of Jupiter, 3 A.M. EST (8h UT)
		This morning, in the hour before dawn, look southeast to see the waning Moon next to Jupiter (−1.8). Venus (−4.6) shines brilliantly above them. In the half hour before dawn, look beneath the group for Mercury (−0.4).
	4	Moon passes 2.8° north of Mercury, 1 P.M. EST (18h UT)
	5	Venus (−4.6) at greatest western elongation (47.0°) at midnight EST (5h UT on Jan. 6). The planet rises more than 3 h before the Sun and is brilliant in the southeast during morning twilight. See p. 210.
		New Moon, 8:28 P.M. EST (1:28 UT on Jan. 6)
	6	**Partial solar eclipse**, greatest at 1:41 UT, when up to 71% of the Sun is obscured as seen from northeastern Siberia. Japan sees the Sun up to 60% obscured.
	12	Moon passes 5.3° south of Mars, 3 P.M. EST (20h UT)
		Mars (0.5), the sole evening planet, rides high in the south as darkness falls. Although relatively faint, the Red Planet stands out among the faint stars of Pisces. Look for Mars above the Moon tonight.
	14	First quarter Moon, 1:45 A.M. EST (6:45 UT)
	15	Venus passes 7.9° north of Antares, 4 P.M. EST (21h UT)
	17	Moon passes 1.6° north of Aldebaran, 2 P.M. EST (19h UT)

(continued)

2019 Almanac (continued)

	21	**Total lunar eclipse,** greatest at 12:12 A.M. EST (5:12 UT; totality is visible throughout the Americas, including Alaska and Hawaii, western Africa, and Europe. See p. 145.

Full Moon, 12:16 A.M. EST (5:16 UT)

22 Venus passes 2.4° north of Jupiter, 1 A.M. EST (6h UT); see p. 210.

Moon passes 2.5° north of Regulus, 9 P.M. EST (2h UT on Jan. 23)

26 Moon passes 7.9° north of Spica, 11 A.M. EST (16h UT)

27 Last quarter Moon, 4:10 P.M. EST (21:10 UT)

29 Mercury in superior conjunction (not visible)

30 Moon passes 8.6° north of Antares, 2 A.M. EST (7h UT)

Moon passes 2.8° north of Jupiter, 7 P.M. EST (0h UT on Jan. 31)

31 Venus (−4.3) and Jupiter (−1.9) straddle the waning crescent Moon this morning. Saturn (0.6) appears beneath them. Look southeast in the half hour before dawn.

Moon passes 0.1° north of Venus, 1 P.M. EST (18h UT)

February   2   Moon passes 0.6° north of Saturn, 2 A.M. EST (7h UT)

The brightest star near the Moon this morning is Saturn (0.6). Look low in the southeast in the half hour before dawn.

4 New Moon,4:04 P.M. EST (21:04 UT)

10 Moon passes 6.1° south of Mars, 11 A.M. EST (16h UT)

Tonight, look for the Moon and fast-fading Mars (0.9) high in the southwest in as darkness falls.

12 First quarter Moon, 5:26 P.M. EST (22:26 UT)

13 Moon passes 1.7° north of Aldebaran, 11 P.M. EST (4h UT on Feb 14)

17 Moon passes 7.0° south of Pollux, 2 A.M. EST (7h UT)

18 Venus passes 1.1° north of Saturn, 9 A.M. EST (14h UT)

Brilliant Venus (−4.2) guides you to Saturn (0.6) this morning. Look southeast in the hour before dawn. Jupiter (−2.0) shines higher farther south.

19 Moon passes 2.5° north of Regulus, 9 A.M. EST (14h UT)

Full Moon,10:54 A.M. EST (15:54 UT)

20 Can you catch Mercury (−1.0) low in the west in the half hour after sunset? See pages 73 and 82.

22 Moon passes 7.8° north of Spica, 8 P.M. EST (1h UT on Feb. 23)

26 Last quarter Moon, 6:28 A.M. EST (11:28 UT)

Moon passes 8.5° north of Antares, 9 A.M. EST (14h UT)

Mercury (−0.5) at greatest eastern elongation (18.1°) at 8 P.M. EST (1h UT on Feb. 27). Look low in the west in the half hour after sunset; see p. 82.

27 Let the Moon guide you through the morning planets over the next three days. This morning, the bright star closest to the Moon is Jupiter (−2.0). Look south-southeast in the hour before dawn.

Moon passes 2.3° north of Jupiter, 9 A.M. EST (14h UT)

March   1   The Moon visits Saturn (0.6) this morning. Look southeast in the hour before dawn. Venus shines below, and Jupiter looks on farther south.

Moon passes 0.3° north of Saturn, 1 P.M. EST (18h UT)

(continued)

2019 Almanac (continued)

	2	The Moon is nearest fading Venus (–4.1) this morning. Look southeast in the hour before dawn, then take in the scene. Three bright planets – Venus, Saturn and Jupiter, in order of altitude – arc out of the southeast. The Summer Triangle hovers in the brightening eastern sky, Scorpius stands on its tail in the south, and Leo is nosing into the northwest.
		Moon passes 1.2° south of Venus, 4 P.M. EST (21h UT)
	6	New Moon, 11:04 A.M. EST (16:04 UT)
	7	Moon passes 8.4° south of Mercury, 8 A.M. EST (13h UT)
	10	Spring forward: U.S. Daylight Saving Time begins this morning.
	11	Moon passes 5.8° south of Mars, 8 A.M. EDT (12h UT)
	13	Moon passes 1.9° north of Aldebaran, 7 A.M. EDT (11h UT)
	14	First quarter Moon, 6:27 A.M. EDT (10:27 UT)
		Mercury in inferior conjunction (not visible)
	16	Moon passes 6.8° south of Pollux, noon EDT (16h UT)
	18	Moon passes 2.6° north of Regulus, 8 P.M. EDT (0h UT on Mar. 19)
	20	Vernal equinox, 5:58 P.M. EDT (21:58 UT)
		Full Moon, 9:43 P.M. EDT (1:43 UT on Mar. 21)
	22	Moon passes 7.6° north of Spica, 8 A.M. EDT (12h UT)
	25	Moon passes 8.3° north of Antares, 6 P.M. EDT (22h UT)
	26	Moon passes 1.9° north of Jupiter, 10 P.M. EDT (2h UT on Mar. 27)
	28	Last quarter Moon, 12:10 A.M. EDT (4:10 UT)
	29	Moon passes 0.1° south of Saturn, 1 A.M. EDT (5h UT)
April	1	Moon passes 2.7° south of Venus, midnight EDT (4h UT on Apr. 2)
	2	This morning's crescent Moon, low in the east-southeast in the half hour before dawn, hovers beneath Venus (–3.9). Look for Mercury (0.7) to the left of this pair.
		Moon passes 3.6° south of Mercury, 7 P.M. EDT (23h UT)
	5	New Moon, 4:50 A.M. EDT (8:50 UT)
	8	This evening, as darkness falls, look for the crescent Moon in the western sky. The Pleiades star cluster lies to the right of the Moon. Above and between them shines dim Mars (1.5). To the left of Mars, look for the horns of Taurus, the celestial bull, now resembling the letter "V." The constellation's brightest star, Aldebaran, now outshines Mars.
	9	Moon passes 4.7° south of Mars, 3 A.M. EDT (7h UT)
		Moon passes 2.1° north of Aldebaran, noon EDT (16h UT)
	11	Mercury (0.3) at greatest western elongation (27.7°) at 4 P.M. EDT (20h UT). Look for it below Venus (–3.9) very low in the east in the half hour before sunup.
	12	First quarter Moon, 3:06 P.M. EDT (19:06 UT)
		Moon passes 6.6° south of Pollux, 7 P.M. EDT (23h UT)
	15	Moon passes 2.8° north of Regulus, 5 A.M. EDT (9h UT)
	16	Mars passes 6.5° north of Aldebaran, 6 P.M. EDT (22h UT)
	18	Moon passes 7.6° north of Spica, 6 P.M. EDT (22h UT)
	19	Full Moon, 7:12 A.M. EDT (11:12 UT)
	22	Moon passes 8.0° north of Antares, 4 A.M. EDT (8h UT)
		Lyrid meteor shower peaks, 8 P.M. EDT (0h UT on Apr. 23)

(continued)

2019 Almanac (continued)

	23	Moon passes 1.6° north of Jupiter, 8 A.M. EDT (12h UT)
		Jupiter (−2.4) is the brilliant "star" near the Moon this morning.
	25	Moon passes 0.4° south of Saturn, 10 A.M. EDT (14h UT)
		Saturn (0.5) accompanies the Moon this morning.
	26	Last quarter Moon, 6:18 P.M. EDT (22:18 UT)
May	2	Moon passes 3.6° south of Venus, 8 A.M. EDT (12h UT)
	3	Moon passes 2.9° south of Mercury, 2 A.M. EDT (6h UT)
	4	New Moon, 6:45 P.M. EDT (22:45 UT)
	6	Moon passes 2.3° north of Aldebaran, 6 P.M. EDT (22h UT)
		Eta Aquariid meteor shower peaks, 10 A.M. EDT (14h UT)
	7	Moon passes 3.2° south of Mars, 8 P.M. EDT (0h UT on May 8)
		You'll find both the young Moon and faint Mars (1.7) at the tips of Taurus the Bull's horns tonight.
	9	Moon passes 6.3° south of Pollux, midnight EDT (4h UT on May 10)
	11	First quarter Moon, 9:12 P.M. EDT (1:12 UT on May 12)
	12	Moon passes 3.0° north of Regulus, 11 A.M. EDT (15h UT)
	16	Moon passes 7.7° north of Spica, 3 A.M. EDT (7h UT)
	18	Full Moon, 5:11 P.M. EDT (21:11 UT)
	19	Moon passes 7.9° north of Antares, 1 P.M. EDT (17h UT)
		Look for the Moon, Jupiter (−2.6) and Antares low in the southeast about 2 h after sunset tonight.
	20	Moon passes 1.7° north of Jupiter, 1 P.M. EDT (17h UT)
	21	Mercury in superior conjunction (not visible)
	22	Moon passes 0.5° south of Saturn, 6 P.M. EDT (22h UT)
		Saturn (0.4) is the bright "star" near the Moon this morning.
	26	Last quarter Moon, 12:34 P.M. EDT (16:34 UT)
June	1	Look for Venus (−3.8) next to the waning crescent Moon this morning, low in the east 20 min before dawn.
		Moon passes 3.2° south of Venus, 2 P.M. EDT (18h UT)
	3	Moon passes 2.3° north of Aldebaran, 2 A.M. EDT (6h UT)
		New Moon, 6:02 A.M. EDT (10:02 UT)
	4	Moon passes 3.7° south of Mercury, noon EDT (16h UT)
		Catch Mercury (−0.8) to the right of the young Moon tonight, low in the west-northwest in the half hour after sunset. Dim Mars (1.8) shines above, midway between the brighter star Pollux and the Moon. See p. 82.
	5	Moon passes 1.6° south of Mars, 11 A.M. EDT (15h UT)
	6	Moon passes 6.2° south of Pollux, 6 A.M. EDT (10h UT)
	8	Moon passes 3.2° north of Regulus, 4 P.M. EDT (20h UT)
	10	First quarter Moon, 1:59 A.M. EDT (5:59 UT)
		Jupiter (−2.6) is at opposition and nearest to Earth, 398.244 million miles away. It rises in the east-southeast at sunset and is visible all night long. Opposition occurs at 11:28 A.M. EDT (15:28 UT).
	12	Moon passes 7.8° north of Spica, 9 A.M. EDT (13h UT)
	15	Moon passes 8.0° north of Antares, 9 P.M. EDT (1h UT on Jun. 16)
		Mercury (−0.1) and Mars (1.8) are less than 2° until the 21st. Look low in the west-northwest in the hour after sunset. Mercury is by far the brighter planet. Binoculars help.

(continued)

2019 Almanac (continued)

	16	Moon passes 2.0° north of Jupiter, 3 P.M. EDT (19h UT)
	17	Full Moon, 4:31 A.M. EDT (8:31 UT)
		Venus passes 4.8° north of Aldebaran, 5 P.M. EDT (21h UT)
	18	Mercury passes 0.2° north of Mars, 11 A.M. EDT (15h UT)
		Tonight, low in the west-northwest in the hour after sunset, look for Mercury (0.1) and faint Mars (1.8). Binoculars help. Turn to the southeast to see brilliant Jupiter (−2.6). About an hour later, the Moon and Saturn (0.2) rise into the southeast together.
		Moon passes 0.4° south of Saturn, midnight EDT (4h UT on Jun. 19)
	21	Mercury passes 5.7° south of Pollux, 1 A.M. EDT (5h UT)
		Look for Mercury (0.3), Pollux, and Mars (1.8) forming a triangle low in the west-northwest in the hour after sunset. Binoculars help.
		Summer solstice, 11:54 A.M. EDT (15:54 UT)
	23	Mars passes 5.6° south of Pollux, 3 A.M. EDT (7h UT)
		Mercury (0.4) at greatest eastern elongation (25.2°) at 7 P.M. EDT (23h UT). It's visible low in the west-northwest in the hour after sunset.
	25	Last quarter Moon, 5:46 A.M. EDT (9:46 UT)
	27	June Boötid meteor shower peaks, 6 P.M. EDT (22h UT)
	30	Moon passes 2.3° north of Aldebaran, noon EDT (16h UT)
July	1	Moon passes 1.6° south of Venus, 6 P.M. EDT (22h UT)
		Catch Venus (−3.9) near the Moon this morning, very low in the east-northeast in the half hour before sunrise.
	2	New moon, 3:16 P.M. EDT (19:16 UT)
		Total solar eclipse, greatest at 19:23 UT; totality courses over Oeno Island in the Pacific, Chile and Argentina. See p. 146.
	3	Moon passes 6.1° south of Pollux, 2 P.M. EDT (18h UT)
		Can you find Mercury (1.4) and Mars (1.8) above tonight's young Moon? Look low in the west-northwest in the half hour after sunset. Binoculars help. Turn to the southeast to see brilliant Jupiter (−2.6). Low in the east-southeast, look for Saturn (0.1) rising.
	4	Moon passes 0.1° north of Mars, 2 A.M. EDT (6h UT)
		Moon passes 3.3° north of Mercury, 5 A.M. EDT (9h UT)
		Earth at aphelion, its farthest point from the Sun this year – 94.5132 million miles away – at 6:11 P.M. EDT (22:11 UT). This is Earth's greatest distance from the sun in the period 2011–2020.
	5	Moon passes 3.2° north of Regulus, 11 P.M. EDT (3h UT Jul. 6)
	7	Mars passes 3.8° north of Mercury, 10 A.M. EDT (14h UT)
	9	First quarter Moon, 6:55 A.M. EDT (10:55 UT)
		Saturn (0.1) is at opposition and nearest to Earth, 839.650 million miles away. It rises in the east-southeast near sunset and is visible all night long. Opposition occurs at 1:07 P.M. EDT (17:07 UT).
		Moon passes 7.9° north of Spica, 2 P.M. EDT (18h UT)
	13	Moon passes 8.0° north of Antares, 3 A.M. EDT (7h UT)
		Moon passes 2.3° north of Jupiter, 4 P.M. EDT (20h UT)
		The brilliant "star" near the Moon this evening and tomorrow morning is Jupiter (−2.5).

(continued)

2019 Almanac (continued)

14	Look for the Moon about midway between Jupiter (−2.5) and Saturn (0.1) tonight. Turn southeast in the hour after sunset.
15	Saturn (0.1) is the bright "star" near the Moon tonight and tomorrow morning.
16	Moon passes 0.2° south of Saturn, 3 A.M. EDT (7h UT)
	Full Moon, 5:38 P.M. EDT (21:38 UT)
	Partial lunar eclipse, greatest at 21:32 UT, visible throughout the Indian Ocean, Africa, southern and central Europe, and Australia. See p. 147.
21	Mercury in inferior conjunction (not visible)
24	Last quarter Moon, 9:18 P.M. EDT (1:18 UT on Jul. 25)
27	Moon passes 2.3° north of Aldebaran, 9 P.M. EDT (1h UT on Jul. 28)
30	Moon passes 4.5° north of Mercury, 10 P.M. EDT (2h UT on Jul. 31)
	Moon passes 6.1° south of Pollux, midnight EDT (4h UT)
	Southern Delta Aquariid meteor shower peaks, 1 A.M. EDT (17h UT)
31	Moon passes 0.6° north of Venus, 5 P.M. EDT (21h UT)
	New Moon, 11:12 P.M. EDT (3:12 UT on Aug. 1)

August	2	Moon passes 3.2° north of Regulus, 8 A.M. EDT (12h UT)
	5	Mercury passes 9.3° south of Pollux, 6 P.M. EDT (22h UT)
		Moon passes 7.8° north of Spica, 9 P.M. EDT (1h UT on Aug. 6)
	7	First quarter Moon, 1:31 P.M. EDT (17:31 UT)
	9	Moon passes 7.9° north of Antares, 9 A.M. EDT (13h UT)
		Moon passes 2.5° north of Jupiter, 7 P.M. EDT (23h UT)
		Mercury (0.0) at greatest western elongation (19.0°) at 7 P.M. EDT (23h UT). The planet is now visible in morning twilight, low in the east-northeast in the hour before dawn. See p. 82.
		Jupiter (−2.4) shines beneath the Moon this evening.
	12	Moon passes 0.0° south of Saturn, 6 A.M. EDT (10h UT)
		Saturn (0.2) lies beneath the Moon this evening.
	13	Perseid meteor shower peaks, 3 A.M. EDT (7h UT)
	14	Venus in superior conjunction (not visible)
	15	Full Moon, 8:29 A.M. EDT (12:29 UT)
	23	Last quarter Moon, 10:56 A.M. EDT (14:56 UT)
	24	Moon passes 2.4° north of Aldebaran, 6 A.M. EDT (10h UT)
	27	Moon passes 6.1° south of Pollux, 11 A.M. EDT (15h UT)
	29	Moon passes 3.2° north of Regulus, 6 P.M. EDT (22h UT)
	30	New Moon, 6:37 A.M. EDT (10:37 UT)
September	2	Moon passes 7.7° north of Spica, 5 A.M. EDT (9h UT)
		Mars in conjunction with the Sun (not visible)
	3	Mercury in superior conjunction (not visible)
	5	Moon passes 7.8° north of Antares, 3 P.M. EDT (19h UT)
		First quarter Moon, 11:10 P.M. EDT (3:10 UT on Sept. 6)
	6	Moon passes 2.3° north of Jupiter, 3 A.M. EDT (7h UT)
	8	Moon passes 0.0° south of Saturn, 10 A.M. EDT (14h UT)
	14	Full Moon, 12:33 A.M. EDT (4:33 UT)
	20	Moon passes 2.7° north of Aldebaran, 1 P.M. EDT (17h UT)

(continued)

2019 Almanac (continued)

	21	Last quarter Moon, 10:41 P.M. EDT (2:41 UT on Sept. 22)
	23	Autumnal equinox, 3:50 A.M. EDT (7:50 UT)
		Moon passes 5.9° south of Pollux, 8 P.M. EDT (0h UT on Sept. 24)
	26	Moon passes 3.3° north of Regulus, 5 A.M. EDT (9h UT)
	27	Moon passes 4.1° north of Mars, 9 P.M. EDT (1h UT on Sept. 28)
	28	New Moon, 2:26 P.M. EDT (18:26 UT)
	29	Moon passes 4.4° north of Venus, 9 A.M. EDT (13h UT)
		Moon passes 7.6° north of Spica, 4 P.M. EDT (20h UT)
		Moon passes 6.2° north of Mercury, 6 P.M. EDT (22h UT)
		Tonight, 20 minutes after sunset, look for the young Moon very low in the west. Venus (−3.9) and Mercury (−0.2) shine beneath it. Binoculars help.
October	2	Moon passes 7.5° north of Antares, 11 P.M. EDT (3h UT on Oct. 3)
	3	Moon passes 1.9° north of Jupiter, 4 P.M. EDT (20h UT)
		As darkness falls tonight, look southwest for the Moon. The bright "star" beneath it is Jupiter (−2.0). Look higher in the south for Saturn (0.5).
	5	First quarter Moon, 12:47 P.M. EDT (16:47 UT)
		Moon passes 0.3° south of Saturn, 5 P.M. EDT (21h UT)
		Saturn (0.5) stands above tonight's Moon. Look south as twilight fades.
	9	Draconid meteor shower activity is expected between 14h and 15h UT; the Americas will be in daylight or bright twilight
	13	Full Moon, 5:08 P.M. EDT (21:08 UT)
	15	Can you spot Venus (−3.8) and Mercury (−0.1) low in the west-southwest 20 min after sunset? Give it a try with binoculars.
	17	Moon passes 2.9° north of Aldebaran, 6 P.M. EDT (22h UT)
	19	Mercury (−0.1) at greatest eastern elongation (24.6°), midnight EDT (4h UT on Oct. 20). Mercury is visible with Venus very low in the west-southwest 20 min after sunset.
	21	Moon passes 5.6° south of Pollux, 3 A.M. EDT (7h UT)
		Last quarter Moon, 8:39 A.M. EDT (12:39 UT)
		Orionid meteor shower peaks, 8 P.M. EDT (0h UT on Oct. 22)
	23	Moon passes 3.5° north of Regulus, 2 P.M. EDT (18h UT)
	26	Moon passes 4.5° north of Mars, 1 P.M. EDT (17h UT)
		This morning, in the hour before dawn, look for faint Mars (1.8) beneath a sliver of Moon.
	27	Moon passes 7.6° north of Spica, 2 A.M. EDT (6h UT)
		New Moon, 11:38 P.M. EDT (3:38 UT on Oct. 28)
	29	Moon passes 3.9° north of Venus, 10 A.M. EDT (14h UT)
		Moon passes 6.7° north of Mercury, 11 A.M. EDT (15h UT)
		Tonight, 20 min. after sunset, look for the young Moon low in the west-southwest. Beneath it shine Venus (−3.8) and Mercury (0.2). As twilight deepens, look southwest for the star Antares, the heart of Scorpius. Jupiter (−1.9) shines brightly above left of the star. Saturn (0.6) gleams higher still in the south-southwest.
	30	Venus passes 2.7° north of Mercury, 4 A.M. EDT (8h UT)
		Moon passes 7.3° north of Antares, 9 A.M. EDT (13h UT)
	31	Moon passes 1.3° north of Jupiter, 10 A.M. EDT (14h UT)

(continued)

2019 Almanac (continued)

November	1	Saturn (0.6) is nearest the Moon tonight.
	2	Moon passes 0.6° south of Saturn, 3 A.M. EDT (7h UT)
	3	Fall back: U.S. Daylight Saving Time ends this morning.
	4	First quarter Moon, 5:23 A.M. EST (10:23 UT)
	8	Mars passes 3.1° north of Spica, 10 A.M. EST (15h UT)
	9	Venus passes 4.0° north of Antares, 6 A.M. EST (11h UT)
	11	Mercury in inferior conjunction. Normally this means that Mercury is not visible, but starting at 7:35 A.M. EST (12:35 UT) the planet can be seen transiting the Sun. Mid-transit occurs at 10:20 A.M. EST (15:20 UT) and Mercury exits the Sun at 1:04 P.M. EST (18:04 UT). The Middle East, Europe, Africa, most of North America, all of Central and South America, and Hawaii will see some portion of the transit. *Seeing this event requires a telescope and suitable eye protection.*
	12	Full Moon, 8:34 A.M. EST (13:34 UT)
	13	Moon passes 3.0° north of Aldebaran, 11 P.M. EST (4h UT on Nov. 14)
	17	Moon passes 5.4° south of Pollux, 7 A.M. EST (12h UT)
		Leonid meteor shower peaks, midnight EST (5h UT on Nov. 18)
	19	Last quarter Moon, 4:11 P.M. EST (21:11 UT)
		Moon passes 3.7° north of Regulus, 7 P.M. EST (0h UT on Nov. 20)
	21	Venus (−3.9) and Jupiter (−1.9) are within 3° of each other until the 26th. Look for them in the southwest in the hour after sunset. Saturn (0.6) shines high above the pair.
	22	Alpha Monocerotid meteor shower peaks, midnight EST (5h UT)
	23	Moon passes 7.7° north of Spica, 11 A.M. EST (16h UT)
	24	The waning crescent Moon visits Mars (1.7) this morning. Mercury (−0.3) gleams below them. Look east-southeast in the hour before dawn; see p. 82.
		Moon passes 4.3° north of Mars, 4 A.M. EST (9h UT)
		Venus passes 1.4° south of Jupiter, 9 A.M. EST (14h UT)
		Venus (−3.9) and Jupiter (−1.8) appear closest tonight. Look southwest in the hour after sunset. Saturn (0.6) shines far above them.
		Moon passes 1.9° north of Mercury, 10 P.M. EST (3h UT Nov. 25)
	26	New Moon, 10:06 A.M. EST (15:06 UT)
		Moon passes 7.2° north of Antares, 6 P.M. EST (23h UT)
	27	Half an hour after the Sun sets tonight, look for the young Moon low in the southwest. It's the anchor for a chain of bright "evening stars" arcing south and up. In order from the Moon, these are the planets Jupiter (−1.8), Venus (−3.9) and Saturn (0.6).
	28	Mercury (−0.6) at greatest western elongation (20.1°), 5 A.M. EST (10h UT). Mercury is visible beneath faint Mars (1.7), low in the east-southeast in the hour before sunrise; see pages 73 and 82.
		Moon passes 0.7° north of Jupiter, 6 A.M. EST (11h UT)
		Moon passes 1.9° north of Venus, 2 P.M. EST (19h UT)
		The Moon stands above bright Venus tonight. Below and to the Moon's right is Jupiter. Above and to the left shines Saturn. Look southwest in the hour after sunset.
	29	Moon passes 0.9° south of Saturn, 4 P.M. EST (21h UT)
		Tonight, as the sky darkens, look southwest toward the Moon. The bright star above our satellite is Saturn (0.6).

(continued)

2019 Almanac (continued)

December	4	First quarter Moon, 1:58 A.M. EST (6:58 UT)
	9	Venus (−3.9) is brilliant in the southwest in the half hour after sunset. Above it shines Saturn (0.6), which becomes more apparent as the sky darkens. Can you still spot Jupiter, below and to their right?
	10	Venus and Saturn are closest tonight. Look west in the half hour after sunset. You'll spot Venus first; look for Saturn above it as twilight fades.
		Venus passes 1.8° south of Saturn, midnight EST (5h UT on Dec. 11)
	11	Moon passes 3.0° north of Aldebaran, 7 A.M. EST (12h UT)
	12	Full Moon, 12:12 A.M. EST (5:12 UT)
	14	Moon passes 5.3° south of Pollux, 1 P.M. EST (18h UT)
		Geminid meteor shower peaks, 2 P.M. EST (19h UT)
	15	Mercury passes 5.1° north of Antares, 11 A.M. EST (16h UT)
	16	Moon passes 3.8° north of Regulus, midnight EST (5h UT on Dec. 17)
	18	Last quarter Moon, 11:57 P.M. EST (4:57 UT on Dec. 19)
	20	Moon passes 7.8° north of Spica, 5 P.M. EST (22h UT)
	21	Winter solstice, 11:19 P.M. EST (4:19 UT on Dec. 22)
	22	Moon passes 3.5° north of Mars, 9 P.M. EST (2h UT on Dec. 23)
		This morning, half an hour before sunrise, look southeast for the waning crescent Moon. Faint Mars (1.6) shines nearby. Mercury (−0.7) is there too, very low in the southeast 20 min before sunrise. Binoculars help.
		Ursid meteor shower peaks, 10 P.M. EST (3h UT on Dec. 23)
	24	Moon passes 7.3° north of Antares, 3 A.M. EST (8h UT)
	25	Moon passes 1.9° north of Mercury, 6 A.M. EST (11h UT)
	26	New Moon, 12:13 A.M. EST (5:13 UT)
		Annular solar eclipse, greatest at 5:18 UT bringing annularity to Saudi Arabia, Oman, India, Sri Lanka, Sumatra, Singapore, Borneo and Guam. See p. 148.
	27	Moon passes 1.2° south of Saturn, 7 A.M. EST (12h UT)
		Jupiter in conjunction with the Sun (not visible)
		Tonight, look for Saturn below right of the young Moon. Brilliant Venus shines above left of the Moon.
	28	Tonight, the hour after sunset, look southwest for the crescent Moon. The dazzling "star" above it is the planet Venus (−3.9).
		Moon passes 1.0° south of Venus, 9 P.M. EST (2h UT on Dec. 29)

2020 Highlights

Mercury seems to be chasing Venus all year, shining beneath the brilliant planet in the evening twilight of February and again in late May, when the two worlds pass within a degree of each other. In July and November, the setup is similar, with Mercury beneath the brighter planet, but this time in the predawn sky.

Venus opens the year as a dazzling evening star and climbs to its best altitude in early April, when it sets 4 h after the Sun. Following its rendezvous with Mercury in late May, Venus plunges toward the solar glare and emerges as a morning star in late June, slowly climbing to its best altitude in early September.

Mars has a predawn encounter with Jupiter and Saturn in late March just when it's bright enough to stand on its own between them. Mars comes to opposition on Oct. 13. That week, early risers will find Mars shining brightly in the west while dazzling Venus shines in the east. Mars remains notable evening planet into March 2021.

Jupiter closes on **Saturn** throughout the year. Their oppositions come less than a week apart, with Jupiter at its brightest on Jul. 14 and Saturn reaching its best on Jul. 20. On Dec. 21, the two giant worlds finally make their closest approach to each other, shining together low in the southwest in the hour after sunset. A view through a small telescope will reveal Jupiter's four Galilean moons, with Saturn serving as a cosmic backdrop.

Meteor showers. The Quadrantid shower peaks with a waxing gibbous Moon, the Perseids with a last quarter Moon. But the Geminid shower is just right, reaching its best activity on the day before new Moon.

Eclipses. The Moon offers up a passel of penumbral eclipses – four in all – but no partial or total ones. The two solar eclipses bring annularity to Africa, the Arabian Peninsula, Pakistan, India, China and Taiwan (June) and totality to Chile and Argentina (December).

2020 Almanac		
January	1	Look for the waxing crescent Moon due south and Venus (−4.0) low in the southwest as the sky darkens tonight.
	2	First quarter Moon, 11:45 P.M. EST (4:45 UT on Jan. 3)
	4	Quadrantid meteor shower peaks, 4 A.M. EST (9h UT)
	5	Earth at perihelion, its closest point from the Sun this year – 91.3982 million miles away – at 2:48 A.M. EST (7:48 UT). In fact, this is Earth's closest to the Sun in the period 2011–2020.
	7	Moon passes 3.0° north of Aldebaran, 5 P.M. EST (22h UT)
	10	Mercury in superior conjunction (not visible)
		Full Moon, 2:21 P.M. EST (19:21 UT)
		Moon passes 5.3° south of Pollux, 10 P.M. EST (3h UT on Jan. 11)
		Penumbral lunar eclipse, greatest at 19:10 UT and visible throughout Asia, Africa, Europe and Australia.
	13	Moon passes 3.8° north of Regulus, 7 A.M. EST (12h UT)
		Saturn in conjunction with the Sun (not visible)
	16	Moon passes 7.7° north of Spica, 10 P.M. EST (3h UT on Jan. 17)
		Mars passes 4.8° north of Antares, 11 P.M. EST (4h UT on Jan. 17)

(continued)

2020 Almanac (continued)

	17	Look for Mars (1.5) in the southeast in the hour before dawn, hovering above its stellar rival, Antares, which is about 1.7 times brighter. The Moon hovers east of Spica this morning and arcs toward Mars over the next few days.
		Last quarter Moon, 7:58 A.M. EST (12:58 UT)
	20	The trio of the Moon, Mars and Antares hovers in the southeast in the hour before dawn.
		Moon passes 7.2° north of Antares, 10 A.M. EST (15h UT)
		Moon passes 2.3° north of Mars, 2 P.M. EST (19h UT)
	22	Moon passes 0.4° south of Jupiter, 10 P.M. EST (3h UT on Jan. 23)
	23	Look for Jupiter (−1.9) above the slender crescent Moon this morning, low in the southeast in the half hour before dawn.
	24	New Moon, 4:42 P.M. EST (21:42 UT)
	25	Moon passes 1.3° south of Mercury, 1 P.M. EST (18h UT)
		Can you spot Mercury (−1.1) to the right of the young crescent Moon, very low in the west-southwest 20 min after sunset tonight? Venus (−4.1) shines brilliantly high above the pair. Watch the Moon climb to and surpass it over the next few days.
	28	Moon passes 4.1° south of Venus, 2 A.M. EST (7h UT)
		Watch Mercury (−1.0) climb toward Venus over the next 12 days. Look low in the west-southwest in the half hour after sunset. See p. 83.
February	1	First quarter Moon, 8:42 P.M. EST (1:42 UT on Feb. 2)
	4	Moon passes 3.1° north of Aldebaran, 2 A.M. EST (7h UT)
	7	Moon passes 5.3° south of Pollux, 8 A.M. EST (13h UT)
	9	Full Moon, 2:33 A.M. EST (7:33 UT)
		Moon passes 3.8° north of Regulus, 5 P.M. EST (22h UT)
	10	Mercury (−0.6) at greatest eastern elongation (18.2°) at 9 A.M. EST (14h UT on Feb. 11). Look low in the west-southwest in the half hour after sunset. Venus (−4.2) shines high above it. See p. 83.
	13	Moon passes 7.6° north of Spica, 5 A.M. EST (10h UT)
		Mars (1.3), Jupiter (−1.9) and Saturn (0.6) – in order, highest to lowest – parade across the predawn sky, low in the southeast in the hour before dawn. The Moon begins its run past them on the 18th.
	15	Last quarter Moon, 5:17 P.M. EST (22:17 UT)
	16	Moon passes 7.1° north of Antares, 3 P.M. EST (20h UT)
	18	Moon passes 0.8° north of Mars, 8 A.M. EST (13h UT)
	19	Moon passes 0.9° south of Jupiter, 3 P.M. EST (20h UT)
	20	Moon passes 1.8° south of Saturn, 9 A.M. EST (14h UT)
	23	New Moon, 10:32 A.M. EST (15:32 UT)
	25	Mercury in inferior conjunction (not visible)
	27	Moon passes 6.3° south of Venus, 7 A.M. EST (12h UT)
March	2	Moon passes 3.3° north of Aldebaran, 11 A.M. EST (16h UT)
		First quarter Moon, 2:57 P.M. EST (19:57 UT)
	5	Moon passes 5.2° south of Pollux, 7 P.M. EST (0h UT on Mar. 6)
	8	Spring forward: U.S. Daylight Saving Time begins this morning.
		Moon passes 3.8° north of Regulus, 5 A.M. EDT (9h UT)
	9	Full Moon, 1:48 P.M. EDT (17:48 UT)
	11	Moon passes 7.4° north of Spica, 4 P.M. EDT (20h UT)

(continued)

2020 Almanac (continued)

14	Moon passes 6.8° north of Antares, 11 P.M. EDT (3h UT on Mar. 15)
16	Last quarter Moon, 5:34 A.M. EDT (9:34 UT)
18	In the hour before dawn, catch brightening Mars (0.9), brilliant Jupiter (−2.1) and Saturn (0.7) low in the southeast. The two brightest "stars" above the crescent Moon are Jupiter and Mars.
	Moon passes 0.7° south of Mars, 4 A.M. EDT (8h UT)
	Moon passes 1.5° south of Jupiter, 6 A.M. EDT (10h UT)
	Moon passes 2.1° south of Saturn, 8 P.M. EDT (0h UT on Mar. 19)
19	Vernal equinox, 11:50 P.M. EDT (3:50 UT on Mar. 20)
20	Jupiter passes 0.7° north of Mars, 2 A.M. EDT (6h UT). Look low in the southeast for the pair in the hour before dawn. Saturn hovers just east of the duo.
21	Moon passes 3.6° south of Mercury, 2 P.M. EDT (18h UT)
23	Mercury (0.1) at greatest western elongation (27.8°) at 10 P.M. EDT (2h UT on Mar. 24). The planet appears very low in the east-southeast 20 minutes before dawn.
24	In the hour before dawn, look southeast to find Jupiter (−2.1), Mars (0.9) and Saturn (0.7) in a 7°-long line.
	New Moon, 5:28 A.M. EDT (9:28 UT)
	Moon at apogee, its farthest point to Earth for the month – 252,706.551 miles – at 11:23 A.M. EDT (15:23 UT). In fact, this is the farthest the Moon gets from Earth in the period covered by this book.
	Venus (−4.5) at greatest eastern elongation (46.1°) at 6 P.M. EDT (22h UT). The planet is brilliant in evening twilight and is visible for more than 4 h after sunset.
28	Moon passes 6.8° south of Venus, 7 A.M. EDT (11h UT)
29	Moon passes 3.6° north of Aldebaran, 6 P.M. EDT (22h UT)
31	Saturn passes 0.9° north of Mars, 7 A.M. EDT (11h UT). Look for the pair in the southeast in the hour before dawn. Jupiter (−2.1), the brightest of the group looks on just to the west.
April 1	First quarter Moon, 6:21 A.M. EDT (10:21 UT)
2	Moon passes 4.9° south of Pollux, 4 A.M. EDT (8h UT)
3	Can you make out the Pleiades star cluster behind the glare of Venus (−4.6) tonight? Look west as the sky darkens for Venus, the brightest object in the sky after the Moon. Take a look through binoculars, too.
4	Moon passes 4.0° north of Regulus, 3 P.M. EDT (19h UT)
7	Full Moon, 10:35 P.M. EDT (2:35 UT on Apr. 8)
8	Moon passes 7.3° north of Spica, 3 A.M. EDT (7h UT)
11	Moon passes 6.6° north of Antares, 8 A.M. EDT (12h UT)
14	The morning sky-show continues as the Moon skirts a chain of planets – in order, by altitude, Jupiter (−2.2), Saturn and Mars (both 0.6). Look south-southeast in the hour before dawn this and the next two mornings.
	Last quarter Moon, 6:56 P.M. EDT (22:56 UT)
	Moon passes 2.0° south of Jupiter, 7 P.M. EDT (23h UT)
15	Moon passes 2.5° south of Saturn, 5 A.M. EDT (9h UT)
16	Moon passes 2.0° south of Mars, 1 A.M. EDT (5h UT)

(continued)

2020 Almanac (continued)

	21	Moon passes 3.1° south of Mercury, 1 P.M. EDT (17h UT)
	22	New Moon, 10:26 P.M. EDT (2:26 UT on Apr. 23)
		Lyrid meteor shower peaks, 3 A.M. EDT (7h UT)
	25	Moon passes 3.8° north of Aldebaran, midnight EDT (4h UT on Apr. 26)
	26	Moon passes 6.1° south of Venus, 11 A.M. EDT (15h UT)
	29	Moon passes 4.7° south of Pollux, 11 A.M. EDT (15h UT)
	30	First quarter Moon, 4:38 P.M. EDT (20:38 UT)
May	1	Moon passes 4.2° north of Regulus, 11 P.M. EDT (3h UT on May 2)
	4	Mercury in superior conjunction (not visible)
	5	Moon passes 7.4° north of Spica, 1 P.M. EDT (17h UT)
		Eta Aquariid shower peaks, 4 P.M. EDT (20h UT)
	7	Full Moon, 6:45 A.M. EDT (10:45 UT)
	8	Moon passes 6.5° north of Antares, 6 P.M. EDT (22h UT)
	12	In the hour before dawn, look south for the waning gibbous Moon. Above it, separated by less than 5°, blaze Jupiter (−2.4) and Saturn (0.5). Mars (0.3, outshining Saturn) stands east of the pair.
		Moon passes 2.3° south of Jupiter, 6 A.M. EDT (10h UT)
		Moon passes 2.7° south of Saturn, 2 P.M. EDT (18h UT)
	14	Last quarter Moon, 10:03 A.M. EDT (14:03 UT)
		Moon passes 2.8° south of Mars, 10 P.M. EDT (2h UT on May 15)
	17	Mercury (−1.0) passes 7.4° north of Aldebaran, 5 A.M. EDT (9h UT). Start looking for the planet low in the west-northwest in the half hour after sunset, about midway between Venus (−4.5) and the horizon. See p. 83.
	21	Venus (−4.3) just over 1.5° from Mercury (−0.6) this evening. Look low in the west-northwest in the half hour after sunset. See p. 83.
	22	Venus passes 0.9° north of Mercury, 4 A.M. EDT (8h UT)
		New Moon, 1:39 P.M. EDT (17:39 UT)
	23	Moon passes 3.8° north of Aldebaran, 6 A.M. EDT (10h UT)
		Moon passes 3.7° south of Venus, 11 P.M. EDT (3h UT on May 24)
	24	Moon passes 2.8° south of Mercury, 7 A.M. EDT (11h UT)
	26	Moon passes 4.5° south of Pollux, 4 P.M. EDT (20h UT)
	29	Moon passes 4.3° north of Regulus, 5 A.M. EDT (9h UT)
		First quarter Moon, 11:30 P.M. EDT (3:30 UT) on May 30
		Prefer observing late at night rather than early morning? Look for Jupiter (−2.6) and Saturn (0.4) low in the southeast just before local midnight, daylight time. They will rise earlier in the evening sky as the year progresses.
June	1	Moon passes 7.5° north of Spica, 10 P.M. EDT (2h UT on Jun. 2)
	3	Venus in inferior conjunction (not visible)
	4	Mercury (0.4) at greatest eastern elongation (23.6°) at 9 A.M. EDT (13h UT)
	5	Moon passes 6.5° north of Antares, 4 A.M. EDT (8h UT)
		Full Moon, 3:12 P.M. EDT (19:12 UT)
		Penumbral lunar eclipse, greatest at 19:25 UT, visible throughout Australia, Asia, Indian Ocean, Africa and Europe

(continued)

2020 Almanac (continued)

<table>
<tr><td>8</td><td>Look south-southwest in the hour before dawn (or southeast about 3 hours after sunset) as brilliant Jupiter (−2.6) and bright Saturn (0.4) accompany the waning gibbous Moon.

Moon passes 2.2° south of Jupiter, 1 P.M. EDT (17h UT)

Moon passes 2.7° south of Saturn, 10 P.M. EDT (2h UT on Jun. 9)</td></tr>
<tr><td>12</td><td>Moon passes 2.8° south of Mars, 8 P.M. EDT (0h UT on Jun. 13). Look for Mars (−0.2) above the Moon, high in the southeast in the hour before dawn.</td></tr>
<tr><td>13</td><td>Last quarter Moon, 2:24 A.M. EDT (6:24 UT)</td></tr>
<tr><td>19</td><td>Venus (−4.4) emerges from twilight and begins its climb into the morning sky. Watch for it in coming mornings very low in the east-northeast in the half hour before sunup. This morning, Venus shines above the slender crescent Moon. Take in the rest of the sky: Peachy Mars (−0.3) shines brightly in the southeast. Jupiter (−2.7) and Saturn (0.3) form a mismatched planetary duo in the southwest. The Summer Triangle, formed by Deneb, Vega and Altair, flies high in the southwest, pointing toward Jupiter.

Moon passes 0.7° north of Venus, 5 A.M. EDT (9h UT)

Moon passes 3.8° north of Aldebaran, 1 P.M. EDT (17h UT)</td></tr>
<tr><td>20</td><td>Summer solstice, 5:44 P.M. EDT (21:44 UT)</td></tr>
<tr><td>21</td><td>New Moon, 2:41 A.M. EDT (6:41 UT)

Annular solar eclipse, greatest at 6:40 UT; annularity tracks from central Africa, the Arabian Peninsula, Pakistan, India, China, Taiwan. See p. 149.</td></tr>
<tr><td>22</td><td>Moon passes 3.9° north of Mercury, 3 A.M. EDT (7h UT)

Moon passes 4.5° south of Pollux, 10 P.M. EDT (2h UT on Jun. 23)</td></tr>
<tr><td>25</td><td>Moon passes 4.4° north of Regulus, 11 A.M. EDT (15h UT)</td></tr>
<tr><td>27</td><td>June Boötid meteor shower peaks, midnight EDT (4h UT)</td></tr>
<tr><td>28</td><td>First quarter Moon, 4:16 A.M. EDT (8:16 UT)</td></tr>
<tr><td>29</td><td>Moon passes 7.5° north of Spica, 4 A.M. EDT (8h UT)</td></tr>
<tr><td>30</td><td>Mercury in inferior conjunction (not visible)</td></tr>
</table>

July

<table>
<tr><td>2</td><td>Moon passes 6.5° north of Antares, 1 P.M. EDT (17h UT)</td></tr>
<tr><td>4</td><td>Earth at aphelion, its farthest point from the Sun this year – 94.5076 million miles away – at 7:35 A.M. EDT (11:35 UT).</td></tr>
<tr><td>5</td><td>Penumbral lunar eclipse, greatest at 12:30 A.M. EDT (4:30 UT), visible throughout the Americas, western Europe, and western and central Africa.

Full Moon, 12:44 A.M. EDT (4:44 UT)

Moon passes 1.9° south of Jupiter, 6 P.M. EDT (22h UT)

An hour after sunset, look southeast for a nearly full Moon flanked by Jupiter (−2.7) and Saturn (0.2), both nearing opposition.</td></tr>
<tr><td>6</td><td>Moon passes 2.5° south of Saturn, 5 A.M. EDT (9h UT)</td></tr>
<tr><td>11</td><td>This morning, the predawn sky offers a sweep of the solar system: In the east, Venus (−4.7) glares near the orange eye of Taurus the Bull. High in the south-southeast, brightening Mars (−0.7) dallies near the waning gibbous Moon. Saturn (0.2) and Jupiter (−2.7) top off the tour low in the southwest.

Moon passes 2.0° south of Mars, 4 P.M. EDT (20h UT)</td></tr>
</table>

(continued)

2020 Almanac (continued)

	12	Venus passes 1.0° north of Aldebaran, 3 A.M. EDT (7h UT)
		Last quarter Moon, 7:29 P.M. EDT (23:29 UT)
	14	Jupiter (−2.8) is at opposition and nearest to Earth, 384.790 million miles away. It rises in the east-southeast near sunset and is visible all night long. Opposition occurs at 3:58 A.M. EDT (7:58 UT).
	16	Moon passes 3.9° north of Aldebaran, 10 P.M. EDT (2h UT on July 17)
	17	Moon passes 3.1° north of Venus, 3 A.M. EDT (7h UT). Can you find Mercury (1.3) below them, low in the east-northeast in the half hour before dawn?
	19	Moon passes 3.9° north of Mercury, midnight EDT (4h UT on Jul. 20)
	20	Moon passes 4.5° south of Pollux, 6 A.M. EDT (10h UT)
		New Moon, 1:33 P.M. EDT (17:33 UT)
		Saturn (0.1) is at opposition and nearest to Earth, 836.110 million miles away. It rises in the east-southeast at sunset and is visible all night long. Opposition occurs at 6:28 P.M. EDT (22:28 UT).
	22	Mercury (0.2) at greatest western elongation (20.1°) at 11 A.M. EDT (15h UT), visible low in the east-northeast half an hour before dawn. See p. 83.
		Moon passes 4.3° north of Regulus, 5 P.M. EDT (21h UT)
	26	Moon passes 7.4° north of Spica, 10 A.M. EDT (14h UT)
	27	First quarter Moon, 8:33 A.M. EDT (12:33 UT)
	29	Moon passes 6.4° north of Antares, 7 P.M. EDT (23h UT)
		Southern Delta Aquariid meteor shower peaks, 7 P.M. EDT (23h UT)
August	1	Moon passes 1.5° south of Jupiter, 8 P.M. EDT (0h UT on Aug. 2)
	2	Mercury passes 6.6° south of Pollux, 2 A.M. EDT (6h UT)
		Moon passes 2.3° south of Saturn, 9 A.M. EDT (13h UT)
	3	Full Moon, 11:59 A.M. EDT (15:59 UT)
	9	Moon passes 0.8° south of Mars, 4 A.M. EDT (8h UT)
	11	Last quarter Moon, 12:45 P.M. EDT (16:45 UT)
		Look for Perseid meteors tonight.
	12	Perseid meteor shower peaks, 9 A.M. EDT (13h UT)
		Venus (−4.4) at greatest western elongation (45.8°) at 8 P.M. EDT (0h UT on Aug 13). The planet is the brightest object in the predawn sky after the Moon.
	13	Moon passes 4.0° north of Aldebaran, 7 A.M. EDT (11h UT)
	15	Moon passes 4.0° north of Venus, 9 A.M. EDT (13h UT)
	16	Moon passes 4.5° south of Pollux, 4 P.M. EDT (20h UT)
	17	Mercury in superior conjunction (not visible)
	18	New Moon, 10:42 P.M. EDT (2:42 UT on Aug. 19)
	22	Moon passes 7.2° north of Spica, 4 P.M. EDT (20h UT)
	25	First quarter Moon, 1:58 P.M. EDT (17:58 UT)
	26	Moon passes 6.2° north of Antares, 1 A.M. EDT (5h UT)
	28	Moon passes 1.4° south of Jupiter, 10 P.M. EDT (2h UT on Aug. 29)
	29	Moon passes 2.2° south of Saturn, 1 P.M. EDT (17h UT)
September	1	Venus passes 8.7° south of Pollux, 1 P.M. EDT (17h UT)
		This morning, in the hours before dawn, look for Venus shining brilliantly in the east. The planet has temporarily joined the "Heavenly G," a huge asterism formed by the brightest stars of numerous constellations. Can you find them all? They are, in order:

(continued)

2020 Almanac (continued)

Capella in Auriga; Castor and Pollux in Gemini; Venus; Procyon in Canis Minor; Sirius in Canis Major – the brightest star in the sky, now far outshone by Venus and Mars; Rigel in Orion; Aldebaran in Taurus; and Betelgeuse in Orion. If this sounds familiar, it's because Venus is in nearly the same position relative to the stars and the Sun as it was on this date in 2012. Look for Mars (−1.8), now rapidly brightening for next month's opposition, high in the southwest.

2	Full Moon, 1:22 A.M. EDT (5:22 UT)
6	Moon passes 0.0° north of Mars, 1 A.M. EDT (5h UT)
9	Moon passes 4.2° north of Aldebaran, 3 P.M. EDT (19h UT)
10	Last quarter Moon, 5:26 A.M. EDT (9:26 UT)
13	Moon passes 4.3° south of Pollux, 1 A.M. EDT (5h UT)
14	Moon passes 4.5° north of Venus, 1 A.M. EDT (5h UT)
15	Moon passes 4.3° north of Regulus, noon EDT (16h UT)
17	New Moon, 7:00 A.M. EDT (11:00 UT)
18	Moon passes 6.4° north of Mercury, 6 P.M. EDT (22h UT). Look for Mercury (−0.1) below the Moon, low in the west-southwest 20 min after sunset.
19	Moon passes 7.0° north of Spica, 1 A.M. EDT (5h UT)
22	Mercury passes 0.3° north of Spica, 5 A.M. EDT (9h UT)
	Moon passes 5.9° north of Antares, 7 A.M. EDT (11h UT)
	Autumnal equinox, 9:31 A.M. EDT (13:31 UT)
23	First quarter Moon, 9:55 P.M. EDT (1:55 UT on Sept. 24)
25	Moon passes 1.6° south of Jupiter, 3 A.M. EDT (7h UT)
	Moon passes 2.3° south of Saturn, 5 P.M. EDT (21h UT)

October	1	Full Moon, 5:05 P.M. EDT (21:05 UT)
		Mercury (−0.0) at greatest eastern elongation (25.8°) at noon EDT (16h UT). The planet is visible low in the west-southwest 20 min after sunset.
		In the hour after sunset, look for Jupiter (−2.4) and Saturn (0.5) in the south, hovering east of the Teapot asterism in Sagittarius. Look east for the rising full Moon, which is followed by an unusual peach-colored "star." That's Mars (−2.5), now nearing its brightest.
	2	Venus passes 0.1° south of Regulus, 8 P.M. EDT (0h UT on Oct. 3)
		Moon passes 0.7° south of Mars, 11 P.M. EDT (3h UT on Oct. 3)
	6	Moon passes 4.5° north of Aldebaran, 10 P.M. EDT (2h UT on Oct. 7)
		Mars (−2.6) is closest to Earth, 38.569 million miles, 10 A.M. EDT (14h UT).
	9	Last quarter Moon, 8:39 P.M. EDT (0:39 UT on Oct. 10)
	10	Moon passes 4.1° south of Pollux, 10 A.M. EDT (14h UT)
	12	Moon passes 4.5° north of Regulus, 10 P.M. EDT (2h UT on Oct. 13)
	13	Mars (−2.6) is at opposition tonight, rising in the east near sunset and visible all night long. Opposition occurs at 7:26 P.M. EDT (23:26 UT). See p. 181.
		Moon passes 4.4° north of Venus, 8 P.M. EDT (0h UT on Oct. 14)
	16	Moon passes 7.0° north of Spica, noon EDT (16h UT)
		New Moon, 3:31 P.M. EDT (19:31 UT)

(continued)

2020 Almanac (continued)

	17	Moon passes 6.8° north of Mercury, 3 P.M. EDT (19h UT)
	19	Moon passes 5.7° north of Antares, 4 P.M. EDT (20h UT)
	21	Orionid meteor shower peaks, 2 A.M. EDT (6h UT)
	22	Moon passes 2.0° south of Jupiter, 1 P.M. EDT (17h UT).

As the sky darkens tonight, look for the waxing crescent Moon. The brightest "star" above it is Jupiter (−2.2); the second-brightest "star" above it is Saturn (0.6). Swing your view eastward to locate Mars (−2.4) — after the Moon, it's now the brightest object in the sky.

Moon passes 2.6° south of Saturn, midnight EDT (4h UT on Oct. 23)

	23	First quarter Moon, 9:23 A.M. EDT (13:23 UT)
	25	Mercury in inferior conjunction (not visible)
	29	Moon passes 3.0° south of Mars, noon EDT (16h UT)
	31	Full Moon, 10:49 A.M. EDT (14:49 UT)
November	1	Fall back: U.S. Daylight Saving Time ends this morning.
	3	Moon passes 4.7° north of Aldebaran, 3 A.M. EST (8h UT)

Start looking for Mercury (0.5) low in the east-southeast in the half hour before sunrise. First, find brilliant Venus (−4.0), which lies 18° above Mercury, then scan below it.

	6	Moon passes 3.8° south of Pollux, 3 P.M. EST (20h UT)
	8	Last quarter Moon, 8:46 A.M. EST (13:46 UT)
	9	Moon passes 4.7° north of Regulus, 6 A.M. EST (11h UT)
	10	Mercury (−0.6) at greatest western elongation (19.1°) at noon EST (17h UT), visible in morning twilight. The waning crescent Moon guides you to these planets on the 12th and 13th; see p. 83.
	12	Moon passes 3.1° north of Venus, 4 P.M. EST (21h UT)
		Moon passes 7.0° north of Spica, 10 P.M. EST (3h UT on Nov. 13)
	13	Moon passes 1.7° north of Mercury, 4 P.M. EST (21h UT)
	15	New Moon, 12:07 A.M. EST (5:07 UT)
		Venus passes 4.1° north of Spica, 8 A.M. EST (13h UT)
	16	Moon passes 5.6° north of Antares, 1 A.M. EST (6h UT)
	17	Leonid meteor shower peaks, 6 A.M. EST (11h UT)
	19	Moon passes 2.5° south of Jupiter, 4 A.M. EST (9h UT)
		Moon passes 2.9° south of Saturn, 10 A.M. EST (15h UT)

Jupiter (−2.1) and Saturn (0.6) are now separated by less than 3.5°. They grow ever closer during the next 4 weeks. Look for the planetary pair in the southwest in the hour after sunset. At the same time, Mars, although fainter than Jupiter, is still a striking peach-colored star in the east-southeast. See p. 210.

	21	First quarter Moon, 11:45 P.M. EST (4:45 UT on Nov. 22)
		Alpha Monocerotid meteor shower peaks, 6 A.M. EST (11h UT)
	25	Moon passes 4.9° south of Mars, 3 P.M. EST (20h UT)
	30	Full Moon, 4:30 A.M. EST (9:30 UT)
		Moon passes 4.7° north of Aldebaran, 9 A.M. EST (14h UT)
		Penumbral lunar eclipse, greatest at 9:43 UT, visible from the Americas, Indonesia and Australia.
December	3	Moon passes 3.8° south of Pollux, 9 P.M. EST (2h UT on Dec. 4)
	6	Moon passes 4.8° north of Regulus, noon EST (17h UT)
	7	Last quarter Moon, 7:37 P.M. EST (0:37 UT on Dec. 8)

(continued)

2020 Almanac (continued)

10	Moon passes 7.1° north of Spica, 7 A.M. EST (12h UT)
12	Moon passes 0.8° north of Venus, 4 P.M. EST (21h UT)
13	Geminid meteor shower peaks, 8 P.M. EST (1h UT on Dec. 14)
14	**Total solar eclipse**, greatest at 16:13 UT; totality tracks through Chile and Argentina. See p. 150.
	New Moon, 11:17 A.M. EST (16:17 UT)
16	Moon passes 2.9° south of Jupiter, 11 P.M. EST (4h UT on Dec. 17)
	Moon passes 3.1° south of Saturn, midnight EST (5h UT on Dec. 17)
	The two giant planets are just 33 arcminutes apart tonight – slightly larger than the Moon's apparent disk – and still closing. See p. 210.
19	Mercury in superior conjunction (not visible)
21	Winter solstice, 5:02 A.M. EST (10:02 UT)
	Jupiter passes 0.1° south of Saturn, 9 A.M. EST (14h UT). In the hour after sunset, look low in the southwest for Jupiter (–2.0) and Saturn (0.6). It's worth a look through binoculars. A small telescope will reveal the disks of the planets, Jupiter's Galilean moons, and Saturn's rings – all in the same field of view. See p. 210.
	First quarter Moon, 6:41 P.M. EST (23:41 UT)
22	Ursid meteor shower peaks, 4 A.M. EST (9h UT)
	Venus passes 5.7° north of Antares, 8 P.M. EST (1h UT on Dec. 23)
23	Moon passes 5.6° south of Mars, 2 P.M. EST (19h UT)
27	Moon passes 4.7° north of Aldebaran, 4 P.M. EST (21h UT)
29	Full Moon, 10:28 P.M. EST (3:28 UT on Oct. 30)
31	Moon passes 3.8° south of Pollux, 3 A.M. EST (8h UT)

Appendix B

Phases of the Moon through 2020

In everyday discussion of the Moon's phases, many people often refer to the first visible evening crescent as "new Moon" or regard the Moon as full within a day or two of the actual date for that phase. Astronomers, not surprisingly, have come to define these terms more precisely. Lunar phases – new, first quarter, full and last quarter – occur when the Moon's apparent angle along the ecliptic differs from the Sun's by 0°, 90°, 180° and 270°, respectively. By this definition, "new Moon" is invisible to us – unless the Moon passes over the Sun's disk, resulting in a solar eclipse. This table gives the instants of lunar phases through 2020. All dates and times given here are in Universal Time (UT). Conversion to Eastern Time is provided in the Timetable of Celestial Events (Appendix A); refer to Table A.1 on page 296 to convert to other U.S. time zones.

2011

New	First quarter	Full	Last quarter
Jan. 4, 9:03	Jan. 12, 11:31	Jan. 19, 21:21	Jan. 26, 12:57
Feb. 3, 2:31	Feb. 11, 7:18	Feb. 18, 8:36	Feb. 24, 23:26
Mar. 4, 20:46	Mar.12, 23:45	Mar. 19, 18:10	Mar. 26, 12:07
Apr. 3, 14:32	Apr. 11, 12:05	Apr. 18, 2:44	Apr. 25, 2:47
May 3, 6:51	May 10, 20:33	May 17, 11:09	May 24, 18:52
Jun. 1, 21:03	Jun. 9, 2:11	Jun. 15, 20:14	Jun. 23, 11:48
Jul. 1, 8:54	Jul. 8, 6:29	Jul. 15, 6:40	Jul. 23, 5:02
Jul. 30, 18:40	Aug. 6, 11:08	Aug. 13, 18:57	Aug. 21, 21:54
Aug. 29, 3:04	Sept. 4, 17:39	Sept. 12, 9:27	Sept. 20, 13:39

(continued)

(continued)

New	First quarter	Full	Last quarter
Sept. 27, 11:09	Oct. 4, 3:15	Oct. 12, 2:06	Oct. 20, 3:30
Oct. 26, 19:56	Nov. 2, 16:38	Nov. 10, 20:16	Nov. 18, 15:09
Nov. 25, 6:10	Dec. 2, 9:52	Dec. 10, 14:36	Dec. 18, 0:48
Dec. 24, 18:06	–	–	–

2012

New	First quarter	Full	Last quarter
–	Jan. 1, 6:15	Jan. 9, 7:30	Jan. 16, 9:08
Jan. 23, 7:39	Jan. 31, 4:10	Feb. 7, 21:54	Feb. 14, 17:04
Feb. 21, 22:35	Mar. 1, 1:21	Mar. 8, 9:39	Mar. 15, 1:25
Mar. 22, 14:37	Mar. 30, 19:41	Apr. 6, 19:19	Apr. 13, 10:50
Apr. 21, 7:18	Apr. 29, 9:57	May 6, 3:35	May 12, 21:47
May 20, 23:47	May 28, 20:16	Jun. 4, 11:12	Jun. 11, 10:41
Jun. 19, 15:02	Jun. 27, 3:30	Jul. 3, 18:52	Jul. 11, 1:48
Jul. 19, 4:24	Jul. 26, 8:56	Aug. 2, 3:27	Aug. 9, 18:55
Aug. 17, 15:54	Aug. 24, 13:54	Aug. 31, 13:58	Sept. 8, 13:15
Sept. 16, 2:11	Sept. 22, 19:41	Sept. 30, 3:19	Oct. 8, 7:33
Oct. 15, 12:02	Oct. 22, 3:32	Oct. 29, 19:49	Nov. 7, 0:36
Nov. 13, 22:08	Nov. 20, 14:31	Nov. 28, 14:46	Dec. 6, 15:31
Dec. 13, 8:42	Dec. 20, 5:19	Dec. 28, 10:21	

2013

New	First quarter	Full	Last quarter
–	–	–	Jan. 5, 3:58
Jan. 11, 19:44	Jan. 18, 23:45	Jan. 27, 4:38	Feb. 3, 13:56
Feb. 10, 7:20	Feb. 17, 20:31	Feb. 25, 20:26	Mar. 4, 21:53
Mar. 11, 19:51	Mar. 19, 17:27	Mar. 27, 9:27	Apr. 3, 4:37
Apr. 10, 9:35	Apr. 18, 12:31	Apr. 25, 19:57	May 2, 11:14
May 10, 0:28	May 18, 4:35	May 25, 4:25	May 31, 18:58
Jun. 8, 15:56	Jun. 16, 17:24	Jun. 23, 11:32	Jun. 30, 4:54
Jul. 8, 7:14	Jul. 16, 3:18	Jul. 22, 18:16	Jul. 29, 17:43
Aug. 6, 21:51	Aug. 14, 10:56	Aug. 21, 1:45	Aug. 28, 9:35
Sept. 5, 11:36	Sept. 12, 17:08	Sept. 19, 11:13	Sept. 27, 3:55
Oct. 5, 0:35	Oct. 11, 23:02	Oct. 18, 23:38	Oct. 26, 23:40
Nov. 3, 12:50	Nov. 10, 5:57	Nov. 17, 15:16	Nov. 25, 19:28
Dec. 3, 0:22	Dec. 9, 15:12	Dec. 17, 9:28	Dec. 25, 13:48

2014

New	First quarter	Full	Last quarter
Jan. 1, 11:14	Jan. 8, 3:39	Jan. 16, 4:52	Jan. 24, 5:19
Jan. 30, 21:39	Feb. 6, 19:22	Feb. 14, 23:53	Feb. 22, 17:15
Mar. 1, 8:00	Mar. 8, 13:27	Mar. 16, 17:08	Mar. 24, 1:46
Mar. 30, 18:45	Apr. 7, 8:31	Apr. 15, 7:42	Apr. 22, 7:52
Apr. 29, 6:14	May 7, 3:15	May 14, 19:16	May 21, 12:59
May 28, 18:40	Jun. 5, 20:39	Jun. 13, 4:11	Jun. 19, 18:39
Jun. 27, 8:08	Jul. 5, 11:59	Jul. 12, 11:25	Jul. 19, 2:08
Jul. 26, 22:42	Aug. 4, 0:50	Aug. 10, 18:09	Aug. 17, 12:26
Aug. 25, 14:13	Sept. 2, 11:11	Sept. 9, 1:38	Sept. 16, 2:05
Sept. 24, 6:14	Oct. 1, 19:33	Oct. 8, 10:51	Oct. 15, 19:12
Oct. 23, 21:57	Oct. 31, 2:48	Nov. 6, 22:23	Nov. 14, 15:15
Nov. 22, 12:32	Nov. 29, 10:06	Dec. 6, 12:27	Dec. 14, 12:51
Dec. 22, 1:36	Dec. 28, 18:31	–	–

2015

New	First quarter	Full	Last quarter
–	–	Jan. 5, 4:53	Jan. 13, 9:46
Jan. 20, 13:14	Jan. 27, 4:48	Feb. 3, 23:09	Feb. 12, 3:50
Feb. 18, 23:47	Feb. 25, 17:14	Mar. 5, 18:05	Mar. 13, 17:48
Mar. 20, 9:36	Mar. 27, 7:43	Apr. 4, 12:06	Apr. 12, 3:44
Apr. 18, 18:57	Apr. 25, 23:55	May 4, 3:42	May 11, 10:36
May 18, 4:13	May 25, 17:19	Jun. 2, 16:19	Jun. 9, 15:42
Jun. 16, 14:05	Jun. 24, 11:03	Jul. 2, 2:20	Jul. 8, 20:24
Jul. 16, 1:24	Jul. 24, 4:04	Jul. 31, 10:43	Aug. 7, 2:03
Aug. 14, 14:53	Aug. 22, 19:31	Aug. 29, 18:35	Sept. 5, 9:54
Sept. 13, 6:41	Sept. 21, 8:59	Sept. 28, 2:50	Oct. 4, 21:06
Oct. 13, 0:06	Oct. 20, 20:31	Oct. 27, 12:05	Nov. 3, 12:24
Nov. 11, 17:47	Nov. 19, 6:27	Nov. 25, 22:44	Dec. 3, 7:40
Dec. 11, 10:29	Dec. 18, 15:14	Dec. 25, 11:11	

2016

New	First quarter	Full	Last quarter
–	–	–	Jan. 2, 5:30
Jan. 10, 1:31	Jan. 16, 23:26	Jan. 24, 1:46	Feb. 1, 3:28
Feb. 8, 14:39	Feb. 15, 7:46	Feb. 22, 18:20	Mar. 1, 23:11
Mar. 9, 1:54	Mar. 15, 17:03	Mar. 23, 12:01	Mar. 31, 15:17

(continued)

(continued)

New	First quarter	Full	Last quarter
Apr. 7, 11:24	Apr. 14, 3:59	Apr. 22, 5:24	Apr. 30, 3:29
May 6, 19:29	May 13, 17:02	May 21, 21:14	May 29, 12:12
Jun. 5, 3:00	Jun. 12, 8:10	Jun. 20, 11:02	Jun. 27, 18:19
Jul. 4, 11:01	Jul. 12, 0:52	Jul. 19, 22:57	Jul. 26, 23:00
Aug. 2, 20:45	Aug. 10, 18:21	Aug. 18, 9:27	Aug. 25, 3:41
Sept. 1, 9:03	Sept. 9, 11:49	Sept. 16, 19:05	Sept. 23, 9:56
Oct. 1, 0:11	Oct. 9, 4:33	Oct. 16, 4:23	Oct. 22, 19:14
Oct. 30, 17:38	Nov. 7, 19:51	Nov. 14, 13:52	Nov. 21, 8:33
Nov. 29, 12:18	Dec. 7, 9:03	Dec. 14, 0:06	Dec. 21, 1:56
Dec. 29, 6:53	–	–	–

2017

New	First quarter	Full	Last quarter
–	Jan. 5, 19:47	Jan. 12, 11:34	Jan. 19, 22:13
Jan. 28, 0:07	Feb. 4, 4:19	Feb. 11, 0:33	Feb. 18, 19:33
Feb. 26, 14:58	Mar. 5, 11:32	Mar. 12, 14:54	Mar. 20, 15:58
Mar. 28, 2:57	Apr. 3, 18:39	Apr. 11, 6:08	Apr. 19, 9:57
Apr. 26, 12:16	May 3, 2:47	May 10, 21:42	May 19, 0:33
May 25, 19:44	Jun. 1, 12:42	Jun. 9, 13:10	Jun. 17, 11:33
Jun. 24, 2:31	Jul. 1, 0:51	Jul. 9, 4:07	Jul. 16, 19:26
Jul. 23, 9:46	Jul. 30, 15:23	Aug. 7, 18:11	Aug. 15, 1:15
Aug. 21, 18:30	Aug. 29, 8:13	Sept. 6, 7:03	Sept. 13, 6:25
Sept. 20, 5:30	Sept. 28, 2:53	Oct. 5, 18:40	Oct. 12, 12:25
Oct. 19, 19:12	Oct. 27, 22:22	Nov. 4, 5:23	Nov. 10, 20:36
Nov. 18, 11:42	Nov. 26, 17:03	Dec. 3, 15:47	Dec. 10, 7:51
Dec. 18, 6:30	Dec. 26, 9:20	–	–

2018

New	First quarter	Full	Last quarter
–	–	Jan. 2, 2:24	Jan. 8, 22:25
Jan. 17, 2:17	Jan. 24, 22:20	Jan. 31, 13:27	Feb. 7, 15:54
Feb. 15, 21:05	Feb. 23, 8:09	Mar. 2, 0:51	Mar. 9, 11:20
Mar. 17, 13:12	Mar. 24, 15:35	Mar. 31, 12:37	Apr. 8, 7:18
Apr. 16, 1:57	Apr. 22, 21:46	Apr. 30, 0:58	May 8, 2:09
May 15, 11:48	May 22, 3:49	May 29, 14:20	Jun. 6, 18:32

(continued)

(continued)

New	First quarter	Full	Last quarter
Jun. 13, 19:43	Jun. 20, 10:51	Jun. 28, 4:53	Jul. 6, 7:51
Jul. 13, 2:48	Jul. 19, 19:52	Jul. 27, 20:20	Aug. 4, 18:18
Aug. 11, 9:58	Aug. 18, 7:48	Aug. 26, 11:56	Sept. 3, 2:37
Sept. 9, 18:01	Sept. 16, 23:15	Sept. 25, 2:52	Oct. 2, 9:45
Oct. 9, 3:47	Oct. 16, 18:02	Oct. 24, 16:45	Oct. 31, 16:40
Nov. 7, 16:02	Nov. 15, 14:54	Nov. 23, 5:39	Nov. 30, 0:19
Dec. 7, 7:20	Dec. 15, 11:49	Dec. 22, 17:49	Dec. 29, 9:34

2019

New	First quarter	Full	Last quarter
Jan. 6, 1:28	Jan. 14, 6:45	Jan. 21, 5:16	Jan. 27, 21:10
Feb. 4, 21:04	Feb. 12, 22:26	Feb. 19, 15:54	Feb. 26, 11:28
Mar. 6, 16:04	Mar. 14, 10:27	Mar. 21, 1:43	Mar. 28, 4:10
Apr. 5, 8:50	Apr. 12, 19:06	Apr. 19, 11:12	Apr. 26, 22:18
May 4, 22:45	May 12, 1:12	May 18, 21:11	May 26, 16:34
Jun. 3, 10:02	Jun. 10, 5:59	Jun. 17, 8:31	Jun. 25, 9:46
Jul. 2, 19:16	Jul. 9, 10:55	Jul. 16, 21:38	Jul. 25, 1:18
Aug. 1, 3:12	Aug. 7, 17:31	Aug. 15, 12:29	Aug. 23, 14:56
Aug. 30, 10:37	Sept. 6, 3:10	Sept. 14, 4:33	Sept. 22, 2:41
Sept. 28, 18:26	Oct. 5, 16:47	Oct. 13, 21:08	Oct. 21, 12:39
Oct. 28, 3:38	Nov. 4, 10:23	Nov. 12, 13:34	Nov. 19, 21:11
Nov. 26, 15:06	Dec. 4, 6:58	Dec. 12, 5:12	Dec. 19, 4:57
Dec. 26, 5:13	–	–	–

2020

New	First quarter	Full	Last quarter
–	Jan. 3, 4:45	Jan. 10, 19:21	Jan. 17, 12:58
Jan. 24, 21:42	Feb. 2, 1:42	Feb. 9, 7:33	Feb. 15, 22:17
Feb. 23, 15:32	Mar. 2, 19:57	Mar. 9, 17:48	Mar. 16, 9:34
Mar. 24, 9:28	Apr. 1, 10:21	Apr. 8, 2:35	Apr. 14, 22:56
Apr. 23, 2:26	Apr. 30, 20:38	May 7, 10:45	May 14, 14:03
May 22, 17:39	May 30, 3:30	Jun. 5, 19:12	Jun. 13, 6:24
Jun. 21, 6:41	Jun. 28, 8:16	Jul. 5, 4:44	Jul. 12, 23:29
Jul. 20, 17:33	Jul. 27, 12:33	Aug. 3, 15:59	Aug. 11, 16:45
Aug. 19, 2:42	Aug. 25, 17:58	Sept. 2, 5:22	Sept. 10, 9:26

(continued)

(continued)

New	First quarter	Full	Last quarter
Sept. 17, 11:00	Sept. 24, 1:55	Oct. 1, 21:05	Oct. 10, 0:39
Oct. 16, 19:31	Oct. 23, 13:23	Oct. 31, 14:49	Nov. 8, 13:46
Nov. 15, 5:07	Nov. 22, 4:45	Nov. 30, 9:30	Dec. 8, 0:37
Dec. 14, 16:17	Dec. 21, 23:41	Dec. 30, 3:28	–

Appendix C

Greatest Elongations of Mercury and Venus through 2020

The tables below list the greatest elongations – the largest angles west or east of the Sun – for Mercury and Venus. Each entry below provides: the date in Universal Time; the planet's maximum angle from the Sun; and its apparent brightness on the astronomical magnitude scale. Smaller magnitude values mean the planet is brighter. In addition, Mercury tends to be brighter in the week before its evening elongations and in the week following its morning elongations. For a more detailed breakdown of the visibility period associated with forthcoming elongations, see Table 3.2 for Venus (p. 67) and Table 3.3 for Mercury (p. 73).

Greatest Elongations of Mercury

Year	West (morning)			East (evening)		
	Date (UT)	Angle	Mag.	Date (UT)	Angle	Mag.
2011	Jan. 9	23.3°	−0.3	Mar. 23	18.6°	−0.3
	May 7	26.6°	0.4	Jul. 20	26.8°	0.4
	Sept. 3	18.1°	−0.3	Nov. 14	22.7°	−0.3
	Dec. 23	21.8°	−0.4	–		
2012	–			Mar. 5	18.2°	−0.4
	Apr. 18	27.5°	0.3	Jul. 1	25.7°	0.4
	Aug. 16	18.7°	−0.1	Oct. 26	24.1°	−0.2
	Dec. 4	20.6°	−0.5	–		
2013	–			Feb. 16	18.1°	−0.6
	Mar. 31	27.8°	0.2	Jun. 12	24.3°	0.4
	Jul. 30	19.6°	0.1	Oct. 9	25.3°	−0.1
	Nov. 18	19.5°	−0.6	–		

(continued)

(continued)

Year	West (morning)			East (evening)		
	Date (UT)	Angle	Mag.	Date (UT)	Angle	Mag.
2014		–		Jan. 31	18.4°	–0.7
	Mar. 14	27.6°	0.1	May 25	22.7°	0.4
	Jul. 12	20.9°	0.3	Sept. 21	26.4°	0.0
	Nov. 1	18.7°	–0.6	–		
2015		–		Jan. 14	18.9°	–0.7
	Feb. 24	26.7°	0.0	May 7	21.2°	0.2
	Jun. 24	22.5°	0.4	Sept. 4	27.1°	0.1
	Oct. 16	18.1°	–0.6	Dec. 29	19.7°	–0.6
2016	Feb. 7	25.6°	–0.1	Apr. 18	19.9°	0.1
	Jun. 5	24.2°	0.5	Aug. 16	27.4°	0.2
	Sept. 28	17.9°	–0.5	Dec. 11	20.8°	–0.5
2017	Jan. 19	24.1°	–0.2	Apr. 1	19.0°	–0.2
	May 17	25.8°	0.4	Jul. 30	27.2°	0.3
	Sept. 12	17.9°	–0.4	Nov. 24	22.0°	–0.4
2018	Jan. 1	22.7°	–0.3	Mar. 15	18.4°	–0.4
	Apr. 29	27.0°	0.4	Jul. 12	26.4°	0.4
	Aug. 26	18.3°	–0.2	Nov. 6	23.3°	–0.3
	Dec.15	21.3°	–0.5	–		
2019		–		Feb. 27	18.1°	–0.5
	Apr. 11	27.7°	0.3	Jun. 23	25.2°	0.4
	Aug. 9	19.0°	0.0	Oct. 20	24.6°	–0.1
	Nov. 28	20.1°	–0.6	Feb. 10	18.2°	–0.6
2020	Mar. 24	27.8°	0.2	Jun. 4	23.6°	0.4
	Jul. 22	20.1°	0.2	Oct. 1	25.8°	0.0
	Nov. 10	19.1°	–0.6	–		

Greatest Elongations of Venus

Year	West (morning)			East (evening)		
	Date (UT)	Angle	Mag.	Date (UT)	Angle	Mag.
2011	Jan. 8	47.0°	–4.6	–		
2012	Aug. 15	45.8°	–4.4	Mar. 27	46.0°	–4.5
2013		–		Nov. 1	47.1°	–4.5
2014	March 22	46.6°	–4.5	–		
2015	Oct. 26	46.4°	–4.5	Jun. 6	45.4°	–4.4
2017	Jun. 3	45.9°	–4.4	Jan. 12	47.1°	–4.6
2018		–		Aug. 17	45.9°	–4.5
2019	Jan. 6	47.0°	–4.6	–		
2020	Aug. 13	45.8°	–4.4	Mar. 24	46.1°	–4.5

Appendix D

Oppositions of Mars, Jupiter and Saturn through 2020

The table below lists information about upcoming oppositions of Mars, Jupiter and Saturn. Each entry provides: the date in Universal Time, the constellation in which the planet appears, the planet's brightness, distance from Earth, and apparent angular size at oppositions through 2020. Distances are given in Astronomical Units (AU), where one AU equals 92.96 million miles or 149.60 million km. For additional details, including distances in miles, see the Timetable of Celestial Events (Appendix A). Note that the smallest distance between Earth and Mars often occurs on a different date than the actual opposition; these dates and distances are noted in the Timetable.

Year	Planet	Date (UT)	Constellation	Mag.	Distance (AU)	Size (arcsec)
2011	Jupiter	Oct. 29	Aries	−2.9	3.970015141	49.7″
	Saturn	Apr. 3	Virgo	0.4	8.613944238	19.3″
2012	Mars	Mar. 3	Leo	−1.2	0.674066787	13.9″
	Jupiter	Dec. 3	Taurus	−2.8	4.068874130	48.5″
	Saturn	Apr. 15	Virgo	0.2	8.852944597	19.1″
2013	Saturn	Apr. 28	Libra	0.1	8.816202998	18.9″
2014	Mars	Apr. 8	Virgo	−1.5	0.620864490	15.1″
	Jupiter	Jan. 5	Gemini	−2.7	4.210645468	46.8″
	Saturn	May 10	Libra	0.1	8.899676184	18.7″
2015	Jupiter	Feb. 6	Cancer	−2.6	4.346243800	45.4″
	Saturn	May 23	Libra	0.0	8.966702667	18.5″
2016	Mars	May 22	Scorpius	−2.1	0.509449196	18.4″
	Jupiter	Mar. 8	Leo	−2.5	4.435364004	44.4″
	Saturn	Jun. 3	Ophiuchus	0.0	9.014904448	18.4″

(continued)

						Size
Year	Planet	Date (UT)	Constellation	Mag.	Distance (AU)	(arcsec)
2017	Jupiter	Apr. 7	Virgo	−2.5	4.455049824	44.3″
	Saturn	Jun. 15	Ophiuchus	0.0	9.042677385	18.4"
2018	Mars	Jul. 27	Capricornus	−2.8	0.386155263	24.3″
	Jupiter	May 9	Libra	−2.5	4.400149556	44.8″
	Saturn	Jun. 27	Sagittarius	0.0	9.048818608	18.4″
2019	Jupiter	Jun. 10	Ophiuchus	−2.6	4.284224065	46.0″
	Saturn	Jul. 9	Sagittarius	0.1	9.032791840	18.4″
2020	Mars	Oct. 13	Pisces	−2.6	0.419239548	22.3″
	Jupiter	Jul. 14	Sagittarius	−2.8	4.139493677	47.6″
	Saturn	Jul. 20	Sagittarius	0.1	8.994700668	18.5″

(continued)

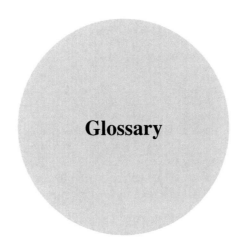

Glossary

Altitude	The elevation angle of an object above the horizon, measured in degrees.
Angular size/distance	The apparent width of an object as seen by an observer, expressed in degrees, arcminutes, or arcseconds. Both the Sun and Moon are about 0.5° or 30 arcminutes across. The same units describe the apparent distance between objects (e.g., Mars is 3° from Venus).
Apparition	The period of observation of a planet, asteroid or comet; an appearance.
Aphelion	The most distant point of a Sun-centered orbit.
Apollo	U.S. space program that included six piloted lunar landings that placing 12 men on the Moon from 1969 through 1972.
Apogee	The most distant point of an Earth-centered orbit.
Arcminute	A unit of angular measure where 60 arcminutes constitutes 1°. It is the smallest angle resolvable to the human eye. Venus is 0.96 arcminute across during its transit in 2012, and Jupiter is 0.83 arcminute across at its 2011 opposition.
Arcsecond	A unit of angular measure where 60 arcseconds constitutes 1 arcminute and 3,600 arcseconds equals 1 degree.

Aspect	The apparent position of the planets or the Moon relative to the Sun, as seen from Earth.
Aurora	Regions of glowing gas in the upper atmosphere whose molecules are stimulated to emit light by collisions with streams of electrons associated with solar activity. Aurora borealis ("northern lights") refers to the phenomenon when seen north of the equator; aurora australis ("southern lights") when seen south of the equator.
Asteroid	A mostly rocky body less than 620 miles (1,000 km) across that orbits the Sun; more accurately called a minor planet. Asteroids orbit the Sun and are most often found in a belt between Mars and Jupiter. Asteroids are the source of most meteorites.
Astronomical unit (AU)	Approximately the average distance between Earth and the Sun, formally 92.955807267 million miles or 149.59787069 million km. Because the AU is defined using a circular orbit and Earth's orbit is elliptical, the value is actually slightly smaller than the mean Earth-Sun distance.
Asterism	A striking pattern of stars, such as the Big Dipper or the Pleiades, that is part of a larger constellation.
Axis	An imaginary line passing through the center of a body, such as a planet, around which that body spins.
Azimuth	The angular distance of an object's direction as measured clockwise along the horizon from a reference point. North is often used as the starting reference of 0° (and also 360°), which makes east 90°, south 180° and west 270°.
Binary star	A system containing two stars in orbit about one another.
Black hole	The most compact object known, usually formed by the gravitational collapse of a star many times the Sun's mass. Not even light can escape a black hole, but gas moving toward it gains enormous energy and emits X-rays and other radiation before falling in. Some of the hottest, fastest-moving gas may escape through particle jets that form very close to the black hole. *Supermassive*: All large galaxies contain central black holes with millions to billions of times the Sun's mass. Our own Milky Way galaxy possesses a supermassive black hole estimated at 4 million solar masses. Astronomers believe that these black holes reached their enormous sizes through collisions and mergers with other galaxies, a process that continues today.

Blue Moon	A popular term for the second full Moon to occur in a calendar month.
Bolide	A fireball that breaks up during its passage through the atmosphere.
Calendar	A system of reckoning time that enumerates days according to their position in cyclic patterns.
Cassini	U.S. mission to Saturn, launched in 1997 and, following successful flybys of Venus (1998, 1999), Earth (1999) and Jupiter (2001), arrived at Saturn in 2004. Cassini deployed a probe named Huygens that entered the atmosphere of Saturn's largest moon, Titan, on Jan. 14, 2005, and returned the first close-up images of the moon as it descended to the surface.
Celestial equator	The projection of the Earth's equator onto the celestial sphere.
Celestial pole	Either of the two points projected onto the celestial sphere by extending extension of the Earth's axis of rotation to infinity. Polaris, the North Star, is currently positioned near the north celestial pole.
Celestial sphere	An immense imaginary sphere whose surface includes all of the stars, planets, and other bodies that is usually centered on an observer on Earth or at Earth's center. It is a convenient tool for mapping the positions of stars and other heavenly bodies and for describing their apparent motions.
Chaos	In its scientific usage, the irregular motion or dynamics of physical systems. The defining characteristics of chaotic systems are periods of order interspersed with randomness and evolution that is extremely sensitive to initial conditions. Chaotic behavior is endemic to most, if not all, physical systems, including the atmosphere and the solar system.
Chromosphere	A distinctly reddish layer in the Sun's atmosphere that marks the transition between the photosphere and the corona, often seen during eclipses.
Comet	A small body made of ice and rock that orbits the Sun, usually much less than 62 miles (100 km) across. As it nears the Sun it usually brightens and develops a gaseous halo, or coma, and a tail of gas and dust. Most comets travel in very elongated orbits that keep them far from the inner solar system.
Conjunction	The alignment of two celestial bodies that occurs when they share similar angles from the Sun as measured along the ecliptic. This is also roughly when the bodies appear closest together in the sky. *Inferior conjunction:* The point in the motions of the

planets Mercury and Venus where they pass between Earth and the Sun. *Superior conjunction:* The point in the motions of Mercury and Venus where they appear in line with the Sun on the far side of their orbits as viewed from Earth. *With the Sun:* The point in the motions of the outer planets where they appear in line with the Sun as viewed from Earth.

Constellation   One of 88 regions into which astronomers have formally divided the sky, based mainly on earlier divisions inspired by historical and mythological figures of Greek and Roman tradition.

Corona   The outermost layer of the Sun's atmosphere. It is visible to the eye only during a total solar eclipse but is now routinely observed by satellites. The corona is the source of the solar wind and, for reasons not entirely understood, is hundreds of times hotter than the Sun's photosphere. Coronal temperatures start at 1.8 million °F (about 1 million °C).

Double star   Two stars that appear close to one another. They can be physically associated (a binary) or simply appear together from the point of view of an observer on Earth.

Dwarf planet   An object in orbit around the Sun that is nearly round, not a planetary moon, and whose orbital vicinity is shared by objects of similar mass. The classification was established in 2006 and includes Pluto and Ceres, the largest asteroid.

Earthshine   A blue-gray light seen during the Moon's crescent phases on the portion not illuminated by the Sun. The source is sunlight reflected by Earth.

Eclipse, lunar An event where the Moon enters into the shadow of the Earth. *Penumbral:* A lunar eclipse where the Moon remains entirely in the brighter portion (penumbra) of Earth's shadow; the event goes largely unnoticed. *Partial:* A lunar eclipse where part of the Moon enters the dark portion (umbra) of Earth's shadow. *Total:* A lunar eclipse where the Moon passes completely into Earth's umbral shadow.

Eclipse, solar An event where the Moon passes in front of the Sun's disk. *Partial:* A solar eclipse where the Moon's darkest shadow (umbra) completely misses Earth, but the brighter penumbra does not. At locations outside of the central track of total, annular, and hybrid eclipses, where the Sun's disk is never completely obscured, the eclipse is also said to be partial. *Total:* A solar eclipse where the Moon entirely covers the Sun. *Annular:* A solar eclipse where the Moon covers all but a thin ring, or annulus, of the Sun. *Hybrid, or annular/total:* A solar eclipse that would otherwise be classified as annular but for a small region along the central track that experiences a brief total eclipse.

Ecliptic	The apparent yearly path of the Sun through the sky. Since this apparent motion is actually a reflection of Earth's movement, the ecliptic also marks the plane of Earth's orbit. The Moon and planets also roughly follow this path.
Electrophonic sounds	Sound produced through the conversion of radio energy at audible wavelengths by the vibration of objects near the observer, such as hair, foliage, or eyeglasses. Meteors, lightning and possibly aurorae can produce sound in this way.
Elongation, greatest	The instant when the angular distance of Mercury or Venus from the Sun is at a maximum as calculated from Earth's center.
Equinox	The date of the year when the Sun's rays illuminate Earth from pole to pole; neither the north pole nor the south pole is angled toward the Sun. This phenomenon occurs on two days of the year, near Mar. 20 and Sept. 23. On these dates, the hours of night nearly equal the hours of daylight (hence the name, meaning "equal night"). The March or vernal equinox is widely regarded as the first day of spring in the Northern Hemisphere; the September or autumnal equinox is thus the first day of autumn.
Farside	The side of the Moon always turned away from Earth.
Fireball	An extremely bright meteor, usually one brighter than magnitude –4.
Galaxy	A vast collection of billions of stars, gas, and dust held together by the gravity of its members. The galaxy where our solar system resides is called the Milky Way.
Gamma-ray burst (GRB)	A brief, intense flash of gamma rays – the most energetic form of light – lasting from milliseconds to several minutes. Some GRBs are associated with supernovae. In March 2008, visible light from GRB 080319B reached naked-eye brightness, even though the dying star was 7.5 billion light-years away.
Galileo	U.S. orbiter that studied Jupiter's atmosphere, moons, and magnetosphere from 1995 to 2003. It deployed a probe into Jupiter's atmosphere and undertook a series of close encounters with the four major moons – Io, Europa, Ganymede and Callisto – and provided evidence for a subsurface ocean at Europa.
Gas giants	The planets Jupiter, Saturn, Uranus and Neptune.

Greatest eclipse	The instant when the axis of the Moon's shadow passes closest to Earth's center. For total eclipses, the instant of greatest eclipse is virtually identical to the instants of greatest duration. For annular eclipses, the instant of greatest duration may occur at either the time of greatest eclipse or near the sunrise and sunset points of the eclipse path.
Horizon	Where the sky meets earth and sea. The ideal or astronomical horizon is a circle on the celestial sphere formed by a plane tangent to an observer's position on Earth's surface.
Inferior conjunction	See conjunction.
International Astronomical Union (IAU)	An international non-governmental organization that promotes all aspects of astronomical science.
Kaguya	A Japanese lunar mission that operated from 2007 to 2009. It deployed two subsatellites to aid in radio studies and was the first lunar probe to carry a high-definition TV camera. The mission ended with the spacecraft's controlled crash on Jun. 10, 2009.
Latitude	*Celestial:* An angle measured north or south from a reference plane defined at 0°; this is often either the ecliptic or the celestial equator. *Geographic:* The angular distance north or south from Earth's equator. The equator is defined at 0° and the poles are at 90° N and 90° S.
Light-year	The distance traveled through space by a beam of light in 1 year. Light travels at 186,282 miles (299,792 km) per second, so a light-year is 5.9 trillion miles (9.5 trillion km), or 63,240 times Earth's distance from the Sun.
Longitude	*Celestial:* An angle measured eastward from the vernal equinox along a reference plane, usually the ecliptic or the celestial equator. *Geographic:* The angular distance east or west between the meridian of a particular place on Earth and that of Greenwich, England, expressed either in degrees or hours of time.
Luna	A series of 24 Soviet robotic lunar missions. In 1959, Luna 3 returned the first image of the Moon's farside.
Lunar Prospector	U.S. mission to the Moon launched on Jan. 6, 1998, that provided evidence for the presence of hydrogen in lunar soil, possibly in the form of water. The mission ended Jul. 31, 1999, with a controlled crash into a crater at the Moon's south pole.

Lunar Reconnaissance Orbiter (LRO) and Lunar CRater Observation and Sensing Satellite (LCROSS)	LRO is a U.S. mission designed to survey the Moon – with particular emphasis on polar regions that appear to hold water in some form – and assess sites for potential human landings. LRO was launched with LCROSS, a companion spacecraft. Once LRO separated, LCROSS and its Centaur rocket body were directed toward a permanently shadowed region of Cabeus crater, which showed strong evidence of water. LCROSS separated from the rocket 11 hours before the Oct. 9, 2009, impact and trailed it as it descended. The spacecraft observed the rocket's crash and flew through the debris before striking the moon itself about four minutes later. LCROSS detected at least 25 gal (95 l) of water in the debris plume, and scientists estimate the area excavated by the impact holds more water than some deserts on Earth.
Lunation	A complete cycle of lunar phases; a lunar or synodic month.
Magnitude	A measure of the relative brightness of stars and other celestial objects. The brighter the object, the lower its assigned magnitude; the brightest objects have negative magnitudes. This logarithmic scale is based on the ancient practice of noting that the brightest stars in the sky were of "first importance" or "first magnitude," the next brightest being "second magnitude," etc. In 1856 Norman Pogson formalized this scale and defined a difference of 5 magnitudes to be exactly a factor of 100 in brightness. See Table 1.1 on page 8 for examples.
Mariner 9	U. S. space mission to Mars, launched in 1971, achieved global imaging of the surface, including the first detailed views of martian volcanoes, Valles Marineris, the polar caps, and the satellites Phobos and Deimos.
Mars Exploration Rovers	U.S. mission consisting of two six-wheeled rovers – named Spirit and Opportunity – that have simultaneously explored opposite sides of Mars since arriving in 2004. Communication with Spirit ended in March 2010. Both rovers have made important discoveries about the history of wet environments on Mars.
Mars Express	Orbiter launched by the European Space Agency that has been operating at Mars since late 2003

Mars Global Surveyor
(MGS)

U.S. orbiter that operated at Mars from 1997 to 2006. Image resolution was several times better than any of those taken by the Viking Orbiter cameras, enabling features just a few meters across to be seen.

Mars Reconnaissance
Orbiter (MRO)

A U.S. mission operating at Mars since 2006. Its high-resolution camera is capable of imaging surface objects the size of a desk.

Mars Odyssey

U.S. orbiter operating at Mars since 2001. Early results indicated large amounts of hydrogen, implying the presence of frozen water in martian soil. In 2010, the mission released the most detailed global map of the Red Planet ever compiled, assembled from more than 20,000 images.

Mars Pathfinder

U.S. lander and small rover that set down on Mars on July 4, 1997, and conducted surface operations through September, when communications were lost. The mission paved the way for the far more ambitious rover missions that followed.

Mars Science Laboratory

A U.S. mission to land and operate a large roving laboratory – named Curiosity – on Mars. It is now slated for launch in late 2011.

Meteor

The streak of light caused by a solid body in orbit about the Sun (a meteoroid) passing through the atmosphere; also called a "shooting star." A meteorite is a meteoroid that strikes the surface of a planet or moon.

Meteor shower

The appearance of many meteors within a few hours that seem to radiate from the same region of the sky. They occur when Earth passes through the dusty debris near a comet's orbit.

Milky Way

A faint band of light around the sky composed of vast numbers of stars too faint to see individually. Also, the name of the galaxy where the solar system resides.

Moon

A general term for any natural satellite orbiting a planet, as well as the name of Earth's natural satellite.

Nautical twilight

See twilight.

Near Earth Asteroid
Rendezvous (NEAR)

U.S. orbiter placed into orbit around asteroid 433 Eros in 2000. On Feb. 12, 2001, at the conclusion of the mission, NEAR became the first spacecraft to land on an asteroid.

Nearside

The side of the Moon that always faces Earth.

Nebula	A cloud of gas and dust, sometimes glowing from the light of nearby stars and sometimes blocking starlight. New stars are born within such clouds.
Nova	A star which suddenly erupts, greatly increasing its brightness.
Nucleus	Of a comet, the solid ice-rock mixture at the center of a comet's gaseous head and tail. Of a spiral galaxy, the dense central portion made of older, redder stars. (plural: nuclei).
Opposition	The point in a planet's orbit where it appears opposite the Sun in the sky. A planet at opposition is visible all night long. Because they orbit closer to the Sun than Earth, Mercury and Venus never reach opposition.
Penumbra	The portion of a shadow cone in which some light from an extended source can be seen.
Perigee	The point where an object in orbit around the Earth is nearest to it.
Perihelion	The point where an object in orbit around the Sun is nearest to it.
Phases	The cycle of varying shape in the sunlit portion of a planet or moon. The Moon, Venus, and Mercury all show phases as seen from Earth.
Phoenix Mars Lander	A U.S. mission that landed in the martian arctic in May 2008. Soil studies and images indicated the presence of frozen water just beneath the surface.
Photosphere	The visible surface of the Sun.
Planet	A body orbiting the Sun that is nearly round and that is the dominant mass in its vicinity. A planet shines by reflecting sunlight.
Plasma	A gas containing roughly equal numbers of positive and negative charges (ions and electrons) that interacts with electric and magnetic fields. Examples include the glowing gas inside fluorescent lights, lightning channels, aurorae, the solar wind, and the Sun.
Radiant	The point in the sky from which shower meteors seem to be emanating.
Retrograde motion	The apparent backward (westward) loop in a planet's motion across the sky. All planets display retrograde motion, but that of Mars is most striking.

Rosetta	A rendezvous and landing mission to comet 67P/Churyumov-Gerasimenko by the European Space Agency. Launched in 2004, the spacecraft flew by asteroids Steins and Lutetia in 2008 and 2010, respectively. Plans call for Rosetta to orbit the comet in Nov. 2014 and deploy a small lander named Philae onto the surface.
Satellite	A natural or artificial body in orbit around a planet.
Scintillation	A tremulous effect of starlight – twinkling – caused by its passage through Earth's turbulent atmosphere. Planets usually don't show this effect, making it easier to identify them.
Sidereal period	The time taken by a planet to complete one revolution around the Sun (or for the Moon to complete an orbit around the Earth) as measured by reference to the background stars.
Solar wind	A stream of electrically charged particles (mainly protons and electrons) moving outward from the Sun with an average speed of a million miles an hour (1.6 million km/h).
Solstice	The date of the year where either Earth's north or south pole is angled most directly toward the Sun. This occurs on two days of the year, near Jun. 21 and Dec. 21. The June solstice, the longest day of the year in the Northern Hemisphere, is when the Sun makes its most northerly arc through the sky. The December solstice, the shortest day of the year, is when the Sun makes its most southerly arc through the sky.
Star	A hot, glowing sphere of ionized gas, usually one that emits energy from nuclear reactions in its core. The Sun is a star.
Sunrise/sunset	Times when the apparent upper limb of the Sun is on the ideal, or astronomical horizon, which does not take into account local geography.
Sunspot	A magnetic disturbance on the Sun. It is cooler than the surrounding area and, consequently, appears darker.
Superior conjunction	See conjunction.
Supernova	An enormous stellar explosion that increases the brightness of a star by more than 100,000 times. The explosion destroys most of the star, but in many cases an ultradense remnant – a neutron star or black hole – may form by compression of the star's central core.
Synodic period	The average time between successive returns of a planet or the Moon to the same apparent position relative to the Sun

	– for example, new Moon to new Moon, or opposition to opposition.
Terminator	The edge of the sunlit portion of the Moon or planets; the line between day and night.
Terrestrial planets	Mercury, Venus, Earth, and Mars.
Tides	Periodic changes in the shape of a planet, moon, or star caused by the gravitational pull of a body near it.
Time zone	A geographic zone offset from Universal Time by a constant number of hours and established as standard, e.g., Eastern Standard Time. Standard time in the U.S. and its territories is observed within eight time zones offset by a whole number of hours. Non-U.S. locations may include offsets with fractional hours.
Train	A dimly visible path in the sky following the passage of a meteor.
Transit	The passage of a planet across the disk of a star. From Earth, only Mercury and Venus can transit the Sun. Planets around other stars also undergo transits and have been discovered because of them.
Twilight	Natural light before sunrise and after sunset provided by sunlight scattering in the upper atmosphere. *Civil:* Begins (morning) and ends (evening) when the Sun's center is 6° below the horizon. Both the brightest stars and the landscape can be seen clearly in good weather. *Nautical:* Begins (morning) and ends (evening) when the Sun's center is 12° below the horizon. In good weather, the horizon itself is indistinct and only the general outlines of ground objects are visible without additional illumination.
Twinkling	See scintillation.
Umbra	The darkest portion of a shadow cone, from which only scattered or refracted light from an extended source can be observed.
Variable star	A star that exhibits significant brightness changes.
Viking	U.S. mission to Mars, composed of two spacecraft, launched in 1975. Viking 1 and Viking 2 both consisted of an orbiter and a lander. Primary mission objectives were to obtain high-resolution images of the Martian surface (55,000 were returned by the orbiters, 1,400 from the landers), characterize the structure and composition of the atmosphere and surface and search for evidence of life.

Voyager	U.S. mission consisting of two spacecraft launched in 1977 to explore Jupiter, Saturn, their moons, rings, and magnetic environments. Voyager 2 went on to explore Uranus and Neptune as well. Both spacecraft are still returning data as they head out of the solar system. As of August 2010, Voyager 1 was 10.6 billion miles (17.1 billion km, or 114.3 AU) from the Sun and Voyager 2 was 8.6 billion miles (13.9 billion kilometers, or 92.9 AU) from the Sun.
White dwarf	A collapsed object formed from a star that has exhausted its nuclear fuel. The Sun will one day become a white dwarf.
Zenith	The point directly overhead, 90° above the ideal horizon.
Zodiac	The band of constellations straddling the ecliptic.

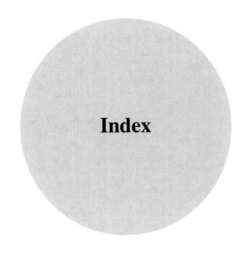

Index